FORTSCHRITTE

DER CHEMIE ORGANISCHER NATURSTOFFE

PROGRESS IN THE CHEMISTRY OF ORGANIC NATURAL PRODUCTS

PROGRÈS DANS LA CHIMIE DES SUBSTANCES ORGANIQUES NATURELLES

HERAUSGEGEBEN VON   EDITED BY   RÉDIGÉ PAR

L. ZECHMEISTER

CALIFORNIA INSTITUTE OF TECHNOLOGY, PASADENA

SECHZEHNTER BAND
SIXTEENTH VOLUME   SEIZIÈME VOLUME

VERFASSER   AUTHORS   AUTEURS

J. BONNER · K. FREUDENBERG · H. KUHN · E. E. VAN TAMELEN
Z. VALENTA · K. WEINGES · K. WIESNER

MIT 27 ABBILDUNGEN   WITH 27 FIGURES   AVEC 27 ILLUSTRATIONS

WIEN · SPRINGER-VERLAG · 1958

ISBN-13: 978-3-7091-8049-5   e-ISBN-13: 978-3-7091-8047-1
DOI: 10.1007/978-3-7091-8047-1

SOFTCOVER REPRINT OF THE HARDCOVER 1ST EDITION 1958

Inhaltsverzeichnis.
Contents. — Table des matières.

Structural Chemistry of Actinomycetes Antibiotics. By E. E. VAN
TAMELEN, Department of Chemistry, The University of Wisconsin,
Madison, Wisconsin 90

Catechine,
andere Hydroxy-flavane und Hydroxy-flavene.

Von KARL FREUDENBERG und KLAUS WEINGES, Heidelberg.

Inhaltsübersicht.

1. Ältere Arbeiten (*19, 20, 69*).

Als „Catechin" wurde ursprünglich der kristallisierte Inhaltsstoff der *Acacia catechu* bezeichnet, der sich als ein Pentahydroxy-flavan (XII) erwiesen hat. Zuerst fanden sich stereoisomere Formen hinzu, später das hydroxylreichere Naturprodukt (XIII) und das hydroxylärmere (XI). Die Zahl der synthetischen Beispiele ist im Wachsen. Hier wird die gesamte Gruppe der Hydroxy-flavane behandelt. Einbezogen werden die Hydroxy-flavene und das ihnen nahestehende Cyanomaclurin (XVIII, S. 15).

Das ursprüngliche Catechin, das 3,5,7,3',4'-Pentahydroxy-flavan (XII), wurde 1821 von F. R. RUNGE aufgefunden. Doch erst 1920 wurde es von FREUDENBERG (*21*) der im Pflanzenreich weitverbreiteten $C_6C_3C_6$-Gruppe zugeordnet, zu der Flavonole, Flavanonole, Flavone, Anthocyanidine, Flavanone, Flavene, Chalkone und Hydrochalkone gehören. Diese Stoffe unterscheiden sich in der Oxydationsstufe ihrer Propaneinheit. Die Catechine stehen in dieser Reihe bei den Flavanonen und Hydrochalkonen. Die vier optisch aktiven und zwei razemischen Formen des Catechins wurden teils in der Natur gefunden, teils durch Umlagerung hergestellt (*42*). In der Natur scheinen primär nur *d*-Catechin und *l*-Epicatechin* vorzukommen, während die anderen Formen — besonders *d,l*-Catechin — erst durch Alterung oder bei der Aufarbeitung der Hölzer entstehen. Die Konstitutionsermittlung wurde durch die Hydrierung des Cyanidins (*30*) in das *d,l*-Epicatechin und damit in *d,l*-Catechin abgeschlossen. Ebenso gelang es, das Pentamethylquercetin in das Pentamethyl-*d,l*-epicatechin zu überführen (*34*). Damit war die Formel der Catechine eindeutig festgelegt und der Zusammenhang mit der α,γ-Diphenylpropangruppe bestätigt.

d-Catechin und *l*-Epicatechin sind in der Natur so verbreitet, daß die Aufzählung der Hölzer und Rinden hier unterbleiben muß. Es sei aber darauf hingewiesen, daß man *d*-Catechin (*41*) in 15—20%iger Ausbeute aus Blockgambir, einem Extrakt aus den Blättern und Zweigen der malaiischen Liane *Uncaria gambir*, und *l*-Epicatechin in zirka 5%iger Ausbeute aus dem Holz der vorderindischen *Acacia catechu* isolieren kann (*42*).

2. Die bekannten Hydroxy-flavane.

In der *Formelübersicht 1* sind diejenigen Hydroxy-flavane synthetischen oder natürlichen Ursprungs angeführt, deren Verhalten gegen Säuren bekannt ist. Die Fähigkeit zur Selbstkondensation im sauren Medium ist durch (—) (fehlend), (+) (langsam) und (++) (schnell) gekennzeichnet. Die mit (n) versehenen Produkte wurden aus Pflanzen isoliert, die mit (s) versehenen synthetisiert. Es ist damit zu rechnen, daß einige der bisher nur synthetisch erhaltenen Verbindungen auch in der Natur vorkommen (*18, 83, 83 a, 88*). Die epimeren Formen sind in der Übersicht nicht berücksichtigt.

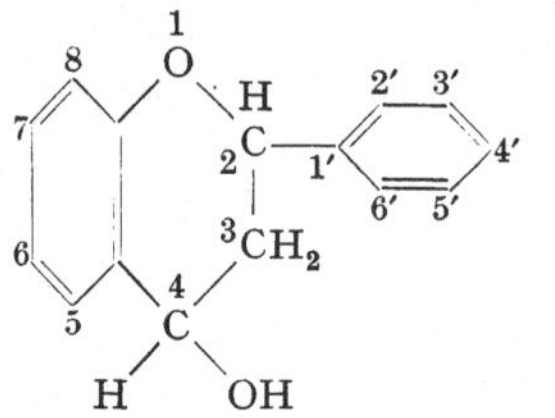

(I.) 4-Hydroxy-flavan (*26, 40*). s. (—)

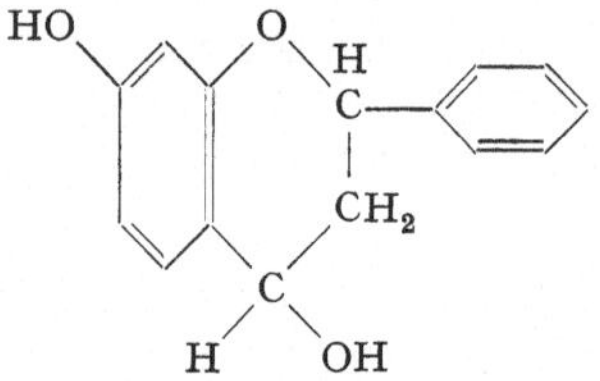

(II.) 4,7-Dihydroxy-flavan (*24*). s. (—)

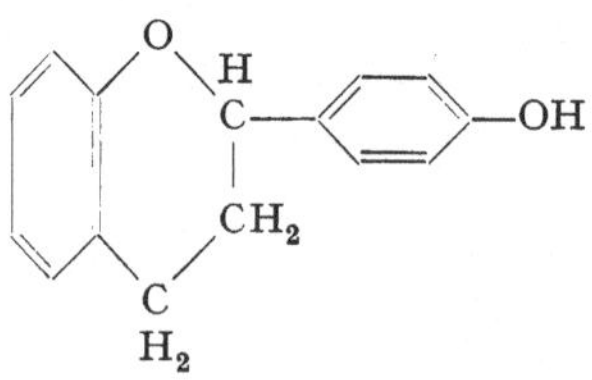

(III.) 4′-Hydroxy-flavan (*47*). s. schwach (+)

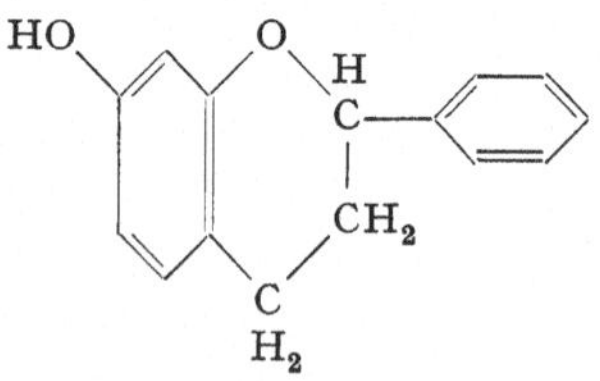

(IV.) 7-Hydroxy-flavan (*79*). s. (—)

---

* Die kleingeschriebenen Präfixe *d* und *l* geben hier lediglich die Drehungsrichtung an.

(V.) 7,4′-Dihydroxy-flavan (*47*). s. (++)

(VI.) 7-Hydroxy-4′-methoxy-flavan (*45*). s. (+)

(VII.) 7-Methoxy-4′-hydroxy-flavan (*45*). s. (+)

(VIII.) 5,7-Dihydroxy-4′-methoxy-flavan (*26*). s. (+)

(IX.) 7,3′,4′-Trihydroxy-flavan (*36*). s. (++)

(X.) 3,7,3′,4′-Tetrahydroxy-flavan (*36*). s. (++)

(XI.) 3,5,7,4′-Tetrahydroxy-flavan
(Epi-afzelechin) (*65*). n. (++)

(XII.) 3,5,7,3′,4′-Pentahydroxy-flavan
(Catechin) (*30*, *41*). s. u. n. (++)

(XIII.) 3,5,7,3′,4′,5′-Hexahydroxy-flavan
(Gallo-catechin) (*71* ,*82* ,*86*). n. (++)

(XIV.) 3,4,7,3′,4′-Pentahydroxy-flavan
(Mollisacacidin) (*62*). s. u. n. (++)

1*

(XV.) 3,4,7,3′,4′,5′-Hexahydroxy-flavan
(Leucorobinetinidin-hydrat) (*44, 83 a, 88*). s. u. n. (++)

(XVI.) 3,4,7,8,3′,4′-Hexahydroxy-flavan
(Melacacidin) (*63*). n. (++)

(XVII.) 3,4,5,7,3′,4′-Hexahydroxy-flavan
(Leucocyanidin-hydrat) (*48, 49*). s. (++)

(XVIII.) Halbketal des 3-Oxo-5,7,2′,4′-tetrahydroxy-
flavans (Cyanomaclurin) (*2, 50, 77*). n. (++)

Formelübersicht 1. Hydroxy-flavane.

Inzwischen ist das 3,7,3′,4′,5′-Pentahydroxy-flavan (Robinetidinol, ++)
synthetisch gewonnen worden (*88*).

Hathway und Seakins (*57*) berichten über weitere synthetische Flavane:
5,7-Dihydroxy-flavan, Schmp. 196°; 3′,4′-Dihydroxy-flavan, Schmp. 132°; 5,7,3′,4′-
Tetrahydroxy-flavan, Schmp. 185°; 5,7-Dihydroxy-3′,4′-dimethoxy-flavan, Schmp.
260°; 3,3′,4′-Trihydroxy-5,7-dimethoxy-flavan (5,7-Dimethyläther des *d*-Catechins),
Schmp. 218—219°. Da das Verhalten gegen Säuren nicht angegeben ist, sind
diese Flavane nicht in der Übersicht angeführt. Das dritte dürfte gegen Säure empfindlich sein. Die Tri-, Tetra- und Penta-methyläther der Catechine (XI), (XII)
und (XIII) sind in der Übersicht nicht angeführt. Sie sind gegen Säure stabil,
ebenso das 5,7,3′,4′-Tetramethoxy-flavan (*32*) und der Methyläther von (VI) (*45*).

Die Hydroxy-flavane (I), (II), (XIV), (XV), (XVI) und (XVII) bilden
eine Untergruppe wegen des Hydroxyls in Stellung 4. Sie sind die
Hydrate der Leucoverbindungen der zugehörigen Flavyliumverbindungen.
Cyanomaclurin (XVIII) steht als Halbketal eines Ketons außerhalb
der Reihe der Catechine. Nur als Acetate bekannt und deshalb nicht
in die Übersicht aufgenommen sind die Acetoxy-flavane (*35*) (XIX)
und (XX), die aus Pelargonidin und Cyanidin hergestellt und neuerdings
aufgeklärt worden sind (*49*) (s. Abschnitte 5 und 6, SS. 9 und 12). Bersin,

(XIX.) R = H. 2,3,5,7,4′-Pentaacetoxy-flavan.
(XX.) R = OAc. 2,3,5,7,3′,4′-Hexaacetoxy-flavan.
(Ac = CH₃·CO)

MÜLLER und SCHWARZ (8) berichten von dem Glycosid eines Hepta-hydroxy-flavans (Pseudocyanidin-hydrats) aus *Crataegus oxyacantha*. Eine andere Gruppe bilden die 3-Hydroxy-flavane oder Catechine.

3. Eigenschaften, Herstellung der Hydroxy-flavane.

Die natürlichen Catechine (XI—XIII) sind farblose, kristalline Verbindungen. Sie sind mehr oder weniger wasserlöslich. Auf dem geringen Unterschied ihrer Wasserlöslichkeit beruht die Trennung der einzelnen aktiven und razemischen Formen (42). In Alkohol und Aceton sind alle leicht, in Äther sind die hydroxylreichen schwer löslich.

Die Ultraviolett-Spektren haben alle zwischen 270—285 mμ ein Maximum, und die Kurven steigen unterhalb von 245 mμ steil an. Zur weiteren Charakterisierung dienen die Acetyl- und Methylderivate, die meist gut kristallisieren.

Zur Papierchromatographie benutzt man als Lösungsmittel in erster Linie Butanol/Eisessig/Wasser (40 : 10 : 50). Für die Leucoanthocyanidin-hydrate hat sich besonders folgendes Gemisch als sehr geeignet erwiesen: 80 ccm *n*-Butanol werden mit 20 ccm Wasser gemischt. Man benutzt die organische Phase und mischt je 50 ccm mit 1 ccm Glykol. Leucocyanidin-hydrat (XVII) darf wegen seiner Empfindlichkeit nicht in sauren Lösungsmitteln chromatographiert werden. Zur Entwicklung dienen: a) diazotierte Sulfanilsäure (17); b) 2%ige alkoholische Eisen-III-chloridlösung; c) methanolische Vanillin-Salzsäure-Lösung: zu der kalten Lösung von 1 g Vanillin in 10 ccm Methanol läßt man langsam, unter Umschütteln 5 ccm konz. Salzsäure zutropfen (die Lösung ist nicht lange haltbar); d) 3%ige, äthanolische *p*-Toluolsulfonsäure-Lösung (83).

Zur Isolierung natürlicher Hydroxy-flavane wird das Pflanzenmaterial erst mit Benzol, dann mit Äther anhaltend extrahiert. Zur Kristallisation dient Wasser. Bei der Synthese müssen Säuren möglichst vermieden werden [s. dagegen HATHWAY (57)]. Folgende Wege sind bisher eingeschlagen worden: Hydrierung von Flavyliumsalzen (30, 36, 47), Hydrierung von Dihydroflavonolen, Leuco-anthocyanidin-hydraten (44, 48, 49, 62, 83a, 88), und Acetaten der Flavylium-pseudobasen (35, 49). Zur Isolierung werden häufig die Acetate hergestellt und mit Baryt, mit NaOCH$_3$ nach ZEMPLÉN oder mit Kaliumacetat in warmem Alkohol verseift.

4. Kondensation durch Säuren.

Aus den Catechinen, Leucoanthocyanidin-hydraten usw., können sich in der Pflanze die löslichen, schwer- und unlöslichen Gerbstoffe und Gerbstoffrote auf zwei Wegen bilden (19). Der eine Vorgang vollzieht sich unter Wirkung von Dehydrasen und verläuft nach Wegnahme von Wasserstoff aus Phenolgruppen spontan, ohne weitere Mitwirkung der Enzyme. Dieser Vorgang ist der Ligninbildung aus den Hydroxyzimt-alkoholen ähnlich. Behandelt man Catechine und Leucoanthocyanidin-hydrate mit Pilzlaccase, so erhält man rotbraune Produkte, die echten

Gerbstoffcharakter haben (49). Ein natürliches Beispiel hierfür ist das Cacaorot, das aus *l*-Epicatechin (28, 39) und außerdem vielleicht aus ähnlichen Stoffen entsteht. Doch liegen hierüber außer Versuchen von Hathway und Seakins (58) noch zu wenige Untersuchungen vor, um bessere Aussagen zu machen.

Der zweite Vorgang ist postmortal und besteht in der Polymerisation unter Einwirkung von Feuchtigkeit und Säure bei langer Zeitdauer. Dieser Vorgang, der sich in der Natur wohl vor allem im Kernholz abspielt, soll im folgenden näher erläutert werden.

Aus der Übersicht geht hervor, daß rasche Kondensation an die Anwesenheit von je einem Hydroxyl in 7- und 4'-Stellung gebunden ist. Somit ist das 7,4'-Dihydroxyflavan (V) das einfachste selbstkondensierende Flavan (47). Langsame Kondensation tritt zwar noch ein, wenn ein Methoxyl in 4'-Stellung und zugleich ein Hydroxyl in 7-Stellung (VI) vorhanden ist (45). Bei weiterer Abdeckung der OH-Gruppen durch Methyl (7,4'-Dimethoxy-flavan) oder bei Weglassen einer dieser Gruppen (II und IV) bleibt die Kondensation ganz aus. Daraus kann der Schluß gezogen werden, daß für die Kondensation durch Säure erforderlich ist: a) Die Gruppierung eines *p*-Hydroxy-benzyläthers (OH in 4' Stellung); b) ein Resorcylsystem (OH in 7). Eines von beiden Hydroxylen darf veräthert sein, aber nicht beide.

Die Hydroxyflavane reagieren also als bifunktionelle Moleküle. Formal besteht der Vorgang in der Öffnung des Benzopyranringes und der Kondensation des C-Atoms 2 mit der Gruppe 6 oder 8 eines anderen Moleküls. Spezieller läßt sich der Mechanismus etwa folgendermaßen darstellen *(Formelübersicht 2)*: Ein Proton lagert sich an das Elektronenpaar des Brückensauerstoffs, wobei sich das Oxoniumion (XXI) bildet. Unter dem aktivierenden Einfluß der *p*-ständigen Hydroxylgruppe wird die Benzylätherbindung unter Bildung eines Carboniumions (XXII) aufgespalten. Dieses reagiert mit seinem positiven C-Atom mit der negativen Stelle eines Resorcylkerns, der einem nicht protonierten Molekül angehört; nach Abspaltung eines Protons bildet sich ein benzoides System (XXIII) zurück. Die Ionenverbindungen (XXI) und (XXII) sind nicht isoliert; (XXIII) stellt das vermutete Zweierstück der Polymerisationsreihe von (V) dar.

Aus allem, was wir aus Modellversuchen wissen, die in den folgenden Abschnitten weiter ausgeführt werden, ist dieser Reaktionsmechanismus gut begründet. Die früher für möglich gehaltene Linearpolymerisation (22) kann ausgeschlossen werden.

Für die Bildung des Oxoniumsalzes (XXI) spricht auch die Beobachtung, daß die Drehung des *d*-Catechins in Dioxan ($[\alpha]_D^{25} = + 18°$) nach Zugabe von wenig mit Chlorwasserstoff gesättigtem Dioxan stark negativ wird. Nach sofortiger Neutralisierung dieser Lösung wird das

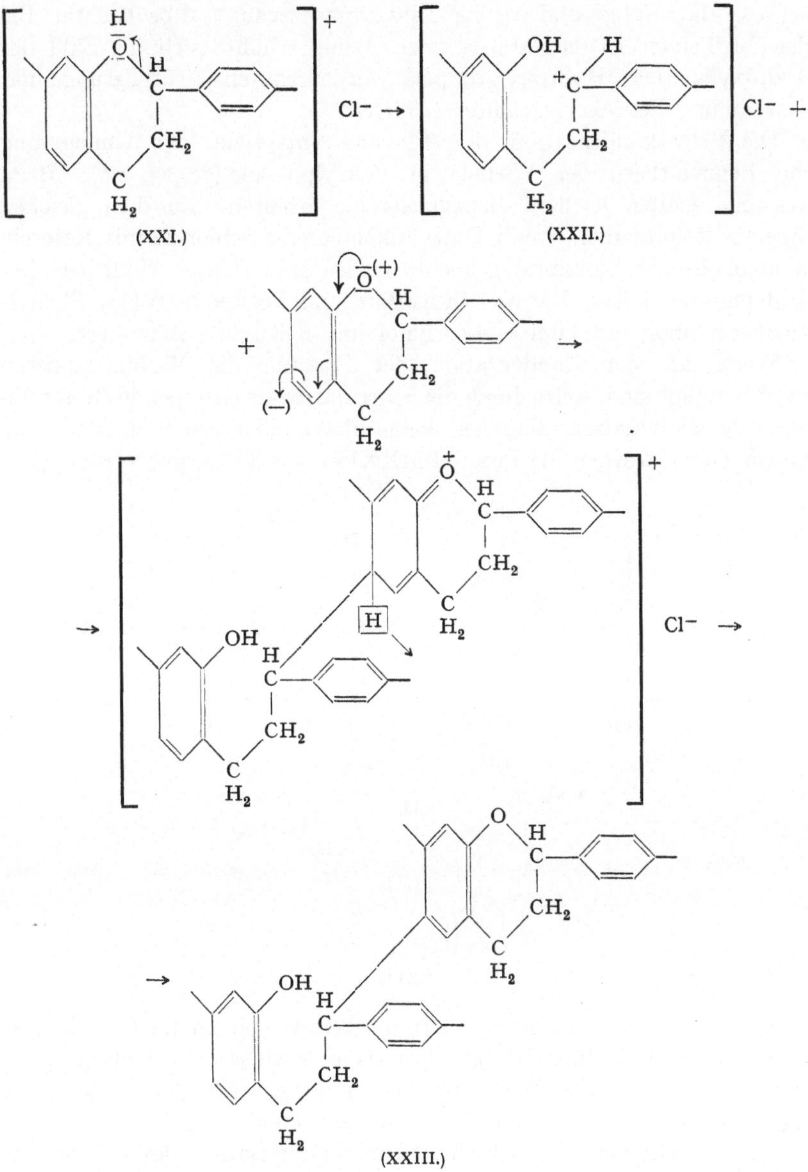

Formelübersicht 2. Selbstkondensation von Hydroxyflavanen.

d-Catechin zurückgewonnen (*26*). *l*-Epicatechin zeigt auch eine starke Linksverschiebung (gleiche Konfiguration am $C_{(2)}$).

Die Öffnung des Pyranringes und damit die Bildung zusätzlichen Hydroxyls wurde bei der Kondensation des Fisetinidols (3,7,3′,4′-Tetrahydroxy-flavan) (*36*) und des *7,4'*-Dihydroxy-flavans (*45*) beobachtet. Bei gleichbleibender Elementarzusammensetzung erhöhte sich die Zahl der Hydroxyle. Die Hydroxylgruppen wurden durch Acetylierung und Bestimmung des Acetylgehaltes ermittelt.

Die Selbstkondensation der Flavane entspricht der Umlagerung von Benzyläthern der Phenole in Benzylphenole (*7*, *13*, *26*). Beim Resorcin werden leicht C-Benzylresorcine erhalten. Zu dem gleichen Ergebnis kommt man, wenn Benzylalkohole oder -chloride mit Resorcin in alkoholischer Salzsäure gekocht werden (*26*). Unter ähnlichen Gesichtspunkten haben Brown, Cummings und Somerfield (*15*) Phenyl-benzyl-carbinol und Diphenyl-carbinol mit Resorcin kondensiert.

Wenn an der Kondensation der Flavane die Methingruppen 6 und 8 beteiligt sind, sollte durch die Sperrung dieser Gruppen im Resorcinkern von Polyhydroxy-flavanen keine Polykondensation eintreten. Zu diesem Zweck wurden die Produkte (XXIV)—(XXVI) synthetisiert (*26*).

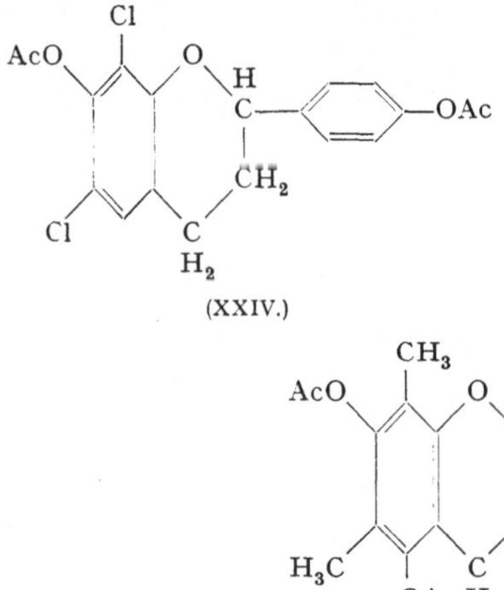

(XXIV.) 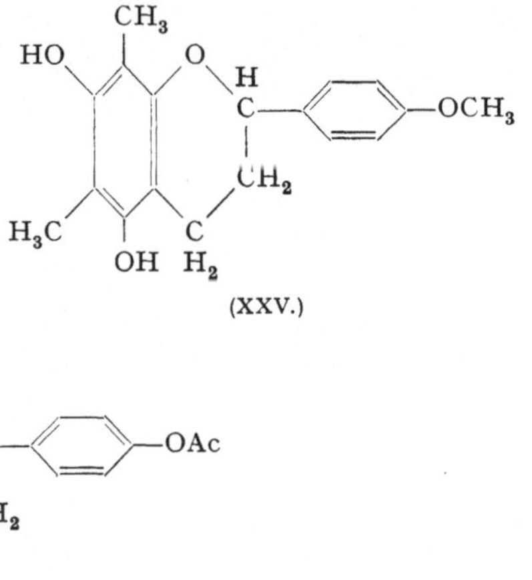   (XXV.)

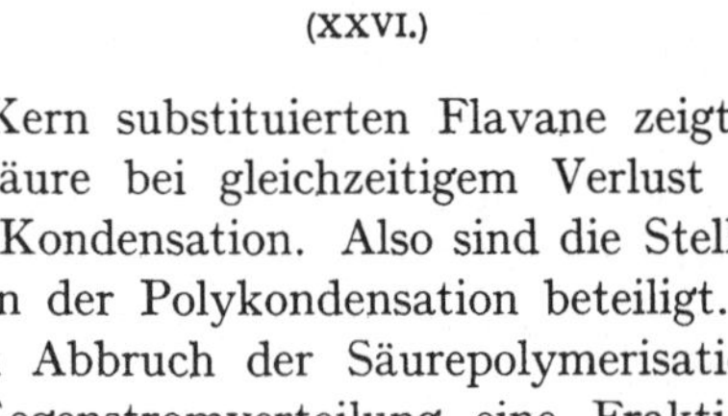

(XXVI.)

Diese drei im Kern substituierten Flavane zeigten auch in heißer, alkoholischer Salzsäure bei gleichzeitigem Verlust der Acetylgruppen keine Neigung zur Kondensation. Also sind die Stellungen 6 oder 8 der Hydroxy-flavane an der Polykondensation beteiligt.

Bei vorzeitigem Abbruch der Säurepolymerisation des *d*-Catechins läßt sich durch Gegenstromverteilung eine Fraktion fassen, die das kristalline Acetat eines dimeren Catechins liefert. Dieses Acetat dreht stark positiv und hat doppeltes Molekulargewicht (*26*). Nach der obigen Ausführung kommt dem Dicatechin die Formel (XXVII) zu, oder eine entsprechende mit Bindung von 2 nach 8. Auch kann Wasser ausgetreten sein.

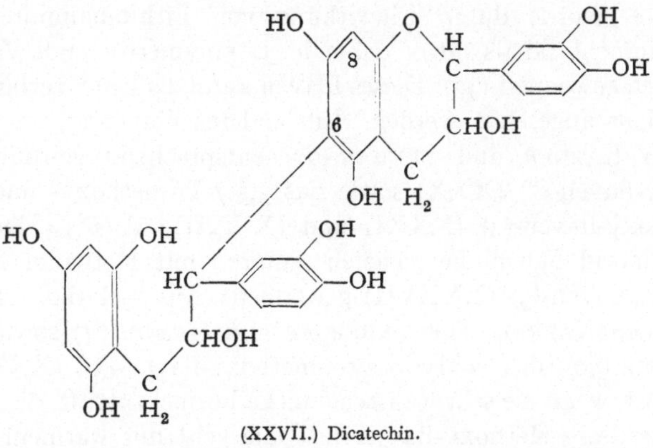

(XXVII.) Dicatechin.

Die optische Aktivität des dimeren Acetats und die Stabilität der kernsubstituierten Polyhydroxy-flavane gegen Säuren sind Argumente für die Kondensation nach Art von (XXIII) und (XXVII).

Die Hydroxyflavane selbst sind keine Gerbstoffe im technischen Sinn. Echte Gerbstoffe erkennt man an ihrer Gelatinefällung:

1 g gute Blattgelatine läßt man in 200 ccm Wasser 24 Stunden lang quellen, dann wird gelinde erwärmt, bis sich die Gelatine löst. Jetzt versetzt man die zu untersuchende Lösung (1%) mit kalter Gelatinelösung. Tritt eine Fällung ein, so liegt ein Gerbstoff vor.

Gerbstoffreaktion zeigt z. B. schon das dimere Catechin. Hier ist gegenüber dem Catechin die Zahl der Phenolhydroxyle vermehrt, und diese Hydroxyle gehören zu einem größeren Molekül. Je größer das Molekül, um so schwerer löslich wird es, wenn es kristallisierte. Da aber die Neigung zur Kristallisation äußerst gering ist, bilden sich übersättigte Lösungen und wird leichte Löslichkeit in Wasser vorgetäuscht. Diese potentielle Unlöslichkeit macht die Substanz zum Gerbstoff (*19*).

Besonders leicht bilden die Leucoanthocyanidin-hydrate (z. B. XIV—XVII) durch Säuren Gerbstoffe und Phlobaphene (*88*). Verbindung (XIV) ist neuerdings auch aus Quebrachoholz isoliert worden (*83a, 88*) und ist der oder einer der Stammkörper des aus diesem Holze gewonnenen Gerbstoffs.

5. Hydroxy-flavene.

Außer den Catechinen (Hydroxyl in 3-) haben wir in den vorangehenden Abschnitten auch diejenigen Hydroxy-flavane erwähnt, die neben anderen Hydroxylen eines in 2- oder 4-Stellung besitzen. Wegen ihrer Beziehung zu den Flavyliumsalzen nehmen sie unter den Hydroxy-flavanen eine Sonderstellung ein. Sie werden im Abschnitt 6 behandelt (S. 12).

Zu ihrem Verständnis müssen die Flavene behandelt werden (*Formelübersicht 3*, S. 11). Als Stammsubstanzen sind Flaven-(2) und Flaven-(3) (XXVIII) und (XXIX) anzusehen. Eines von beiden, wahrscheinlich das letztere, wurde durch Einwirkung von Lithiumaluminiumhydrid auf Flavyliumchlorid als autoxydable, leicht polymerisierende Verbindung kristallinisch gewonnen (47). Dieses Flaven kann als Leucoverbindung des Flavyliumions angesehen werden. Ein anderes Flaven-(3) ist das schon früher von Karrer und Seyhan (60) entsprechend gewonnene 3,4'-Dimethoxy-flaven-(3) (XXX) sowie das 3,5,7-Trimethoxy- und 3,5,7,4'-Tetramethoxy-flaven-(3) (XXXI) und (XXXII). Aus 3,4'-Dimethoxy-flavyliumchlorid haben die gleichen Autoren mit Methanol das 2,3,4'-Trimethoxy-flaven-(3) (XXXVII) gewonnen. Aus 3-Methoxy-flavylium-chlorid haben Karrer, Trugenberger und Hamdi (61) sowie Karrer und Fatzer (59) das 2-Hydroxy-3-methoxy-flaven-(3) (XXXIII) bereitet. Es ist, wenn die sehr wahrscheinliche Formel zutrifft, die kristalline Pseudobase des 3-Methoxy-flavyliums. Sie geht mit warmem Methanol oder Äthanol in ihre Äther über (2-Methoxy- bzw. Äthoxy-3-methoxy-flaven-(3) (XXXIV) und (XXXV). Ihnen wurde später (47) das zugehörige Acetat (XXXVI) hinzugefügt, das mit Pyridin/Acetanhydrid aus 3-Methoxy-flavyliumchlorid entsteht.

Dieses Verfahren haben bereits 1935 Freudenberg, Karimullah und Steinbrunn (35) auf Flavyliumsalze angewendet. Aus Pelargonidin entsteht das 3,5,7,4'-Pentaacetoxy-flaven-(3) (XXXVIII), aus Cyanidin das 2,3,5,7,3',4'-Hexaacetoxy-flaven-(3) (XXXIX). Die Konstitution dieser Verbindungen wird weiter unten bewiesen. Sie sind die Acetate der Pseudobasen der zugehörigen Anthocyanidine. Durch Hydrierung gehen sie in die entsprechenden Acetoxy-flavane (XIX) und (XX) über.

Aus der Klasse der Flavene-(2) scheint bislang nur ein Vertreter bekannt zu sein, das 5,7,3',4'-Tetramethoxy-flaven-(2) (XL). Man kann es zum 5,7,3',4'-Tetramethoxy-flavan hydrieren (29, 31). Mit Chlorwasserstoff gibt das Tetramethoxy-flaven-(2) unter gleichzeitiger Oxydation (3) das zugehörige Tetramethoxy-flavyliumchlorid.

Man kann, wie oben geschehen, das Flaven-(3) (XXIX) als die Leuco-verbindung des Flavyliumchlorids und das 2-Hydroxy-flaven-(3) (XLI und z. B. XXXIII) als eine Pseudobase ansprechen. In diesem Fall zeichnet sich der Unterschied zwischen Leuco- und Pseudoverbindung am Kohlen-stoffatom 2 ab. Bei den Flavenen-(2) (XXVIII) könnte man mit dem gleichen Recht von Leucoverbindungen sprechen, wenn das Kohlenstoff-atom 4 mit zwei Wasserstoffatomen besetzt ist, und von einer Pseudobase, wenn es mit einem Wasserstoffatom und einem Hydroxyl versehen ist (XLII).

Wenn wir die Bezeichnung Leuco- und Pseudo- verwenden, so beziehen wir sie jedoch, wie schon oben geschehen, ausschließlich auf das Kohlenstoffatom 2. Andernfalls würde die Nomenklatur zu unübersichtlich. Am besten ist es, auf die Stammverbindung Flavan oder Flaven zurückzugreifen und den Sitz der Hydroxyle anzugeben. Leuco-

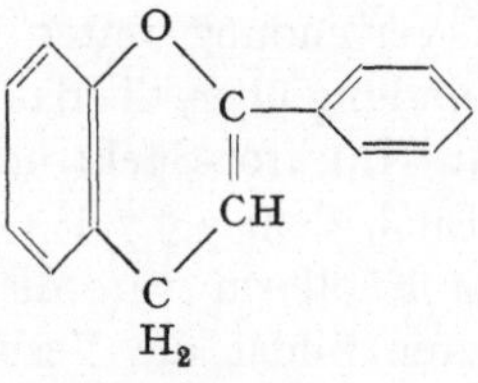

(XXVIII.) Flaven-(2).

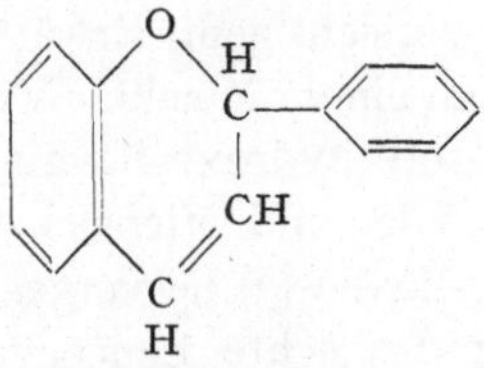

(XXIX.) Flaven-(3).

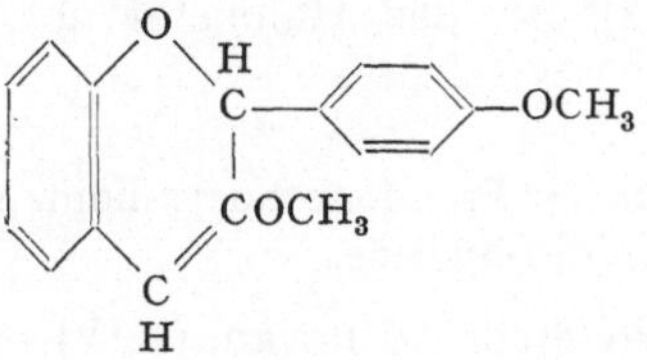

(XXX.) 3,4'-Dimethoxy-flaven-(3).

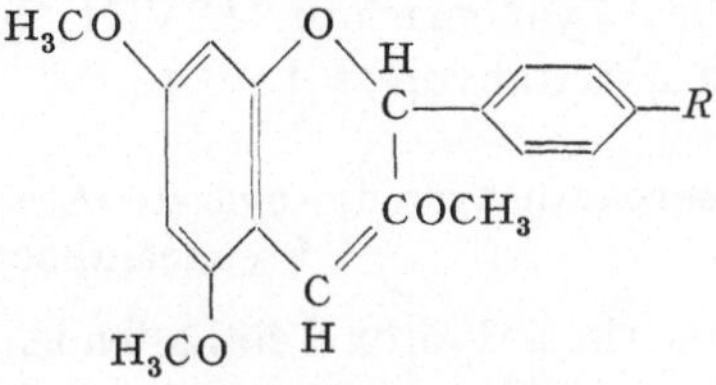

(XXXI.) $R = H$. 3,5,7-Trimethoxy-flaven-(3).
(XXXII.) $R = OCH_3$. 3,5,7,4'-Tetramethoxy-flaven-(3)
(Tetramethyl-leucopelargonidin).

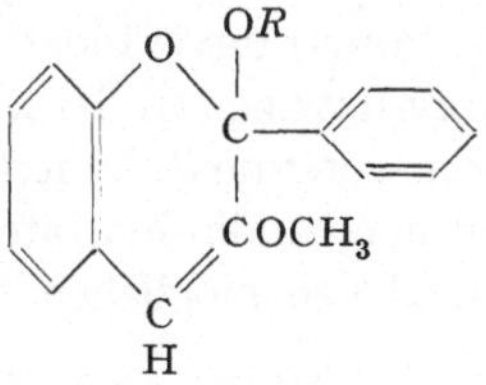

(XXXIII.) $R = H$. 2-Hydroxy-3-methoxy-flaven-(3).
(XXXIV.) $R = CH_3$. 2,3-Dimethoxy-flaven-(3).
(XXXV.) $R = C_2H_5$. 2-Äthoxy-3-methoxy-flaven-(3).
(XXXVI.) $R = COCH_3$. 2-Acetoxy-3-methoxy-flaven-(3).

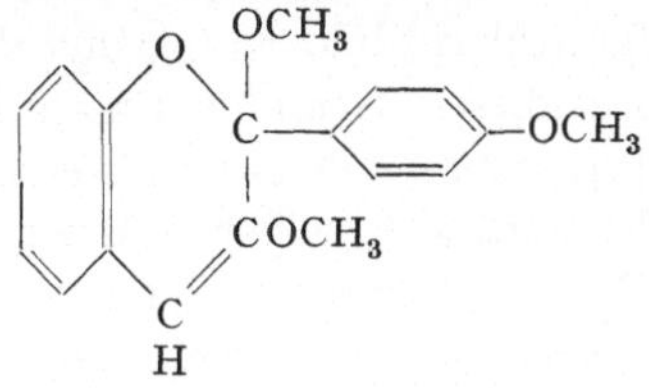

(XXXVII.) 2,3,4'-Trimethoxy-flaven-(3).

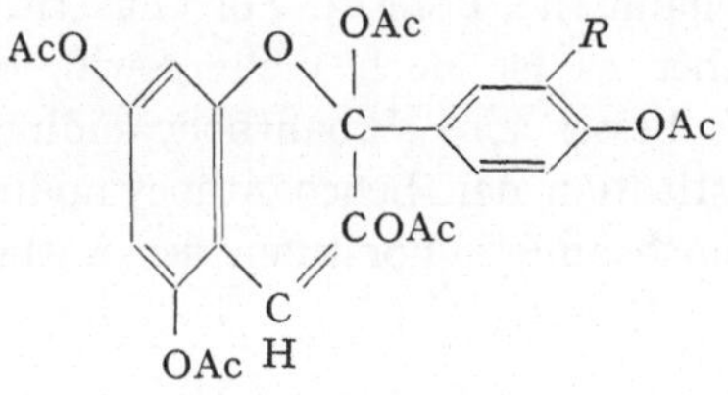

(XXXVIII.) $R = H$. 2,3,5,7,4'-Pentaacetoxy-flaven-(3)
(Acetat des Pseudopelargonidins).
(XXXIX.) $R = OCOCH_3$. 2,3,5,7,3',4'-Hexaacetoxy-flaven-(3)
(Acetat des Pseudocyanidins).

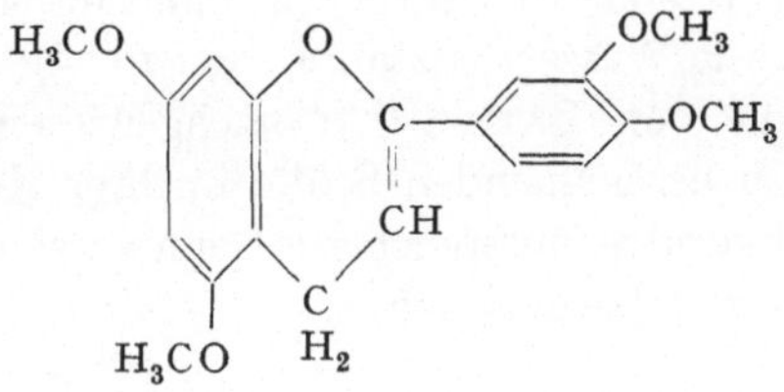

(XL.) 5,7,3',4'-Tetramethoxy-flaven-(2).

Formelübersicht 3. Flavene.

cyanidin ist unter Beachtung dieser Beschränkung das 3,5,7,3',4'-Penta-
hydroxy-flaven-(3). Das oben bereits angeführte 3,4,5,7,3',4'-Hexa-
hydroxy-flavan (XVII) ist dementsprechend als Leucocyanidin-hydrat
zu bezeichnen (S. 4).

Es ist verwirrend, eine solche Verbindung Leucocyanidin zu nennen. Auch
Cyanidiol oder ähnliches schafft nur Verwirrung.

Bei vorsichtigem Erwärmen geht diese Verbindung unter Verlust von genau einem Molekül Wasser in eine Verbindung über, die das 3-Oxo-5,7,3′,4′-tetrahydroxy-flavan zu sein scheint (Infrarot-Spektrum zeigt C=O-Bande) und offensichtlich aus einem Enol, dem 3,5,7,3′,4′-Pentahydroxy-flaven-(3) hervorgegangen ist. In der Enolform wäre diese Verbindung das echte Leucocyanidin. Mit Säuren bildet die Verbindung das rote Cyanidinsalz und mit Alkalien die blaue Phase.

Das Cyanomaclurin (XVIII, S. 4) ist das Halbketal des 3-Oxo-5,7,2′,4′-tetrahydroxy-flavans.

6. Leucoanthocyanidin-hydrate, Acetate der Pseudoanthocyanidine und der Pseudoanthocyanidin-hydride.

Die ein 4-Hydroxyl enthaltenden Polyhydroxy-flavane (XIV)—(XVII) (SS. 3—4) bilden mit Säuren neben den oben erwähnten Polykondensationsprodukten die entsprechenden Anthocyanidine (*11, 44, 48, 62, 63, 64, 85*), während das 4,7-Dihydroxy-flavan (II, S. 2) nur das 7-Hydroxy-flavyliumchlorid bildet (*24*), und das 4-Hydroxy-flavan (I) unverändert bleibt. Aus diesem Grund will man diesen Polyhydroxy-flavanen in der lebenden Pflanzenzelle den Charakter einer Anthocyanidin-vorstufe (*80*) zuerkennen. Es wird sich zeigen müssen, ob den Leucoanthocyanidin hydraten diese Rolle zukommt und ihnen unter den verschiedenen möglichen anderen Formen allein vorbehalten ist.

Bate-Smith (*4—6*) hat sich mit der systematischen Verbreitung von „Leucoanthocyanidinen" im Pflanzenreich beschäftigt. Er kocht hierzu Pflanzenextrakte in 2 *n*-Salzsäure und chromatographiert die entstandenen Anthocyanidine im sogenannten „Forestal Solvent" (Eisessig/Wasser/Salzsäure, 30 : 10 : 3). Aber es ist noch nicht gewiß, ob die von Bate-Smith nachgewiesenen, besser als „Proanthocyanidine" zu bezeichnenden Stoffe in ihrer Konstitution den Leucoanthocyanidinhydraten entsprechen. Denn es sind noch andere Vorstufen der Anthocyanidine möglich.

Tabelle 1. Natürliche Dihydro-flavonole.

Name	Literatur
Dihydrofisetin (3,7,3′,4′-Tetrahydroxy-flavanon-4)	Fustin (*72*)
Dihydrorobinetin (3,7,3′,4′,5′-Pentahydroxy-flavanon-4)	— (*33*)
Dihydrogalangin (3,5,7-Trihydroxy-flavanon-4)	Pinobanksin (*68*)
Dihydrokämpferol (3,5,7,4′-Tetrahydroxy-flavanon-4)	Katsuranin (*68*)
Dihydromorin (3,5,7,2′,4′-Pentahydroxy-flavanon-4)	— (*16*)
Dihydroquercetin (3,5,7,3′,4′-Pentahydroxy-flavanon-4)	Taxifolin (*67, 78*) Distylin*
Dihydromyricetin (3,5,7,3′,4′,5′-Hexahydroxy-flavanon-4) . . .	Ampelopsin (*73*)

  * Diese Bezeichnung ist überflüssig.

Die Synthese der Leucoanthocyanidin-hydrate (Polyhydroxy-flavan-3,4-diole) geht von den entsprechenden Dihydro-flavonolen (Flavanonolen) aus, die in der Natur verbreitet sind. Seit der Isolierung des Dihydro-fisetins (Fustin; 3,7,3',4'-Tetrahydroxy-flavanon-4) durch OYAMADA (72) hat sich auch in dieser Gruppe die Zahl sehr vergrößert, wie *Tabelle 1* zeigt.

Aus den Infrarot-Spektren der Dihydroflavonole geht hervor, daß diejenigen, die in der 5-Stellung keine OH-Gruppe besitzen, die Bande der Ketogruppe 4 deutlich zeigen. Wenn das Hydroxyl 5 vorhanden ist, rückt die Bande nach längeren Wellen und verbreitert sich (Wasserstoffbindung). Inzwischen scheint sich zu bestätigen, daß sich die Dihydroflavonole ohne Wasserstoffbindung leicht zu Leucoanthocyanidin-hydraten (XIV), (XV) hydrieren lassen (44, 62), während die ein 5-Hydroxyl tragenden nur nach Aufhebung der Brückenbindung durch die Benzylierung der Phenolgruppen, anschließender Reduktion mit LiAlH$_4$ und vorsichtiger Abhydrierung der Benzylgruppen in das Leucoanthocyanidin-hydrat (XVII, S. 4) überführt werden können (48, 49).

Wie erwähnt, ist (XVII) äußerst empfindlich gegen Säuren, von denen es zur Selbstkondensation veranlaßt wird; es darf daher nicht mit Lösungsmitteln chromatographiert werden, die Säure enthalten. Ob die Kondensation der des Catechins entspricht, wissen wir nicht. In jedem Falle entstehen polymere Kondensate, die im Chromatogramm nicht oder wenig wandern und dennoch die Anthocyanidinfärbung zeigen, weil am Ende eines jeden kondensierten Moleküls eine zur Flavyliumsalzbildung befähigte Gruppe steht.

Aber noch aus einem anderen Grunde ist die Darstellung der Leucoanthocyanidin-hydrate von Interesse. FREUDENBERG, KARIMULLAH und STEINBRUNN (35, s. auch 84) erhielten 1935 bei der Acetylierung von Anthocyanidinen mit Essigsäureanhydrid und Pyridin Acetate, die eine Acetoxygruppe mehr enthielten, als nach der Zahl der Hydroxylgruppen im ursprünglichen Anthocyanidin zu erwarten war. Es wurden damals mehrere Möglichkeiten der Konstitution diskutiert: a) Ein Acetat wie (XXXVIII) und (XXXIX), abgeleitet vom Flaven-(3), bei dem das zusätzliche Hydroxyl in 2-Stellung steht. Die Stammsubstanz ist die Flavylium-pseudobase (XLI). b) Eine von der isomeren Pseudobase (XLII)

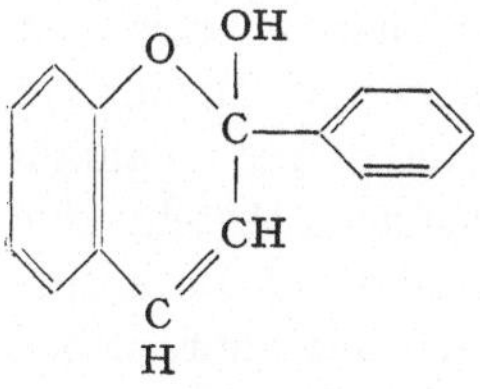

(XLI.) 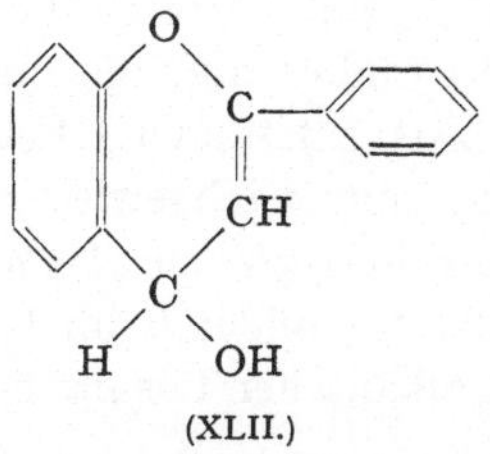   (XLII.) 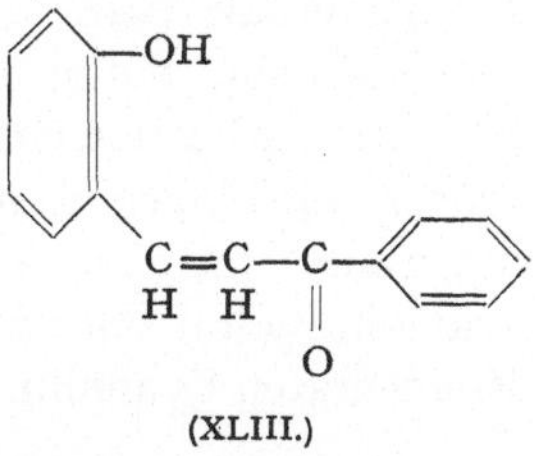   (XLIII.)

abgeleitete Flavenstruktur, bei der das zusätzliche Hydroxyl in 4-Stellung steht. c) Ein Chalkon (XLIII) mit möglicher *cis-trans*-Isomerie.

Eine Entscheidung konnte damals nicht getroffen werden. Ähnliche Ergebnisse erhielten Brockmann und Junge (*14*) bei der Acetylierung der Anhydrobasen des 7-Hydroxy-, des 5,7-Dihydroxy- und des 7,8-Dihydroxy-2,4-diphenylbenzopyranols. Auch sie konnten keine Entscheidung treffen.

Diese Acetate ließen sich nicht zu den entsprechenden Catechinen hydrieren, sondern lieferten teils amorphe, teils kristalline Hydrierungsprodukte, wobei die Zusammensetzung der kristallinen Produkte einem Mehrgehalt von einem Mol Wasserstoff entsprach. Es mußten also die Acetate der Pseudoanthocyanidin-hydride (XLIV), der Leucoanthocyanidin-hydrate (I) oder der Hydrochalkone (XLV) entstanden sein.

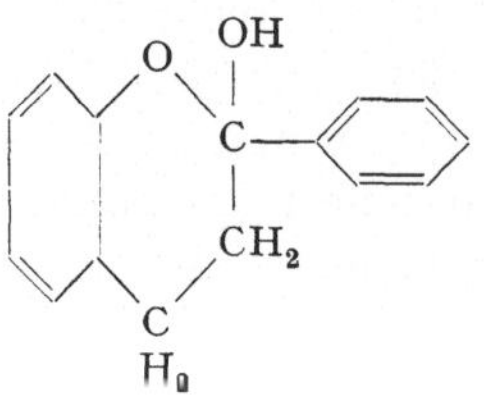

(XLIV.) 2-Hydroxy-flavan. 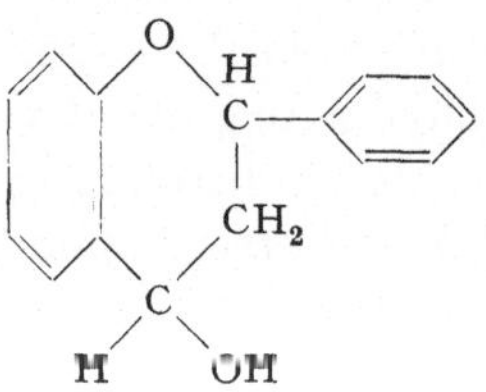   (I.) 4-Hydroxy-flavan. 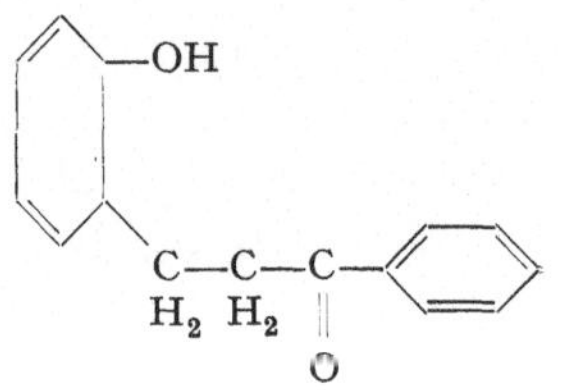   (XLV.) 2′-Hydroxy-hydrochalkon

Auf Grund ihrer Infrarot-Spektren konnte zunächst Hydrochalkon (C=O-Bande) und Flavan unterschieden werden (*49*). Dabei ergab sich, daß die hydrierten Acetate aus 4′-Hydroxy-, 4′-Methoxy-, 4′-Hydroxy-7-methoxy- und 7,4′-Dimethoxy-flavyliumsalz eine Hydrochalkon-Struktur besitzen, während die aus Cyanidin- und Pelargonidinchlorid Flavan-Struktur haben. Das hydrierte Acetat aus Cyanidin erwies sich verschieden von dem aus Dihydroquercetin (Taxifolin) erhaltenen Acetat des Leucocyanidin-hydrats (XVII, S. 4), so daß man für das hydrierte Acetat aus Cyanidin die Struktur des 2,3,5,7,3′,4′-Hexahydroxy-flavans (XX, S. 4) annehmen muß.

Ein wesentlicher Unterschied vom Leucocyanidin-hydrat zeigt sich — außer den Infrarot-Spektren und Schmelzpunkten — bei der Behandlung dieser beiden Acetate mit Säure. Die Lösungen der *Acetate (XIX) und (XX) werden mit Säure blau.* Bei Zugabe von Alkali zu dieser blauen Lösung erhält man zunächst eine farblose Lösung; nach kurzer Zeit tritt die blaue Farbe des Cyanidins auf, die nun in saurem Medium in die rote Farbe übergeht. Das entstandene Pelargonidin und Cyanidin wurden im Papierchromatogramm nachgewiesen. Demnach besteht eine mit Säure *sich blau färbende Vorstufe der Anthocyanidine.* Leucoanthocyanidin-hydrat oder sein Acetat bildet beim Erhitzen mit Säure neben Kondensaten Cyanidin. In alkalischer Lösung erwärmt, gibt es die blaue

Phase des Cyanidins. Wenn Leucocyanidin-hydrat (XVII, S. 4) dagegen *auf dem Papier* mit Alkali besprüht und auf 80° erwärmt wird, nimmt es eine rotviolette Farbe an, die mit Säure in *Blau übergeht*. Also auch hier ein blaues Kation. Cyanomaclurin verhält sich in Lösung und auch auf dem Papier wie das Leucocyanidin-hydrat.

Zusammenfassend läßt sich über die Acetate aus Flavyliumsalzen aussagen: Cyanidin- und Pelargonidin-chlorid, also Flavyliumsalze, die in 3- und 5-Stellung eine Hydroxylgruppe haben, ergeben bei ihrer Acetylierung Acetate der Pseudobasen vom Typus (XLI), die bei der Hydrierung die Acetate (XIX) und (XX, S. 4), abgeleitet vom 2-Hydroxy-flavan (XLIV), ergeben. Die zweite Gruppe bilden die Flavyliumsalze, die weder in 5- noch in 3-Stellung eine OH-Gruppe haben. Sie bilden Acetoxy-chalkone und nach der Hydrierung Acetoxy-hydrochalkone. Eine dritte Gruppe besteht aus Acetaten, die bisher keine kristallinen Hydrierungsprodukte lieferten und deren Struktur deshalb noch nicht feststeht; doch kann man aus den Ultraviolett-Spektren mit einer gewissen Wahrscheinlichkeit auf eine Chalkonstruktur schließen (47). In diese Gruppe fallen die Flavyliumsalze, die einen Resorcylkern haben: 3,7,3',4'-Tetrahydroxy-flavylium-chlorid (Fisetinidin-chlorid), 7-Hydroxy-, 7-Hydroxy-4'-methoxy- und 7,3',4'-Trihydroxy-flavylium-chlorid.

7. Oxydations- und Reduktionsübergänge in der $C_6C_3C_6$-Reihe.

In der Gruppe $C_6C_3C_6$ sind die verschiedenen Oxydationsstufen durch einen in vitro realisierbaren Reduktions-Oxydations-Mechanismus verknüpft. In *Formelübersicht 4* (S. 16) sind die vollzogenen Übergänge der meistuntersuchten und in der Natur am wichtigsten Quercetinreihe angegeben.

Die für die Catechinchemie interessierenden Übergänge sind schon in den vorangehenden Kapiteln näher ausgeführt worden, lediglich die Oxydation des Catechins (XII) zum Cyanidin muß noch erwähnt werden. Appel und Robinson (1) konnten aus dem Tetramethyl-d-catechin durch Brom und anschließende Abspaltung der Methylgruppen mit HJ kristallines Cyanidin erhalten. *d*-Taxifolin wird über *d*-Leuco-cyanidin-hydrat zu *d*-Catechin reduziert, woraus die Konfiguration des Taxifolins hervorgeht (48, 49).

Der Übergang von Cyanidin zu Leucocyanidin ist noch nicht bewerkstelligt worden; dagegen haben Karrer und Seyhan (60) das Tetramethyl-pelargonidin in das Tetramethyl-leuco-pelargonidin (XXXII) verwandelt.

8. Cyanomaclurin.

Cyanomaclurin (XVIII, S. 4) ist bisher das einzige natürliche Leucoanthocyanidin. Auf der gleichen Oxydationsstufe der C_3-Kette

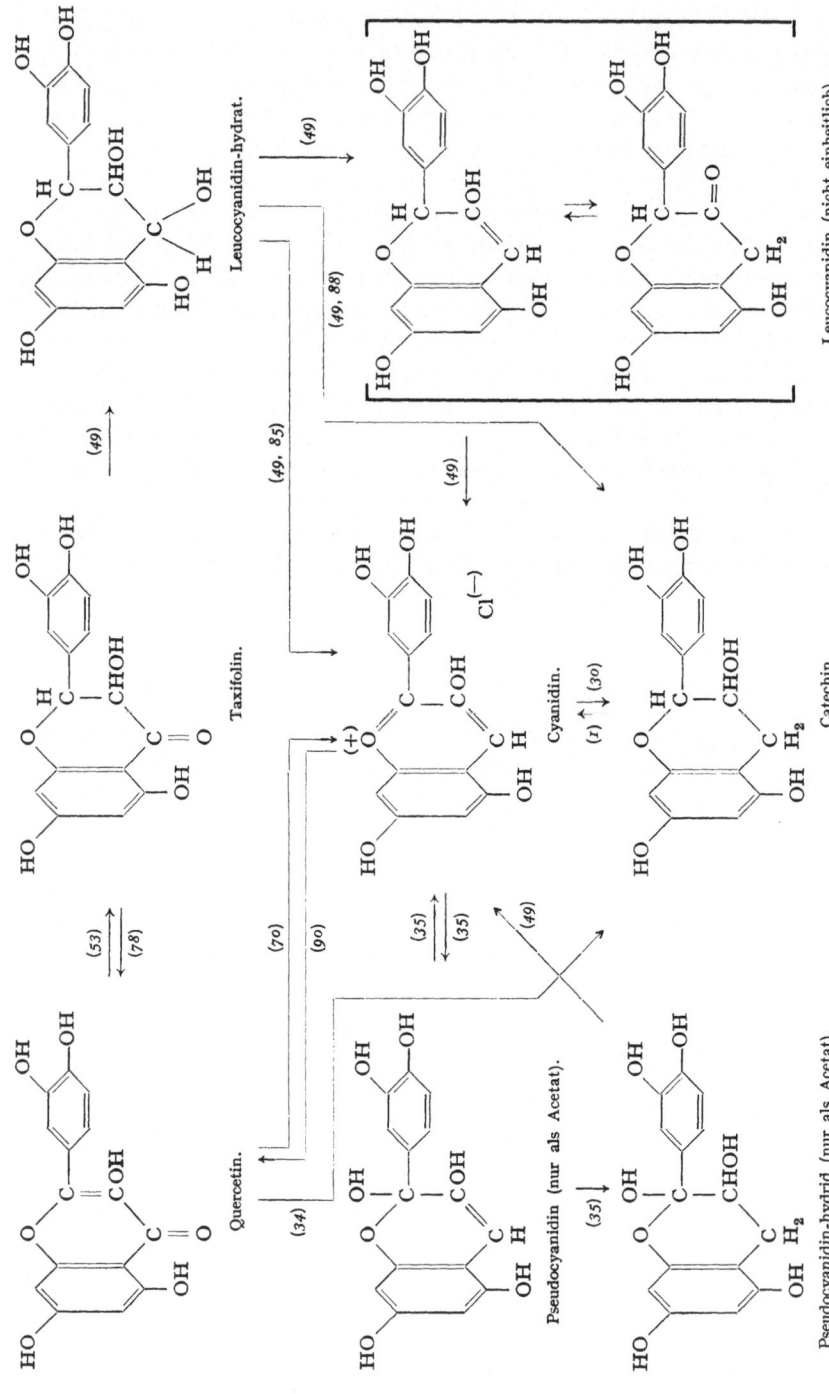

Formelübersicht 4. Oxydations- und Reduktionsübergänge. (Die Zahlen sind Literaturhinweise.)

stehen die Naturstoffe Mollisacacidin (XIV), Leucorobinetinidin-hydrat (XV) und Melacacidin (XVI, S. 3, 4). Das Cyanomaclurin geht mit Alkali und anschließender Säurebehandlung in das 3,5,7,2′,4′-Pentahydroxy-flavyliumchlorid (Morinidin) über, das auf diesem Wege von APPEL und ROBINSON (2) in kristalliner Form isoliert wurde. Bei direkter Säure-behandlung entsteht aus Cyanomaclurin ein Polymerisat. Es zeigt also ähnliche Eigenschaften wie die Leucoanthocyanidin-hydrate.

PERKIN (76, 77) isolierte das Cyanomaclurin aus dem Holz des vorder-indischen Jackbaumes *(Artocarpus integrifolia)* in optisch aktiver Form. Als Begleiter konnte jetzt das hierzu gehörige Razemat gefunden werden (49, 50), das auf Grund der verschiedenen Wasserlöslichkeit vom aktiven Cyanomaclurin getrennt werden kann.

Durch die Überführung des Cyanomaclurins in das Morinidin war der Phloroglucin- und Resorcylkern eindeutig nachgewiesen und lediglich der heterocyclische Ring bedurfte einer Aufklärung. APPEL und ROBIN-SON (2) erörterten die Keto- (XLVI) und die Halbketal-Struktur (XVIII, S. 4) und gaben letzterer den Vorzug.

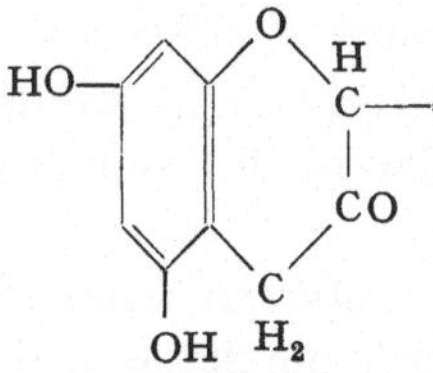

(XLVI.)

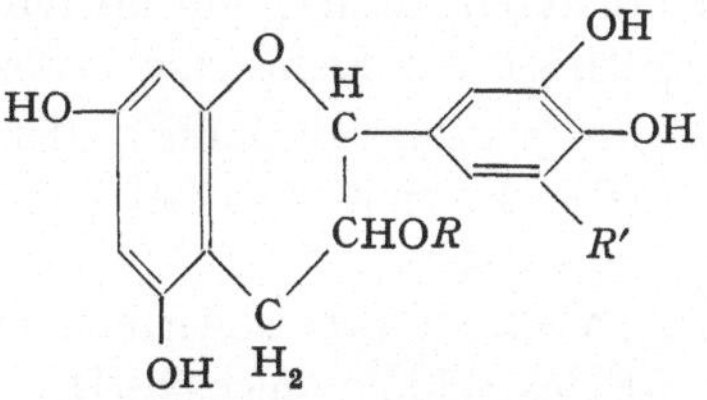

(XLVII.) R = Galloyl, R' = H.
(XLVIII.) R = Galloyl, R' = OH.
(XLIX.) R = Glucosyl, R' = H.

Die Tatsache (49, 50), daß das Infrarot-Spektrum des kristallinen Tetra-acetyl- und kristallinen Tetramesyl-cyanomaclurins keine Carbonyl-bande zeigt, beweist, daß die Ketoformel ausgeschlossen werden muß. Das kristalline Trimethylmonoacetat beweist die Sonderstellung einer Hydroxylgruppe. Demnach trifft die von APPEL und ROBINSON (2) aufgestellte Halbketalform zu. Die zu erwartende Mutarotation fällt aus, weil der starre 5-Ring nur in *cis*-Stellung an den teilweise starren 6-Ring angebaut sein kann.

Es ist sehr wahrscheinlich, daß weitere 3-Oxoflavane in der Natur vorkommen, die sich aus den Leucoanthocyanidin-hydraten durch Wasserabspaltung und Umlagerung bilden können, aber voraussichtlich empfindlich gegen Sauerstoff sind.

9. Catechin-3-gallat und Catechin-3-glucosid.

Den Catechinen kommt noch eine andere Bedeutung für die Gerb-stoffchemie zu, die zu den Depsiden EMIL FISCHERS zurückführt. Im

Tee fand Tsujimura (*86*) das 3-Galloyl-*l*-epicatechin (XLVII). Später isolierten Bradfield und Penney (*12*), gleichfalls aus Tee, den entsprechenden Gallussäureester (XLVIII) des *l*-Epigallocatechins. Letztere verwendeten zur Auftrennung der im Tee enthaltenen Catechine (3-Galloyl-*l*-epicatechin, 3-Galloyl-*l*-epigallocatechin, *l*-Epicatechin und *l*-Epigallocatechin) eine Silicagelsäule. In einfacherer Weise läßt sich das Gemisch heute durch Gegenstromverteilung mit einem Phasensystem Äther/Wasser trennen, wobei man die Auftrennung chromatographisch verfolgen kann.

Daß die Gallussäure mit dem Hydroxyl 3 verestert ist, hatte schon Tsujimura (*86*) wahrscheinlich gemacht. Es konnte nun durch die Synthese des Heptaacetyl-3-galloyl-*l*-epicatechins bestätigt werden.

Freudenberg, Rein und Porter (*43*) synthetisierten das Acetat des noch nicht in der Natur gefundenen 3-Galloyl-*d*-catechins (XLVII). Diese Synthese wurde auf das Epicatechin übertragen (*25*) und das kristalline Heptaacetyl-3-galloyl-*l*-epicatechin mit dem aus Tee isolierten und acetylierten 3-Galloyl-*l*-epicatechin verglichen. Die Infrarot-Spektren und Analysen stimmen gut überein. Doch zeigt das synthetische Produkt im polarisierten Licht eine etwas niedrigere Drehung ($[\alpha]_D^{25} = + 91°$; 2% in Benzol) gegenüber dem natürlichen, acetylierten Produkt (+ 102°). Es muß also bei der Synthese eine teilweise Razemisierung eingetreten sein.

Gachokidse (*51*) isolierte aus den Schoten von *Gleditschia triacanthos* das *l*-Epicatechin-3-glucosid (IL). Das Glucosid wurde durch das Acetyl- und Methylderivat charakterisiert. Die Hydrolyse des methylierten Glucosids führte zu dem bekannten Tetramethyl-*l*-epicatechin. Dieser Umstand erlaubt es anzunehmen, daß die Glucose in 3-Stellung mit *l*-Epicatechin verbunden ist.

Ein vermeintlicher, in der Natur vorkommender Catechin-3-methyläther ist zu streichen. Er verdankt sein Auftreten in der Literatur einem Versehen im Referat in Chem. Abstr. 38, 5364 (1944). In der Originalabhandlung von Paris (*75*) wird der Äther nicht erwähnt.

10. Stereochemie der Catechine.

Das Catechin besitzt zwei asymmetrische Zentren (C-Atome 2 und 3) und kommt demnach in vier optisch aktiven und zwei Razemformen vor, die alle realisiert worden sind (*41*, *42*). Die relative und absolute Konfiguration konnte erst in neueren Arbeiten von Freudenberg (*23*, *24*, *27*), Hardegger (*56*) und Birch (*10*) bestimmt werden.

Die erste Konfigurationsbestimmung der Catechine ist von Freudenberg 1955 gegeben worden (*23*, *24*). Aus den gegenseitigen Umlagerungen der Catechine und Epicatechine geht hervor, daß beide im Verhältnis der Epimerie stehen. Es ist das Kohlenstoffatom 2 oder 3, das sich umlagert, wenn d-Catechin in d-Epicatechin übergeht. Dabei wird eine erhebliche Verschiebung der Drehung nach rechts beobachtet, die größer ist, als bei der Umlagerung eines sekundären Carbinols zu erwarten wäre. Anderseits weiß man, z. B. aus den Untersuchungen der Mandelsäure (37) und des Ephedrins (38), daß Phenylcarbinole und ihre Äther bedeutende Drehungsbeiträge geben. Dieser Umstand weist auf die Epimerisierung am Kohlenstoffatom 2 hin. Für die Umlagerung dieses Asymmetriezentrums spricht auch seine Labilität, denn ein Phenylcarbinol oder sein Äther wird sehr viel leichter umgelagert, als ein Carbinol zwischen aliphatischen Gruppen. Aus diesem Grunde muß $C_{(2)}$ als Sitz der Epimerisierung von d-Catechin zu d-Epicatechin angesehen werden (vgl. S. 7).

Die Tetramethyläther des Catechins und Epicatechins verhalten sich bei der Wasserabspaltung, die sich von Hydroxyl 3 nach Wasserstoffatom 2 vollzieht, sehr verschieden. Das Catechinderivat erleidet dabei eine tiefgreifende Umlagerung, während die Wasserabspaltung aus dem Epicatechin-tetramethyläther glatt verläuft. Während FREUDENBERG und HARDER (32) daraus keine Schlüsse auf die Konfiguration zogen und die Frage offen ließen, hat WAWZONEK (87) die richtige Konsequenz gezogen, daß im Epicatechin die Wasserstoffatome 2 und 3 in cis-Stellung stehen.

Rechtsverschiebung erleiden Phenylcarbinole, wenn sie wie im (—)-Ephedrin (38) [das in (+)-Pseudoephedrin übergeht] und in D-(—)-Mandelsäure (37) [die in L-(+)-Mandelsäure übergeht] gelagert sind. Sowohl im Ephedrin wie in der Mandelsäure hat die mehr links drehende Gruppierung, der das (+)-Catechin entspricht, die hier angeschriebene Konfiguration. Um diese Anordnung auf das (+)-Catechin anwenden zu können, müssen wir das Hydroxyl nach oben, das zweite Kohlenstoffatom der Seitenkette nach unten stellen. Durch doppelten Austausch erhalten wir die Gruppierung A (S. 20).

Der Äthersauerstoff dieser Gruppierung und die drei Kohlenstoffatome sind Glieder eines Ringes. Das Wasserstoffatom am CarbinolKohlenstoff muß in trans-Stellung zum oberen stehen. Damit ergibt sich für das (+)-Catechin die absolute Konfiguration B oder die angeschriebene Formel für (+)-Catechin und (+)-Epicatechin. Neuerdings haben HARDEGGER, GEMPELER und ZÜST (56) durch Ozonisierung von Catechin, Veresterung der entstandenen Säuren und Reduktion mit Lithiumaluminiumhydrid den 2-Desoxy-D-adonit C erhalten. Die beiden endständigen, primären Carbinolgruppen entstammen den Benzolsystemen des Catechins. Durch Umstellung wird daraus die Projektionsformel D, durch welche die frühere Ableitung der Konfiguration des d-Catechins bestätigt wird.

Eine dritte Ableitung haben BIRCH, CLARK-LEWIS und ROBERTSON gegeben (10). Sie wenden das Verfahren von PRELOG an. (—)-Epi-

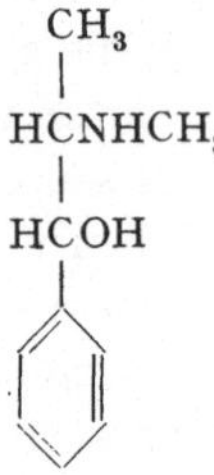

(—)-Ephedrin (*38*). 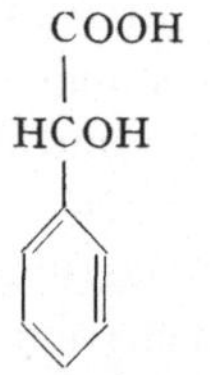   D-(—)-Mandelsäure (*37*). 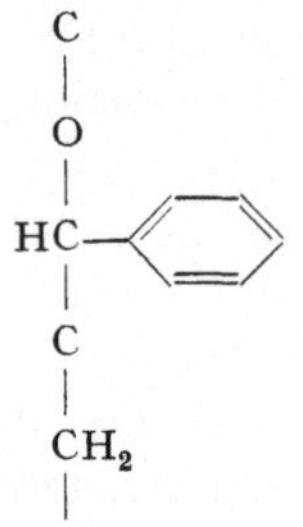   *A* 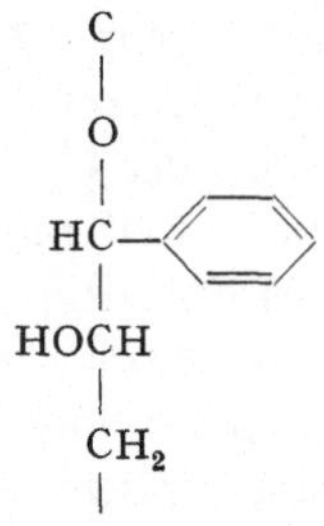   *B*

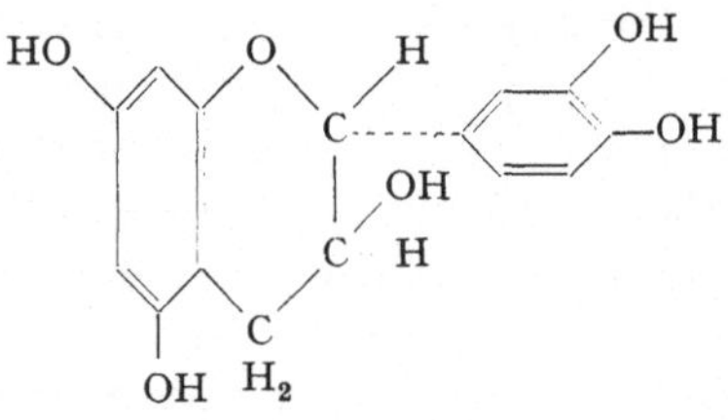

(+)-Catechin. 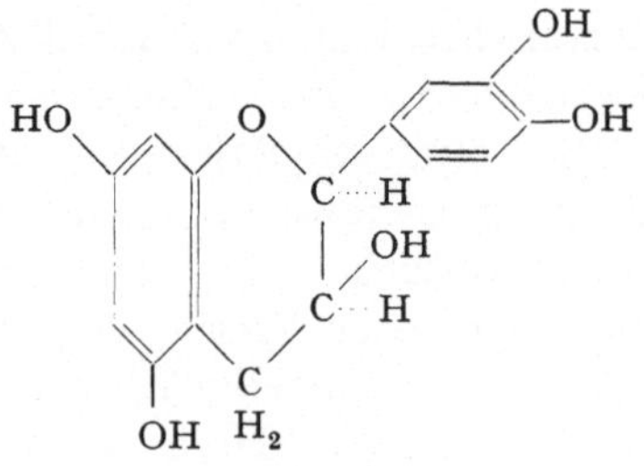   (+)-Epicatechin.

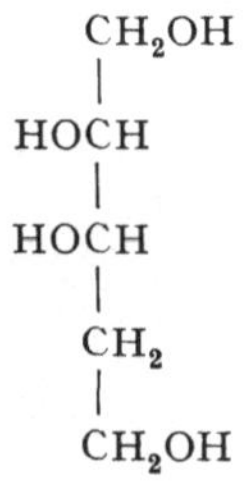

C 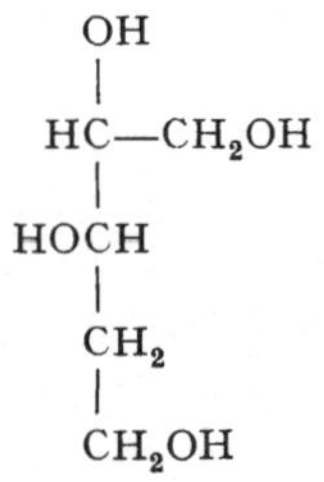   *D*   *E*

2-Desoxy-*D*-adonit.

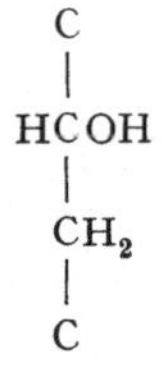

catechin-tetramethyläther wird mit Phenylglyoxylsäure verestert. Durch Methylmagnesiumhalogenid wird der Ester in einen solchen der Atrolactinsäure verwandelt. Nach der Verseifung erhält man neben razemischer Atrolactinsäure einen erheblichen Anteil an (—)-Atrolactinsäure, deren Konfiguration bekannt ist (*46*). Diese asymmetrische Synthese rührt daher, daß sich das Methylmagnesiumhalogenid dem Ester von derjenigen Seite her nähert, die räumlich am besten geeignet ist. Überlegungen, die mit der Raumerfüllung zusammenhängen, ergeben für die Carbinolgruppe des (—)-Epicatechin-tetramethyläthers die Konfiguration *E*. Ferner haben die britischen Autoren durch Abbau der Methyläther mit Natrium in Ammoniak gezeigt, daß (+)-Catechin und (+)-Epicatechin in der Carbinolgruppe gleichkonfiguriert sind. Damit ist die absolute Konfiguration der Catechine erneut bestätigt.

Catechin einer *D*- oder *L*-Gruppe zuzuordnen, ist zwecklos.

11. Biogenese der $C_6C_3C_6$-Gruppe.

Ein in vitro ausgeführter Übergang eines Produktes in ein anderes ist zwar eine Stütze für die biogenetische Auffassung, doch den entscheidenden Beweis muß der biologische Versuch bringen. Man kann annehmen, daß die Catechine in engem Zusammenhang mit den anderen in der Natur vorkommenden Stoffen der $C_6C_3C_6$-Gruppe stehen. Aus diesem Grunde muß zunächst die Biogenese der gesamten Gruppe betrachtet werden, um später vielleicht Schlüsse auf die biogenetischen Zusammenhänge der einzelnen Oxydationsstufen zu ziehen. Seit der Möglichkeit, ^{14}C-markierte Verbindungen dem pflanzlichen Stoffwechsel zuzusetzen und sie im Organismus zu verfolgen, sind einige beweiskräftige Versuche zur Biogenese dieser Stoffklasse ausgeführt worden.

Über die Biosynthese der Phenylpropankörper (C_6C_3) ist viel und erfolgreich gearbeitet worden. In jedem Fall führte der Weg über die Shikimisäure zur Prephensäure, die unter Aromatisierung weiter in Brenztraubensäure, Zimtsäure oder Phenylalanin umgewandelt wird. Durch Fütterung von Phenylalanin-2-^{14}C an Buchweizen *(Fagopyrum esculentum)* konnten GEISSMAN und SWAIN *(54)* nachweisen, daß der Ring *B* des erhaltenen radioaktiven Quercetins über die Kaffeesäure (3,4-Dihydroxyzimtsäure) aus dem Phenylalanin gebildet worden ist. Auch andere Versuche sprechen für die Auffassung, daß der Ring *B* durch Hydroxyzimtsäuren gebildet wird *(52)*.

Die Frage nach der Bildung des Ringes *A* wurde von GRISEBACH *(55)* sowie GEISSMAN und SWAIN *(54)* durch Gabe von Acetat-(1 oder 2)-^{14}C an Pflanzen geklärt. Der Abbau der erhaltenen aktiven C_{15}-Körper (Quercetin und Cyanidin) zeigte, daß das Acetat-^{14}C in spezifischer Weise in den Ring *A* eingebaut wird. Die frühere Ansicht *(66)*, daß der Ring *A* aus meso-Inosit entsteht, konnte neuerdings widerlegt werden *(89)*. Meso-Inosit-[14]C führt in der Pflanze zu keiner nachweisbaren Radioaktivität der isolierbaren C_{15}-Körper.

Die experimentellen Befunde stützen die von BIRCH *(9)* angegebene Hypothese zur Biogenese des Flavangerüstes, die auf folgendem Wege ablaufen soll:

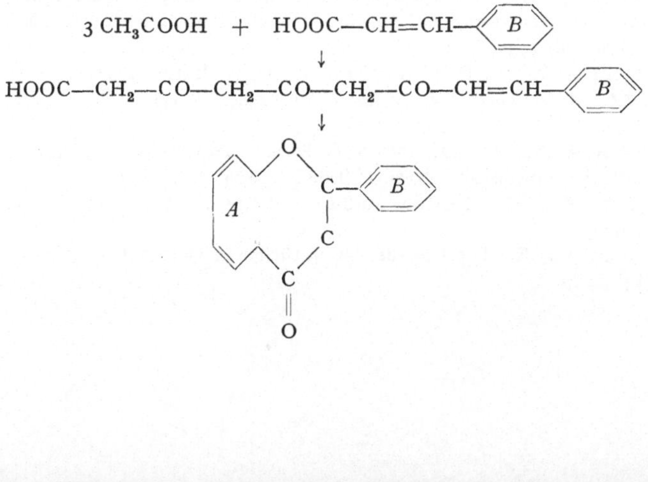

Während Robinson (*81*) der Auffassung ist, daß zunächst eine gemeinsame C_{15}-Vorstufe (z. B. Chalkon oder eine bisher unbekannte C_{15}-Verbindung) aufgebaut wird und sich aus dieser die weiteren Oxydationsstufen bilden, vertritt Paech (*74*) die Hypothese, daß die Oxydationsstufe schon bei der Bildung der Phenylpropankörper festgelegt ist. Doch fehlt für beide Auffassungen die experimentelle Unterbauung.

Literaturverzeichnis.

1. Appel, H. and R. Robinson: The Transformation of *d*-Catechin into Cyanidin Chloride. J. Chem. Soc. (London) **1935**, 426.

2. — — The Constitution of Cyanomaclurin. J. Chem. Soc. (London) **1935**, 752.

3. Baker, W.: Some Substances derived from the Anhydrocatechin Tetramethyl Ethers. J. Chem. Soc. (London) **131**, 1596 (1929).

4. Bate-Smith, E. C.: Colour Reactions of Flowers attributed to (a) Flavanols and (b) Carotenoid Oxides. J. exp. Botan. **4**, 1 (1953).

5. — Leuco-Anthocyanins. 1. Detection and Identification of Anthocyanidins Formed from Leuco-Anthocyanins in Plant Tissues. Biochemic. J. **58**, 122 (1954).

6. Bate-Smith, E. C. and N. H. Lerner: Leuco-Anthocyanins. 2. Systematic Distribution of Leuco-Anthocyanins in Leaves. Biochemic. J. **58**, 126 (1954).

7. Behaghel, O. und H. Freiensehner: Umlagerung von Phenol-benzyläthern bei höherer Temperatur. Ber. dtsch. chem. Ges. **67**, 1368 (1934).

8. Bersin, T., A. Müller und H. Schwarz: Über Inhaltsstoffe von *Crataegus oxyacantha* L. Arzneimittel-Forsch. **5**, 490 (1955); Schweiz. Apotheker-Ztg. **95** 403 (1957).

9. Birch, A. J.: Biosynthetic Relations of Some Natural Phenolic and Enolic Compounds. Fortschr. Chem. organ. Naturstoffe **14**, 186 (1957).

10. Birch, A. J., J. W. Clark-Lewis and A. V. Robertson: The Relative and Absolute Configurations of Catechins and *epi*-Catechins. J. Chem. Soc. (London) **1957**, 3586.

11. Bottomley, W.: The Conversion of Melacacidin into 3:3′:4′:7:8-Pentahydroxyflavylium Chloride. Chem. and Ind. **1954**, 516.

12. Bradfield, A. E. and M. Penney: The Catechins of Green Tea. Part II. J. Chem. Soc. (London) **1948**, 2249.

13. Braun, J. v. und H. Reich: Synthesen in der fettaromatischen Reihe. XVI: Gechlorte Amine und Aminosäuren. Liebigs Ann. Chem. **445**, 225 (1925).

14. Brockmann, H. und H. Junge: Über Benzopyryliumverbindungen. II. Mitt.: Acylierung und Methylierung. Ber. dtsch. chem. Ges. **77**, 44 (1944).

15. Brown, B. R., W. Cummings and G. A. Somerfield: Polymerisation of Flavans. Part I. The Condensation of Methoxybenzyl Alcohols with Phenols. J. Chem. Soc. (London) **1957**, 3757.

16. Carruthers, W. R., R. H. Farmer and R. A. Laidlaw: Dihydromorin from East African Mulberry (*Morus lactea* Mildbr.). J. Chem. Soc. (London) **1957**, 4440.

17. Evans, R. A., W. H. Parr and W. C. Evans: Paper Partition Chromatography of Phenolic Substances. Nature (London) **164**, 674 (1949).

18. Forsyth, W. G. C.: "Leuco-cyanidin" and Epicatechin. Nature (London) **172**, 726 (1953).

19. Freudenberg, K.: Die Chemie der natürlichen Gerbstoffe. Berlin: Springer-Verlag. 1920.

20. FREUDENBERG, K.: Die natürlichen Gerbstoffe. In: G. KLEIN, Handbuch der Pflanzenanalyse. Wien: Springer-Verlag. 1932. Hiermit größtenteils übereinstimmend das Kapitel über Gerbstoffe in: Tannin, Cellulose, Lignin. Berlin: Springer-Verlag. 1933.

21. — Über Gerbstoffe. V. Phloroglucin-Gerbstoffe und Catechine. Konstitution des Gambir-Catechins. Ber. dtsch. chem. Ges. 53, 1416 (1920).

22. — Polykondensationsprinzip der Catechingerbstoffe und des Lignins. Angew. Chem. 67, 84 (1955).

23. — Catechine und verwandte Verbindungen. Angew. Chem. 67, 728 (1955).

24. — Catechins and Related Substances. Sci. Proc. Roy. Dublin Soc. 27, 153 (1956).

25. — Beiträge zur Chemie der Catechine und Flavane. Festschrift Arthur Stoll, S. 199. Basel: Birkhäuser. 1957.

26. FREUDENBERG, K. und J. M. ALONSO DE LAMA: Zur Kenntnis der Catechingerbstoffe. Liebigs Ann. Chem. 612, 78 (1958).

27. FREUDENBERG, K., O. BÖHME und L. PURRMANN: Raumisomere Catechine. II. Ber. dtsch. chem. Ges. 55, 1734 (1922).

28. FREUDENBERG, K., R. F. B. COX and E. BRAUN: The Catechin of the Cacao Bean. J. Amer. Chem. Soc. 54, 1913 (1932).

29. FREUDENBERG, K., H. FIKENTSCHER und M. HARDER: Abbau- und Aufbauversuche am Catechin. Liebigs Ann. Chem. 441, 157 (1925).

30. FREUDENBERG, K., H. FIKENTSCHER, M. HARDER und O. TH. SCHMIDT: Die Umwandlung des Cyanidins in Catechin. Liebigs Ann. Chem. 444, 135 (1925).

31. FREUDENBERG, K., H. FIKENTSCHER und W. WENNER: Die Konstitution des Catechins. Liebigs Ann. Chem. 442, 309 (1925).

32. FREUDENBERG, K. und M. HARDER: Synthesen von Abkömmlingen des Catechins. Liebigs Ann. Chem. 451, 213 (1927).

33. FREUDENBERG, K. und L. HARTMANN: Inhaltsstoffe der Robinia pseudacacia. Liebigs Ann. Chem. 587, 207 (1954).

34. FREUDENBERG, K. und A. KAMMÜLLER: Übergänge aus der Gruppe der Flavone in die des Catechins. Liebigs Ann. Chem. 451, 209 (1927).

35. FREUDENBERG, K., KARIMULLAH und G. STEINBRUNN: Umwandlung der Anthocyanidine und Catechine. Liebigs Ann. Chem. 518, 37 (1935).

36. FREUDENBERG, K. und P. MAITLAND: Der Quebracho-Gerbstoff. Liebigs Ann. Chem. 510, 193 (1934).

37. FREUDENBERG, K. und L. MARKERT: Konfiguration der Mandelsäure. Ber. dtsch. chem. Ges. 58, 1753 (1925).

38. FREUDENBERG, K. und F. NIKOLAI: Die Konfiguration des Ephedrins. Liebigs Ann. Chem. 510, 223 (1934).

39. FREUDENBERG, K. und L. OEHLER: Die Catechine der Colanuß. Liebigs Ann. Chem. 483, 140 (1930).

40. FREUDENBERG, K. und K. L. ORTHNER: Die Reduktion des Flavanons. Ber. dtsch. chem. Ges. 55, 1748 (1922).

41. FREUDENBERG, K. und L. PURRMANN: Raumisomere Catechine. III. Ber. dtsch. chem. Ges. 56, 1185 (1923).

42. — — Raumisomere Catechine. IV. Liebigs Ann. Chem. 437, 274 (1924).

43. FREUDENBERG, K., H. G. REIN und J. PORTER: Synthese des 3-Galloyl-catechinacetats. Liebigs Ann. Chem. 603, 177 (1957).

44. FREUDENBERG, K. und D. G. ROUX: Umwandlung des Dihydrorobinetins in das zugehörige Anthocyanidin. Naturwiss. 41, 450 (1954).

45. FREUDENBERG, K., J. H. STOCKER und J. PORTER: Das Kondensationsprinzip der Catechin-gerbstoffe. Chem. Ber. 90, 957 (1957).

46. Freudenberg, K., J. Todd und R. Seidler: Die Konfiguration des tertiären Kohlenstoffatoms. Atrolactinsäure, Mandelsäure und verwandte Verbindungen. Liebigs Ann. Chem. 501, 199 (1933).

47. Freudenberg, K. und K. Weinges: Zur Kenntnis der Catechine und Catechingerbstoffe. Liebigs Ann. Chem. 590, 140 (1954).

48. — — Leucocyanidin-hydrat. Angew. Chem. 70, 51 (1958).

49. — — Leuko- und Pseudoverbindungen der Anthocyanidine. Liebigs Ann. Chem. 613, 61 (1958).

50. Freudenberg, K., K. Weinges und J. M. Alonso: Catechine und verwandte Verbindungen. Angew. Chem. 69, 679 (1957).

51. Gachokidse, A. M.: Untersuchung des Glucosids aus Gleditschia triacanthos. J. Appl. Chem. (USSR) 19, 1197 (1946) [Chem. Zbl. 1946, I, 952].

52. Geissman, T. A. and E. Hinreiner: Theories of the Biogenesis of Flavonoid Compounds. Botan. Rev. 18, 77 (1952).

53. Geissman, T. A. and H. Lischner: Flavanones and Related Compounds. VII. The Formation of 4,6,3',4'-Tetrahydroxy-2-benzylcoumaranone-3 by the Sodium Hydrosulfite Reduction of Quercetin. J. Amer. Chem. Soc. 74, 3001 (1952).

54. Geissman, T. A. and T. Swain: Biosynthesis of Flavonoid Compounds in Higher Plants. Chem. and Ind. 1957, 984.

55. Grisebach, H.: Zur Biogenese des Cyanidins. I. Mitt.: Versuche mit Acetat-[1-^{14}C] und Acetat-[2-^{14}C]. Z. Naturforsch. 12 b, 227 (1957).

56. Hardegger, E., H. Gempeler und A. Züst: Die absolute Konfiguration des (+)-Catechins. Helv. Chim. Acta 40, 1819 (1957).

57. Hathway, D. E. and J. W. T. Seakins: Autoxidation of Polyphenols. Part III. Autoxidation in Neutral Aqueous Solution of Flavans Related to Catechin. J. Chem. Soc. (London) 1957, 1562.

58. — — Enzymic Oxydation of Catechin to a Polymer Structurally Related to some Phlobatannins. Biochemic. J. 67, 239 (1957).

59. Karrer, P. und W. Fatzer: Herstellung von Tocopherol-ähnlichen Verbindungen und Flavonolen aus Benzo-pyryliumsalzen. — Die Oxydation von Benzopyryliumsalzen zu Flavonolen. — Über ein Dipheno-spiranderivat mit konstitutionellen Beziehungen zu den Tocopherolen. Helv. Chim. Acta 25, 1129, 1138, 1140 (1942).

60. Karrer, P. und M. Seyhan: Über die Reduktion der Pyryliumsalze mit Lithiumaluminiumhydrid. Helv. Chim. Acta 33, 2209 (1950).

61. Karrer, P., C. Trugenberger und G. Hamdi: Umwandlungsprodukte einfacher Benzopyryliumverbindungen. Helv. Chim. Acta 26, 2116 (1943).

62. Keppler, H. H.: The Isolation and Constitution of Mollisacacidin, a New leucoAnthocyanidin from the Heartwood of Acacia mollisima Willd. J. Chem. Soc. (London) 1957, 2721.

63. King, F. E. and W. Bottomley: The Chemistry of Extractives from Hardwoods. Part XVII. The Occurrence of a Flavan-3:4-diol (Melacacidin) in Acacia melanoxylon. J. Chem. Soc. (London) 1954, 1399.

64. King, F. E. and J. W. Clark-Lewis: The Constitution and Synthesis of Leucoanthocyanidins. J. Chem. Soc. (London) 1955, 3384.

65. King, F. E., J. W. Clark-Lewis and W. F. Forbes: The Chemistry of Extractives from Hardwoods. Part XXV. (—)-epiAfzelechin, a New Member of the Catechin Series. J. Chem. Soc. (London) 1955, 2948.

66. Kursanow, A. L.: Synthese und Umwandlung der Gerbstoffe in der Teepflanze. Berlin: VEB. Verlag Volk und Gesundheit. 1954.

67. Kurth, E. F. and F. L. Chan: Extraction of Tannin and Dihydroquercetin from Douglas-fir Bark. J. Amer. Leather Chem. Assoc. 48, 20 (1953).

68. LINDSTEDT, G.: Constituents of Pine Heartwood. XXI. The Structure of Pinobanksin. Acta Chem. Scand. 4, 772 (1950).
69. MASON, F. A.: The Catechin Problem. J. Soc. Chem. Ind. 47, 269 T (1928).
70. MIRZA, R. and R. ROBINSON: Conversion of Flavonols into Anthocyanidins. Nature (London) 166, 997 (1950).
71. OSIMA, Y. und H. ITO: Untersuchungen über pflanzliche Gerbmittel aus Formosa. Bull. Agric. Chem. -Soc. (Japan) 15, 108 (1939) [Chem. Zbl. 1939, II, 3919].
72. OYAMADA, T.: Über die Konstitution des Fustins. Liebigs Ann. Chem. 538, 44 (1939).
73. KOTAKE, M. und T. KUBOTA: Über Inhaltsstoffe von Ampelopsis meliaefolia KUDO (Haku-Tya). Liebigs Ann. Chem. 544, 253 (1940).
74. PAECH, K.: Biochemie und Physiologie der sekundären Pflanzenstoffe, S. 189. Berlin: Springer-Verlag. 1950.
75. PARIS, R.: Sur une Combrétacée africaine, le «kinkéeliba». Bull. sci. pharmacol. 49, 181 (1942).
76. PERKIN, A. G.: Cyanomaclurin. J. Chem. Soc. (London) 87, 716 (1905).
77. PERKIN, A. G. and F. COPE: The Constituents of Artocarpus integrifolia. J. Chem. Soc. (London) 67, 937 (1895).
78. PEW, J. C.: A Flavanone from Douglas-Fir Heartwood. J. Amer. Chem. Soc. 70, 3031 (1948).
79. ROBERTSON, A., V. VENKATESWARLU and W. B. WHALLEY: The Pigments of "Dragon's Blood" Resin. Part V. Some Flavans. J. Chem. Soc. (London) 1954, 3137.
80. ROBINSON, G. M. and R. ROBINSON: A Survey of Anthocyanins. III. Notes on the Distribution of Leuco-anthocyanins. Biochemic. J. 27, 206 (1933).
81. ROBINSON, R. and G. M. ROBINSON: The Colloid Chemistry of Leaf and Flower Pigments and the Precursors of the Anthocyanins. J. Amer. Chem. Soc. 61, 1605 (1939).
82. ROUX, D. G.: d-Gallocatechin from the Bark of Casuarina equisetifolia LINN. Nature (London) 179, 158 (1957).
83. — Some Recent Advances in the Identification of Leucoanthocyanins and the Chemistry of Condensed Tannins. Nature (London) 180, 973 (1957).
83a. ROUX, D. G. und K. FREUDENBERG: Über Leuko-robinetinidin-hydrat und Leuko-fisetinidin-hydrat. Liebigs Ann. Chem. 613, 56 (1958).
84. SHRINER, R. L.: Chemistry of Flavylium Salts; Reactions with Amines. In: Roger Adams Symposium, p. 103. New York: J. Wiley and Sons. 1954.
85. SWAIN, T.: Leucocyanidin. Chem. and Ind. 1954, 1144.
86. TSUJIMURA, M.: Über aus grünem Tee isoliertes Teetannin. Scient. Pap. Inst. Physic. Chem. Res. (Japan) 14, 63 (1930) [Chem. Zbl. 1930, II, 3032].
87. WAWZONEK, S.: In: N. R. ELDERFIELD, Heterocyclic Compounds, vol. II, p. 358. New York: J. Wiley and Sons. 1951.
88. WEINGES, K.: Über Catechine und ihre Herstellung aus Leukoanthocyanidin-hydraten. Liebigs Ann. Chem. (1958) (im Druck).
89. WEYGAND, F., W. BRUCKER, H. GRISEBACH und E. SCHULZE: Stoffwechsel-untersuchungen mit meso-Inosit-^{14}C. Z. Naturforsch. 12 b, 222 (1957).
90. WILLSTÄTTER, R. und H. MALLISON: Über die Verwandtschaft der Anthocyane und Flavone. Sitzber. preuß. Akad. Wiss. 1914, 769.

(Eingelaufen am 2. Januar 1958.)

Recent Progress in the Chemistry of the Aconite-Garrya Alkaloids.

By KAREL WIESNER and ZDENEK VALENTA, Fredericton, N. B., Canada.

Contents.

Acknowledgement. We wish to thank the *National Research Council* in Ottawa and the *Eli Lilly & Co.*, Indianapolis, Indiana, U. S. A. for the support of the investigations of the garrya-aconite alkaloids which were performed at the University of New Brunswick during the past six years.

I. Introduction.

Alkaloids derived from various species of *Aconitum* and *Delphinium* have been the subject of a large number of experimental studies. Among the early investigations the meticulous and accurate work of W. A. JACOBS and his collaborators overshadows that of anybody else. It was JACOBS who coined the name "Aconite alkaloids" for bases of both genera. This is well justified not only by the close similarity in the chemical behaviour of both types of bases, but also by occurrences of identical alkaloids in both genera. It was also JACOBS who suspected, in his early studies, the diterpenoid nature of these compounds. However, in spite of the large amount of experimental work, no satisfactory structures have been proposed to accommodate the available mass of chemical data. Some progress in this direction was achieved only when it was recognized that a third genus, *Garrya*, yields especially simple and easily available bases, which undoubtedly belong, according to their chemical properties, to the class of aconite alkaloids.

The aconite alkaloids may be divided into two large groups: First, the non-toxic group of penta- or hexacyclic alkaloids with few substituents and a $C_{(20)}$ skeleton. One of the distinguishing chemical properties of this group is the formation of phenanthrene hydrocarbons on dehydrogenation in good yield. This group includes the garrya alkaloids, atisine, napelline, and many other alkaloids of as yet unknown constitution.

The second group is a highly toxic class of bases distinguished by many substituents (chiefly methoxy, hydroxy, and acyloxy groups) and a C_{19}, probably hexacyclic, skeleton. Also in this group progress has been very recently achieved due mainly to the crystallographic work of PRZYBYLSKA (50) which provided a structural framework for the chemical studies of EDWARDS and MARION (18, 19) on lycoctonine and COOKSON (3–8) on delpheline. Some of the lycoctonine chemistry, however, is still baffling.

Since extensive reviews of all the early experimental work are available (52), the present survey will deal only with such cases of garrya or aconite alkaloids in which either the complete structure has been deduced or at least a plausible if not rigidly proved formula may be proposed. In addition, the biogenetic origin and interrelations of aconite alkaloids will be discussed.

II. Garrya Alkaloids.

1. The Structure of Veatchine and Garryine.

In a search for quinine substitutes Oneto (45) has investigated several species of Garrya for alkaloid content. The richest source of alkaloids found by this author was the bark of Garrya veatchii Kellog, from which two bases, veatchine and garryine, were isolated.

In a reinvestigation of this plant material (59) the properties of the two compounds were confirmed and the isolation method was improved by showing that garryine and veatchine may be easily separated by a short countercurrent distribution. It was shown that both compounds in the anhydrous form have the empirical formula $C_{22}H_{33}O_2N$ and thus are isomeric with some aconite alkaloids.

While veatchine crystallized only as the anhydrous compound, garryine crystallized as a monohydrate which easily lost water to give the oily anhydrous garryine. Both garryine (pK = 8.7) and veatchine (pK = 11.5) were shown to owe their strong basicity to the ability to give salts of the quaternary Schiff type. This was borne out by the finding that both alkaloids gave the same dihydroveatchine by lithium aluminum hydride reduction of an unsaturation associated with the nitrogen, and the same tetrahydroveatchine by catalytic hydrogenation of this unsaturation and an isolated terminal methylene group. Both dihydro-veatchine (pK = 6.9) and tetrahydroveatchine (pK = 6.8) were found to be much weaker bases than the parent alkaloids. While dihydro- and tetrahydroveatchine clearly showed the presence of two hydroxyls in a Zerewitinoff determination and by giving a basic O-diacetyl derivative on acetylation, veatchine and garryine appeared to possess only a single hydroxyl. Both alkaloids also have a N-alkyl group, and the clarification of the precise nature of the unsaturation associated with the nitrogen was intimately connected with the identification of this group.

Environment of the Nitrogen. It was realised (60) that in order to accommodate the manifold reactions in the vicinity of the nitrogen, which will be described in the sequel, veatchine must be represented by the partial structure (I), garryine by (II), garryine hydrate by (III), and dihydroveatchine by (IV).

Pyrolysis of veatchine or garryine in the presence of selenium gave two isomeric compounds $C_{20}H_{29}ON$ formulated by the partial structures (V) and (VI). Compound (V) showed in the infrared a Δ_1-piperideine double bond (band at 1650 cm.$^{-1}$) and a five-membered ketone (1735 cm.$^{-1}$) which must have originated from a ketonisation of an exocyclic allylic alcohol. In compound (VI) only the Δ_1-piperideine double bond created by the degradation of the oxazolidine ring was present; the rest of the molecule remained undisturbed.

It was possible to prove these assumptions rigorously as follows.

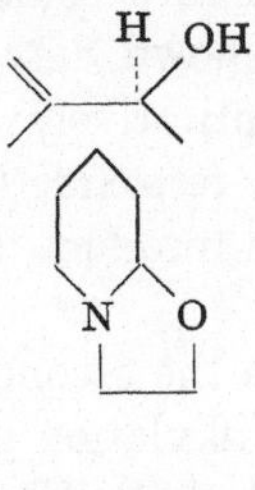

(I.) Veatchine. 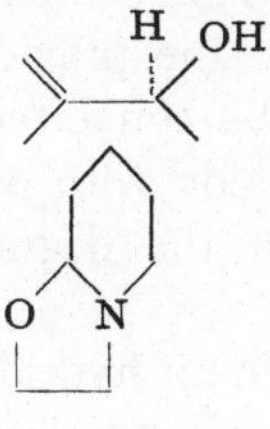   (II.) Garryine. 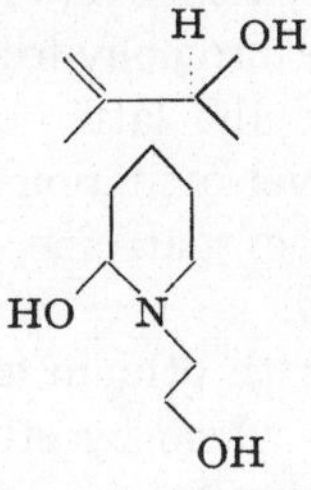   (III.) Garryine hydrate. 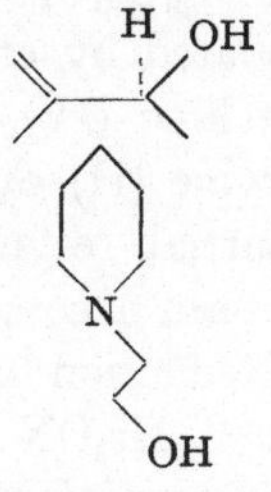   (IV.) Dihydroveatchine.

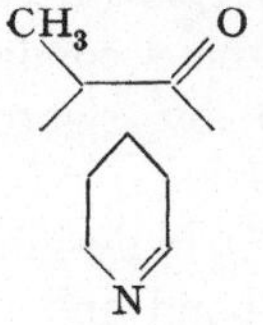

(V.) 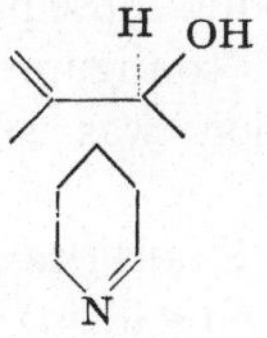   (VI.) 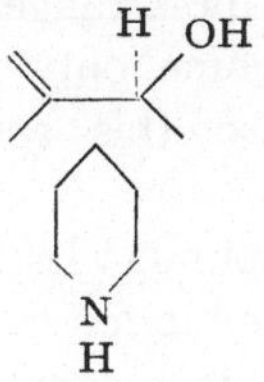   (VII.) 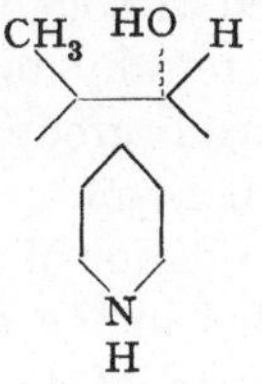   (VIII.)

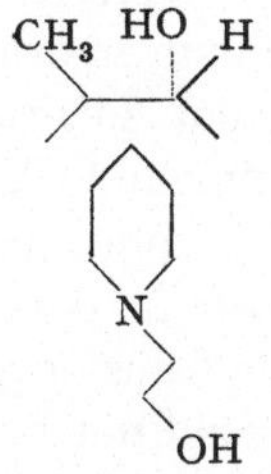

(IX.) 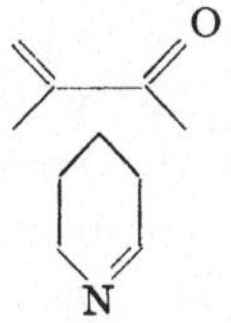   (X.) 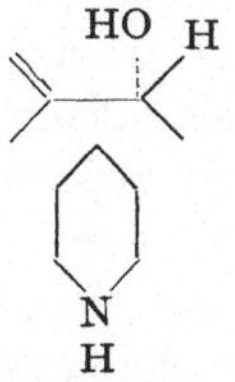   (XI.) 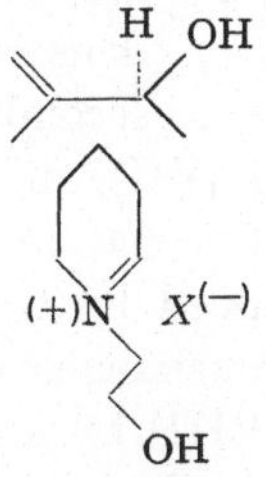   (XII.)

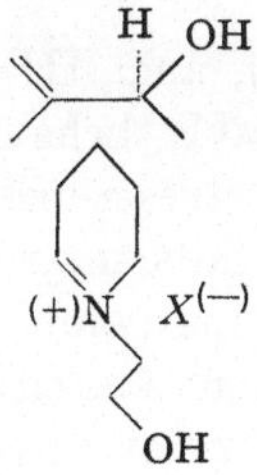

(XIII.) 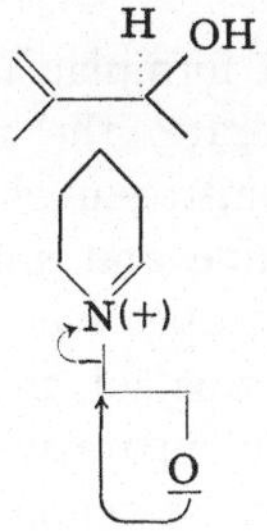   (XIV.) 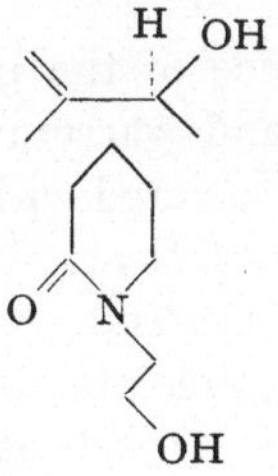   (XV.) Oxogarryine. 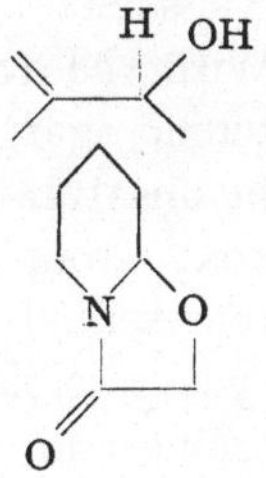   (XVI.) Oxoveatchine A.

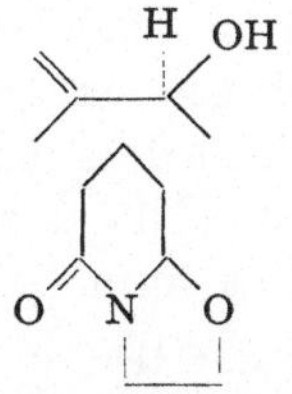

(XVII.) Oxoveatchine B.

Reduction of (VI) with lithium aluminum hydride gave the secondary base $C_{20}H_{31}ON$ represented by (VII). This compound in turn could be alkylated by ethylene bromohydrine to the previously known dihydro-veatchine (IV). Since the latter can be converted quantitatively into garryine (II) by a novel oxidation reaction with osmium tetroxide (60), a partial resynthesis of garryine from the degradation product (VI) has been accomplished.

Reduction of (V) with lithium aluminum hydride gave the secondary base $C_{20}H_{33}ON$ (VIII) which by ethylene bromohydrine alkylation gave an isomer of tetrahydroveatchine (IX). It was assumed that (V) was formed without skeletal rearrangement and, consequently, (IX) is different from tetrahydroveatchine only in the configuration of the secondary hydroxyl group. Also this assumption was supported by rigorous evidence (56).

Oxidation of compound (VII) by the Sarett reagent (CrO_3 in pyridine) resulted in the reintroduction of a Δ_1-piperideine double bond and the oxidation of the allylic alcohol to an α,β-unsaturated ketone (U. V. λ_{max} 236 mμ, log $\varepsilon = 4$) to give a compound $C_{20}H_{27}ON$ represented by (X). This compound, when treated with lithium aluminum hydride, gave a secondary basic alcohol $C_{20}H_{31}ON$ which was not identical with (VII) and must be represented as the epimeric compound (XI). Hydrogenation of (XI) gave, finally, compound (VIII), previously obtained by reduction of (V). Since (VII) has been shown to possess the unchanged garrya skeleton (see above) and now has been transformed to (VIII) by a series of oxidations and reductions which could not have involved a rearrangement, also (VIII) and (V) must possess the original garrya skeleton.

With the acceptance of the partial formulae (I), (II), and (III) for veatchine, garryine and garryine hydrate, the remarkable behaviour of the unsaturation associated with the nitrogen which involves a masked hydroxy group becomes clear. Veatchine and garryine salts may now be represented by (XII) and (XIII), respectively. At this point it is not yet possible to appreciate the reasons for the different basicity of veatchine and garryine as well as the comparative stability of garryine hydrate (III) which has no counterpart in veatchine. The explanation of these properties is intimately connected with the complete structure and stereochemistry of the garrya alkaloids and will be dealt with later.

The pyrolytic cleavage of the oxazolidine ring to the pyrolysis bases (V) and (VI) may be regarded as displacement proceeding *via* the quaternary zwitter-ionic form (XIV), which could be in thermal equilibrium with the oxazolidine form (I).

Finally, it should be pointed out that it is not possible at this stage to distinguish whether the Δ_1-piperideine double bond in the pyrolysis bases mentioned is in the position corresponding to garryine or to veatchine, as both these compounds give the same products (V) and (VI). This is understandable since veatchine may be easily isomerized to garryine in alkaline medium and presumably the two possible pyrolysis bases, differing from the isolated compounds (V) and (VI) by the orientation of the Δ_1-piperideine double bond, would be susceptible to isomerisation to (V) and (VI).

Through the entire discussion the partial structures used have involved a six-membered nitrogen ring, and no evidence as yet has been advanced to support this particular feature. Studies on mild permanganate oxidation of garryine and veatchine have been instrumental in a further clarification of the nitrogen environments (59, 60).

The oxidation of garryine with permanganate gave a good yield of oxogarryine $C_{22}H_{33}O_3N$, which has according to its infrared spectrum (1618 cm.$^{-1}$) a six-membered, or less likely, a larger lactam ring. The formation of this oxidation product is easily rationalized if we consider that garryine exists in water-containing media predominantly as the carbinolamine (III, p. 29). Oxogarryine consequently must be formed by oxidation of the carbinolamine group to the corresponding lactam and is best represented by (XV). In the case of veatchine the oxazolidine form is more stable and it appears that the equilibrium concentration of a veatchine-carbinolamine is exceedingly low (for rationalisation, see below). Consequently, in a mild permanganate oxidation of veatchine, the oxazolidine (I) is the substrate and the products isolated are two isomeric lactams $C_{22}H_{31}O_3N$ (oxoveatchines A and B), represented by the partial structures (XVI) and (XVII). In agreement with these formulations compound (XVI) shows a carbonyl peak in the infrared corresponding to a five-membered lactam ring (1700 cm.$^{-1}$). Compound (XVII) on the other hand has a lactam carbonyl frequency (1630 cm.$^{-1}$) similar to the one shown by (XV).

Structure of the Skeleton. At this point we shall digress from the description of studies concerned with the exploration of the nitrogen environment and discuss some experiments which helped in the clarification of the complete structure. This is advantageous, since one can fully understand all the peculiarities of the garrya oxazolidine system only in terms of the complete structure and stereochemistry of veatchine and garryine.

Dehydrogenation of veatchine or garryine with selenium (59) gave a good yield of 1-methyl-7-ethylphenanthrene (XVIII) and a compound $C_{16}H_{15}N$, which according to its ultraviolet spectrum was clearly an azaphenanthrene. A biogenetically very plausible explanation of these results and of the fact that veatchine and garryine showed the presence of one C-methyl group in the KUHN-ROTH determination is the assumption of structure (XX) for dihydroveatchine. This possibility had been

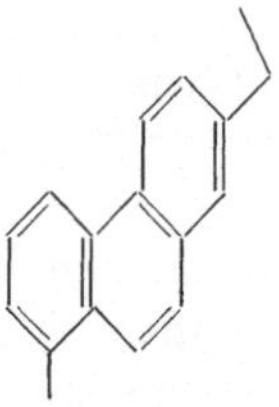

(XVIII.) 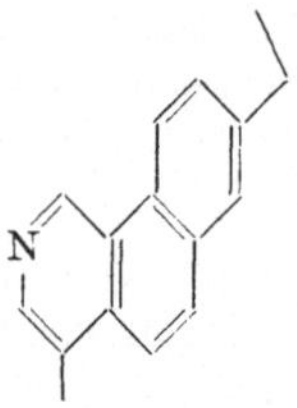 (XIX.) 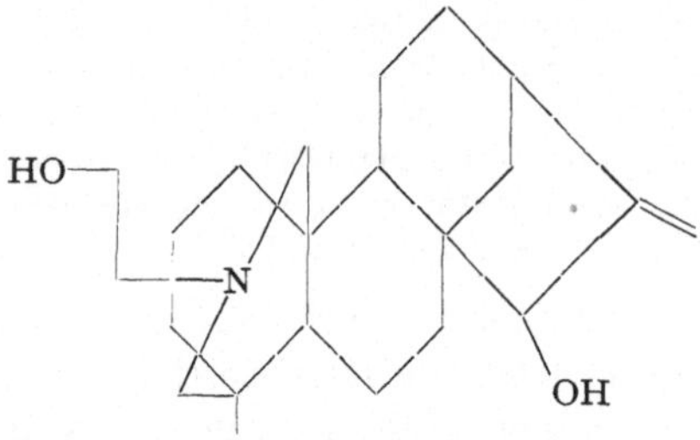 (XX.) Dihydroveatchine.

considered, but was initially discarded for two reasons: (a) The ultra-violet spectrum of the veatchine azaphenanthrene was practically super-imposable on the spectrum of a disubstituted phenanthridine. (b) The chemical properties of some acidic oxidation products of garryine and veatchine (cf. p. 34), closely paralleled by the behavior of similar compounds derived from the aconite alkaloid atisine, seemed to exclude structures of the type represented by (XX).

However, both these difficulties were invalidated and the dihydro-veatchine structure (XX) accepted in 1953 (57). An X-ray crystallographic study of the veatchine azaphenanthrene performed by HUGHES and NATHAN (29a) at the California Institute of Technology showed that the substitution of this compound was probably identical with the one of the veatchine-phenanthrene (XVIII), that is 1-methyl-7-ethyl. Simultaneously it was shown (57) that the methiodide of the veatchine azaphenanthrene had an ultraviolet spectrum very different from the spectrum of a phenanthridine methiodide. Also the products of a permanganate oxidation performed on "natural" and model methiodides were very different. While the model disubstituted phenanthridine gave a N-methyl phenanthridone dicarboxylic acid in good yield, the "natural" azaphenanthrene gave benzene 1,2,3,4-tetracarboxylic acid as the only major product.

Since also at this time the unusual behaviour of the veatchine oxidation acids was rigorously explained and shown to be compatible with structure (XX), no more obstacles seemed to prevent the acceptance of this structure for dihydroveatchine and (XIX) for the veatchine azaphenanthrene. Structure (XIX) was further confirmed by a synthesis (2) and crystallographically by HUGHES and NATHAN (29a), who by further refinements have been able to locate the nitrogen atom.

The acceptance of (XX) for dihydroveatchine clearly implies, in view of the preceding discussion, structures (XXI) and (XXII) for veatchine and garryine or vice versa. While it is possible to deduce on theoretical grounds that (XXI) is the correct representation of veatchine, we shall enter into these arguments only after presenting some classical evidence which points in the same direction. In the next paragraph, however, we shall consider some degradations which corroborate the structure and mode of attachment of the five-membered ring in veatchine (XXI).

Structure of the Oxoveatchine Dicarboxylic Acids. A vigorous oxidation (57, 56) of (XXI) by potassium permanganate gave two lactam carboxylic acids, (XXIII) and (XXIV). Their infrared spectra clearly indicated a five-membered lactam in (XXIII) and a six-membered one in (XXIV). Thus, the relationship of the two isomeric dicarboxylic acids is the same as of the two oxoveatchines (XVI) and (XVII). Both acids by treatment with diazomethane gave dimethylesters. These compounds show in the infrared the absence of a hydroxyl group, a feature which corroborates the presence of the oxazolidine ring.

Saponification of the dimethylesters with alkali yielded the corresponding monoesters. This finding is in agreement with the tertiary nature of one of the two carboxyls. Finally, the anhydrides of the two dicarboxylic acids were prepared, and the infrared spectra of these compounds in the carbonyl region were characteristic of a substituted glutaric anhydride.

The selenium dehydrogenation of (XXIV) yielded puzzling results. The first compound isolated was a small amount of pimanthrene (XXV). This product seems to be in direct disagreement with the proposed structure (XXIV) and rather indicates the partial structure (XXVI) for the dicarboxylic acid. It is clear that such a partial structure would require a complete revision of all our preceding deductions. This is the finding mentioned above which prevented for a time the acceptance of the garrya skeleton (XX). A completely analogous situation in the chemistry of atisine has also led to the postulation of an incorrect skeletal structure (32). The problem was finally resolved (57, 56) when it was convincingly shown that one of the methyls of (XXV) has its origin in the reduction of a carboxyl in (XXIV), during the selenium dehydrogenation.

Reduction of (XXIV) by lithium aluminium hydride gave the triol (XXVII) which gave by dehydrogenation also (XXV) in a much better yield *(Chart 1)*. Furthermore, the acidic fraction from the dehydrogenation of (XXIV) yielded an acid which on spectroscopic grounds has been assigned the structure (XXVIII). Thus, the partial structure (XXVI) may be definitely dismissed, and all the chemical and spectroscopic properties of the two lactam dicarboxylic acids are in agreement with the structures (XXIII) and (XXIV).

There is a further point implicit in the acceptance of (XXI) and (XXII) for veatchine and garryine which may be checked rather easily (56). While the salts of both alkaloids are undoubtedly quaternary SCHIFF

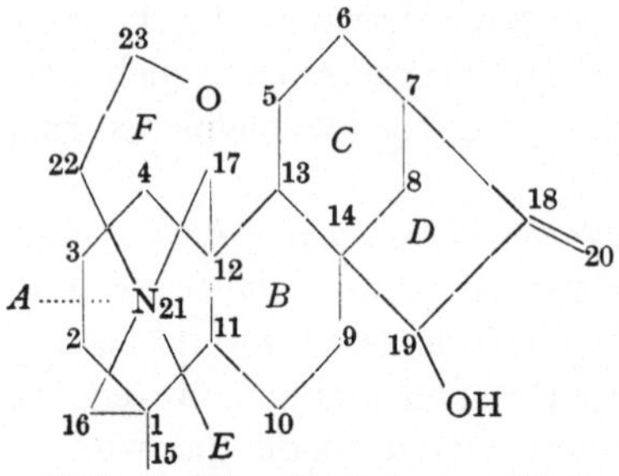

(XXI.) Veatchine. [Numbering according to DJERASSI et al. (12).]

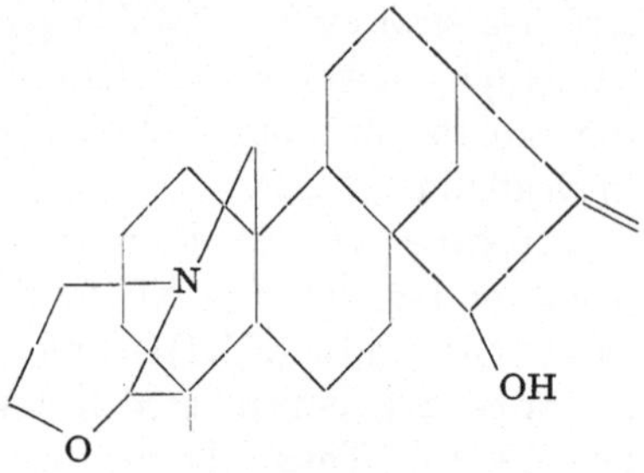

(XXII.) Garryine.

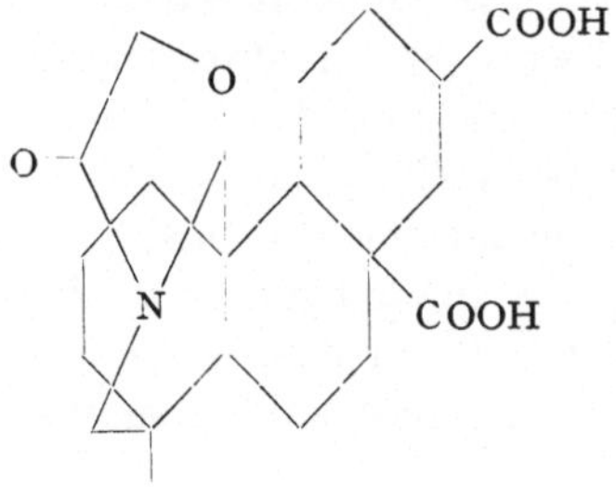

(XXIII.) Oxoveatchine dicarboxylic acid A.

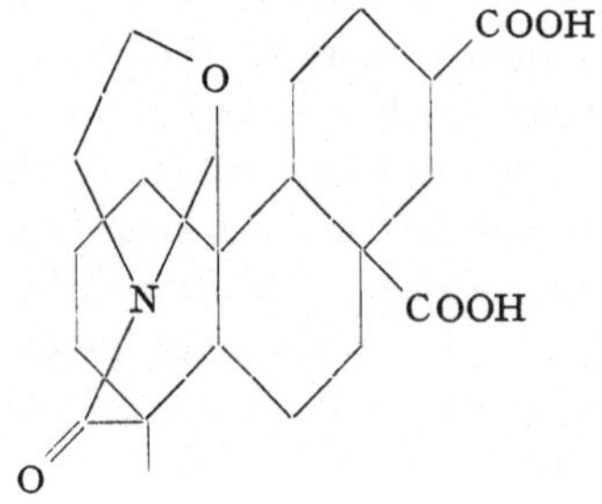

(XXIV.) Oxoveatchine dicarboxylic acid B.

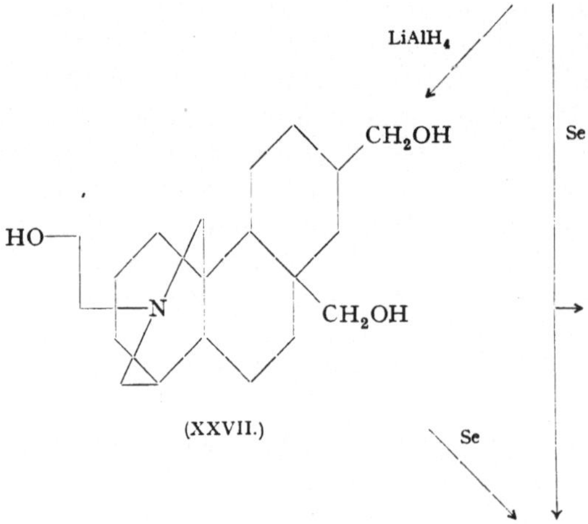

LiAlH₄

Se

(XXVII.)

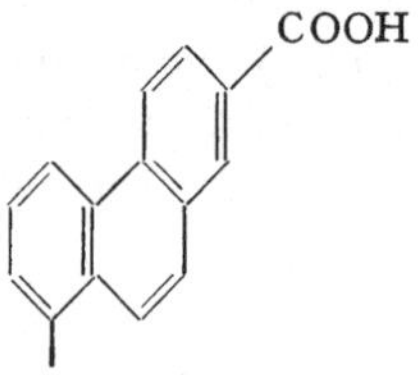

(XXVIII.)

Se

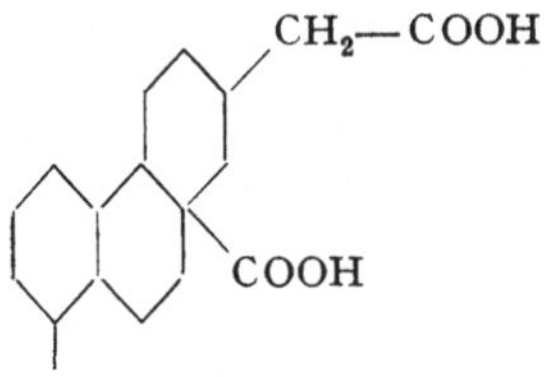

(XXVI.)

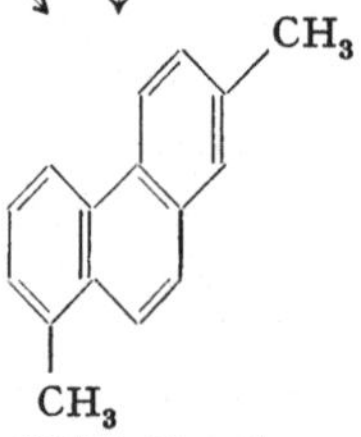

(XXV.) Pimanthrene.

Chart 1.

salts (XII and XIII), the free bases can exist in either the oxazolidine forms (I–XXI; II–XXII) or in the open carbinolamine forms (see partial formula III, p. 29), which have the elementary composition of monohydrates of veatchine and garryine. The tertiary vinylamine isomers of veatchine and garryine represented by the partial structures (XXIX a) and (XXIX b) are clearly impossible within the framework of the complete structures (XXI) and (XXII). Thus, the ability of veatchine and garryine, according to the structures (XXI) and (XXII), to exist as anhydrous bases $C_{22}H_{33}O_2N$ is entirely due to the presence in the molecule of a hydroxyl group, which can form an internal ether. If this is the case, then a derivative of veatchine or garryine which retains the essential structural features of these two compounds and lacks hydroxy groups should exist as free base exclusively in the carbinolamine form and be incapable of dehydration. Such a derivative represented by one of the two structures (XXX a) and (XXX b) was easily prepared by forming the methiodide of the pyrolysis base (V, p. 29) and liberating the free

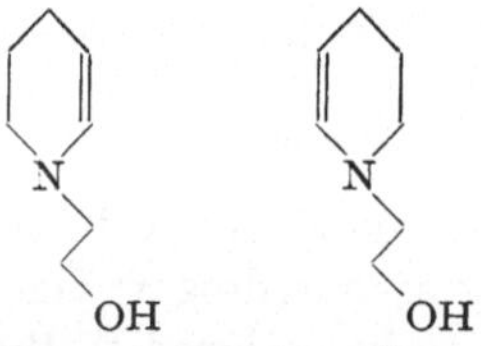

(XXIX.)

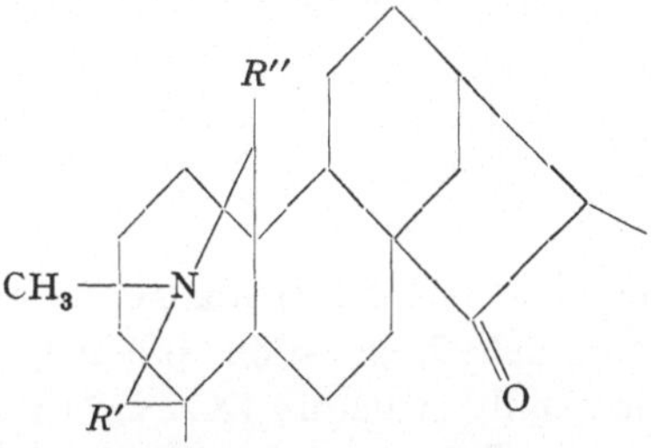

(XXX a.) ($R'' = H$; $R' = OH$)
(XXX b.) ($R' = H$; $R'' = OH$).

base with concentrated sodium hydroxide. The carbinolamine (XXX) indeed had the expected properties. It analysed for $C_{21}H_{33}O_2N$, showed the presence of a hydroxyl in the infrared spectrum and, most importantly, was completely stable to dehydration. Thus, this study again did not reveal any discrepancy between the proposed structures and the observed chemical properties of compound (XXX).

Structure Assignment to Veatchine and Garryine. The task which remained to be done at this point was to decide which of the two available structures (XXI) and (XXII) represented veatchine and which garryine. The reaction of garryine with methylmagnesium iodide gave methyldihydrogarryine which has either the structure (XXXI) (if garryine is XXI) or (XXXII) (if garryine is XXII) (58).

Dehydrogenation of methyldihydrogarryine by selenium gave as expected a good yield of 1-methyl-7-ethyl-phenanthrene (XVIII; p. 32) and a base $C_{17}H_{17}N$. The U. V. spectrum of this base paralleled the

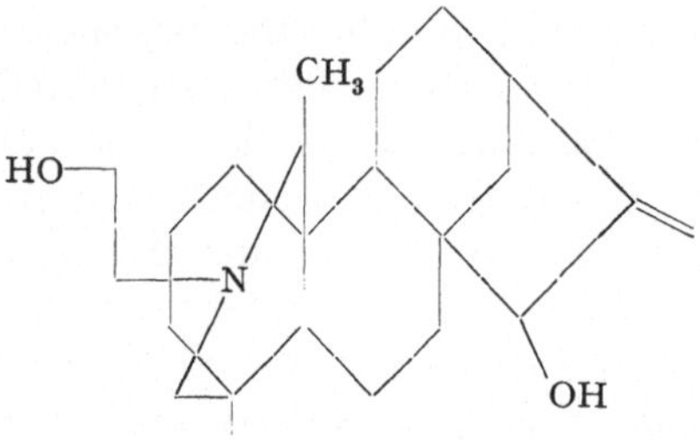

(XXXI.) Methyldihydroveatchine.

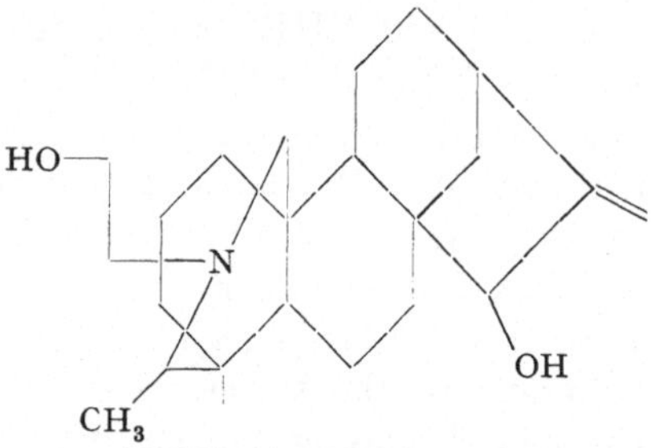

(XXXII.) Methyldihydrogarryine.

spectrum of the azaphenanthrene (XIX), except that it was slightly shifted to longer wavelengths by the influence of an additional substituent. Consequently, the base may be represented by either the structure (XXXIII) (if garryine is XXI) or (XXXIV) (if garryine is XXII).

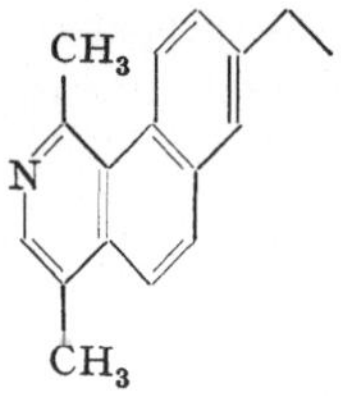

(XXXIII.)

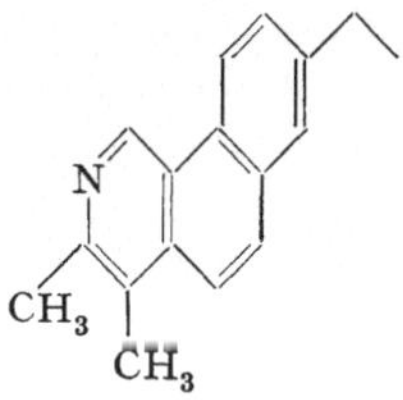

(XXXIV.)

Compound (XXXIII) was synthesised and found not to be identical with the dehydrogenation product of methyldihydrogarryine. Consequently, only structure (XXXIV) remains to be assigned to the latter compound. This in turn enables us to assign (XXXII) to methyldihydrogarryine, (XXII) to garryine, and (XXI) to veatchine.

It was a logical extension of this study to prepare in an analogous manner methyldihydroveatchine which must now be represented by (XXXI) and should give (XXXIII) by dehydrogenation. However, (XXXI) was obtainable only in poor yield and the small amount available did not give any detectable azaphenanthrene on dehydrogenation. The poor reactivity of veatchine with the GRIGNARD reagent is reminiscent of the instability of a veatchine carbinolamine as compared to the quite stable garryine carbinolamine. Both these findings are a consequence of the stereochemical situation in veatchine and garryine which will be discussed in the sequel.

2. The Stereochemistry of the Garrya Skeleton and its Influence on the Properties of Garrya Alkaloids.

While no rigorous evidence on the stereochemistry of the garrya skeleton exists to date, it would appear that the two steric arrangements (XXXV) and (XXXVI) deserve serious consideration.

In order to differentiate between these two possibilities the two epimeric alcohols represented by (XXXV) or (XXXVI) ($R' = C_2H_5$, $R'' = $ OH, $R''' = $ H; or $R' = C_2H_5$, $R'' = $ H, $R''' = $ OH) and their acetates have been prepared. It is easily seen from models, that if structure (XXXVI) were correct, there would be considerably more

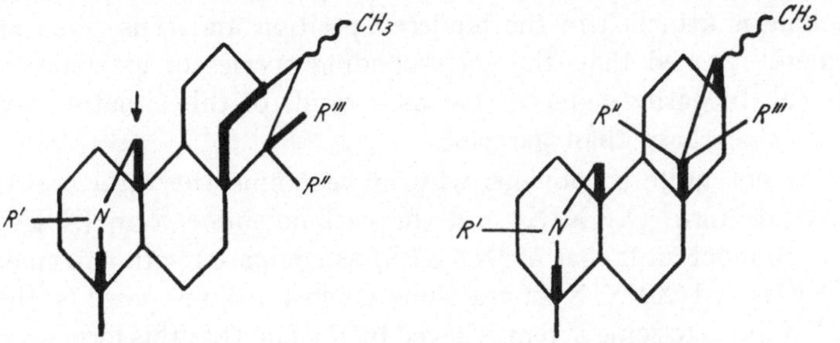

(XXXV.) (XXXVI.)

interaction between a hydroxy or acetoxy group and the nitrogen in one of the two epimers than in the other. In (XXXV) such an interaction would be negligible in both epimers. The basicities of all four compounds were determined and found identical within experimental error (pK $= 7.4$). This finding leaves us with (XXXV) as the more probable steric arrangement (58). The same skeletal stereochemistry was deduced in an ingenious biogenetic hypothesis by WENKERT (54).

A correct steric structure of the garrya skeleton must be able to provide us with a plausible explanation of the remarkable differences in the properties of veatchine and garryine. It is clear that structurally both these compounds are symmetrical and that consequently their different behaviour must have stereochemical causes.

It becomes immediately apparent on a model that in the accepted garrya skeleton one methylene group flanking the nitrogen is exceedingly strongly hindered (see arrow in XXXV). It is now possible to interpret the differences in behaviour of veatchine and garryine in terms of this steric hindrance (58).

A solution of veatchine or garryine in aqueous media contains an equilibrium mixture of the species (XXXVII), (XXXVIII), and (XXXIX) represented by the following partial structures:

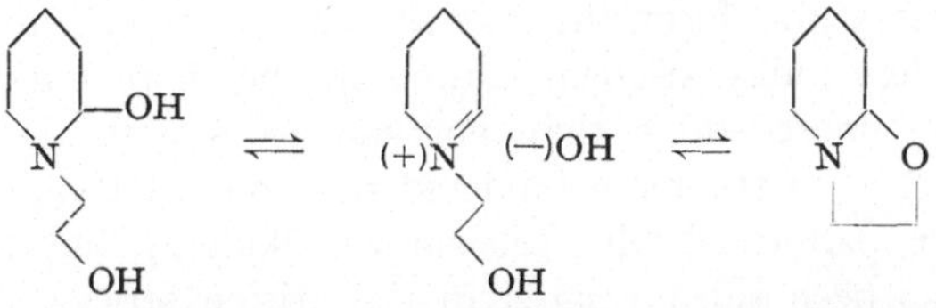

(XXXVII.) (XXXVIII.) (XXXIX.)

Many experiments devised to trap or detect a fourth aldehyde tautomer have failed and this species may consequently be neglected in the equilibrium.

It is clear that in veatchine the species (XXXVIII) is much more favoured in the equilibrium mixture than in garryine. The alcoholic or oxazolidine oxygen in the species (XXXVII) and (XXXIX) of veatchine is attached in the hindered position and thus these products are more strained than the corresponding species of garryine.

It can be easily deduced that as a result of this situation veatchine is a stronger base than garryine.

It is not quite so obvious why in veatchine the ratio between the oxazolidine form (XXXIX) and the carbinolamine form (XXXVII) is shifted so much in favour of (XXXIX) as compared with garryine. Both (XXXVII) and (XXXIX) of veatchine are strained, but possibly, the strain in (XXXIX) is to some extent relieved by the fact that this form may partly have the character of a zwitter-ion and be represented by (XXXIXa).

(XXXIX a.)

The different behaviour of veatchine and garryine in the reaction with methylmagnesium iodide is explainable in the same manner, i. e., by the steric hindrance of the carbon atom in veatchine which is being attacked by the Grignard reagent and also by the strain in the resulting methyldihydroveatchine.

Another interesting reaction is the conversion of veatchine to garryine by hot alkali *(Chart 2)*. In this medium their quaternary forms are undoubtedly in equilibrium with an ylid ion. Because of the high concentration of alkali the bulk of veatchine and garryine will be present in the equilibrium as the forms (XXXVII) and (XXXIX) and only a small amount as the quaternary bases (XXXVIII) and the ylid. It is therefore obvious that the equilibrium must be shifted toward the sterically favourable forms (XXXVII) and (XXXIX) of garryine and not to the corresponding strained forms of veatchine (58).

Chart 2 is also valid for the pair atisine-isoatisine which has the identical structure of the relevant nitrogenous portion (p. 47).

It must be emphasized that the greater stability of garryine is only true in a basic medium.

For the salts of veatchine and garryine which exist exclusively in the quaternary Schiff form, the reverse situation should hold. Veatchine salts have a less bulky trigonal carbon in the more hindered position and should be more stable than garryine salts with the more bulky tetragonal carbon in the more hindered position. Relevant experiments have not been performed with the garrya alkaloids, but this rationalisation (58) has been substantiated in the atisine series.

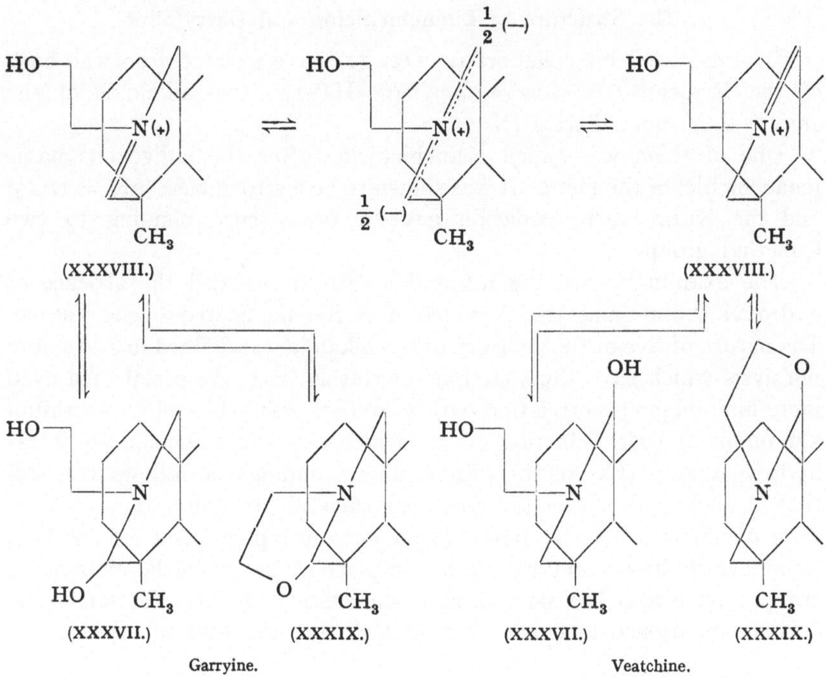

Chart 2. Base-catalysed Isomerisation of Veatchine.

It has been observed by DJERASSI and co-workers (*12*) that the isomerisation of garryfoline to isogarryfoline (which differ from veatchine and garryine only by the configuration of the secondary hydroxyl) proceeds also without the agency of an outside base in refluxing alcohol. This observation also holds for the conversion of veatchine to garryine.

The explanation of DJERASSI et al. involved the conversion of the oxazolidine into the zwitter-ionic form (XIV, p. 29). In this form the alcoxide ion may act as internal base and detach a proton adjacent to the nitrogen. This permits the shift of the double bond and recyclisation of the oxazolidine ring in the garryine position. The medium obviously must be sufficiently basic by the presence of the alkaloids themselves to maintain the bulk of the alkaloids in the carbinolamine or oxazolidine form. This is, according to our previous discussion, a necessary condition. DJERASSI's mechanism is plausible, especially since a partially zwitter-ionic character of the veatchine oxazolidine fits well with several other observations (see above).

The manifold behaviour of garryine and veatchine seems now to be adequately accounted for in terms of the proposed structures and skeletal stereochemistry, and the only missing information is the configuration of the secondary alcoholic group.

The discussion of this feature will be, however, deferred (p. 42), since it will be of advantage to describe first the experiments performed by DJERASSI and his group on the alkaloid garryfoline.

3. The Structure of Cuauchichicine and Garryfoline.

Djerassi and his collaborators (*12, 11*) have isolated from the bark of the Mexican tree *Garrya laurifolia* Hartw. two alkaloids of the empirical formula $C_{22}H_{33}O_2N$.

One of them was called cuauchichicine after the indigenous name (cuauchichic) of the plant. It was shown to be a strong base (pK = 11.15) and the Kuhn-Roth oxidation gave a result corresponding to two C-methyl groups.

The examination of the infrared spectrum revealed the absence of hydroxyl groups and the presence of a five-membered cyclic ketone. The nature of five of the six rings of the alkaloid was defined in a selenium pyrolysis which gave the veatchine pyrolysis base (V, p. 29). An even more unambiguous correlation with veatchine was achieved by a lithium aluminum hydride reduction of cuauchichicine which yielded epi-tetra-hydroveatchine (IX, p. 29). From these findings it follows (*11, 12*) that cuauchichicine has the identical skeletal structure as veatchine, from which it differs by possessing a keto group in place of the $C_{(19)}$ secondary hydroxyl and a C-methyl in place of the exocyclic methylenic group. Since also the skeletal stereochemistry must be the same, the authors mentioned assigned to cuauchichicine the structure (XL).

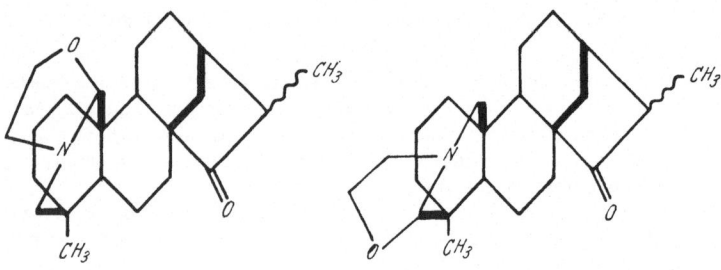

(XL.) Cuauchichicine.   (XLI.) Isocuauchichicine.

The attachment of the oxazolidine ring at $C_{(17)}$ follows from the basicity of cuauchichicine. In this property the compound resembles veatchine and not garryine. Cuauchichicine is isomerised in basic medium easily to isocuauchichicine (pK = 8.6) which corresponds to garryine and must be represented by (XLI).

The second alkaloid, called garryfoline was isolated by a modified method, when it was realized that it is labile to mineral acid (*12*). The method involved partitioning of the alcoholic extract between dilute acetic acid and methylene dichloride followed by countercurrent distribution. Garryfoline is an isomer of cuauchichicine and veatchine. It has a similar basicity (pK = 11.8) as both these compounds and like veatchine has one hydroxyl, one C-methyl and one exocyclic methylene group.

The action of mineral acid on garryfoline at room temperature gave a good yield of cuauchichicine (XL). With lithium aluminum hydride garryfoline yielded *F*-dihydrogarryfoline, formulated as the $C_{(19)}$ epimer of dihydroveatchine (XX, p. 32). This compound in turn by treatment with acid gave *F*-dihydrocuauchichicine, also obtainable by catalytic hydrogenation of cuauchichicine and represented by (XLII)*.

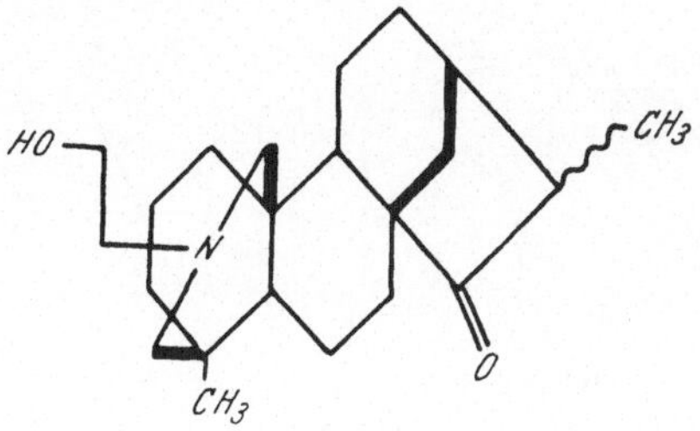

(XLII.)

Finally, lithium aluminum hydride reduction of (XLII) gave epi-tetrahydroveatchine (IX, p. 29). These transformations are, according to DJERASSI, consistent only with the formulation of garryfoline as 19-epi-veatchine (cf. XXI, p. 34).

This conclusion was corroborated by EDWARDS in the laboratory of the writers in the following manner (*17*). Veatchine was oxidised by chromium trioxide in pyridine to the α,β-unsaturated ketone (XLIII), which gave *F*-dihydrogarryfoline by lithium aluminum hydride reduction of the keto-group and the oxazolidine ring. The con-

(XLIII.)

figuration of the secondary hydroxyl in *F*-dihydrogarryfoline must, consequently, be the sole difference between this compound and dihydroveatchine and both products must possess the structural formula (XX, p. 32).

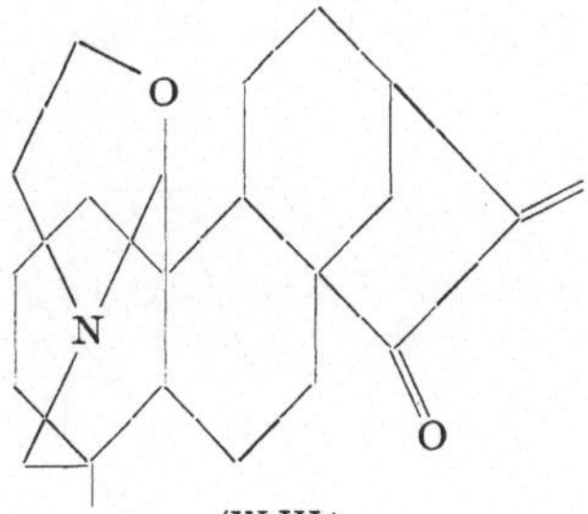

While garryfoline suffers isomerisation of the allylic alcohol group to a ketone by acid treatment at room temperature, such a reaction does not take place with veatchine, and as a matter of fact has been realised in the veatchine series only on selenium pyrolysis with a simultaneous loss of the side-chain.

---

* In the "*F*-dihydro" garrya derivatives reduction of ring *F* has occurred (see p. 34).

The difference in the acid catalysed ketonisation of the two epimers
garryfoline and veatchine must be a consequence of the stereochemistry
of the $C_{(19)}$ hydroxyl and will be discussed in the sequel.

4. Configuration of the Secondary Hydroxyl Group in Veatchine and Garryfoline.

From the preceding paragraph it is clear that garryfoline and veatchine
are epimers and must be represented by the two partial steric structures
(XLIV) and (XLV) or vice versa. The remaining problem is to decide

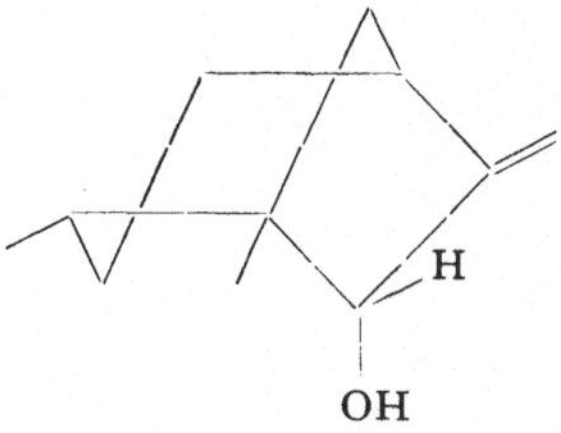

(XLIV.) 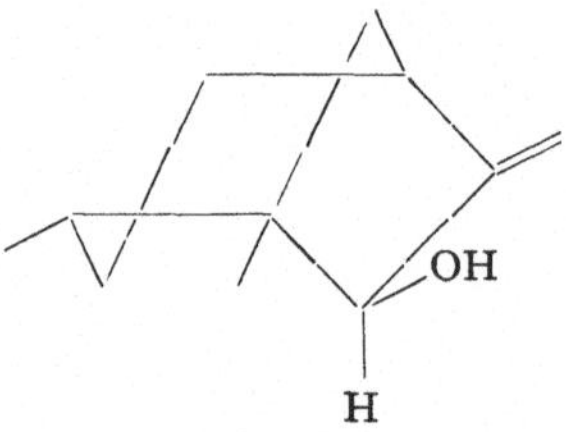   (XLV.)

which of these two structures belongs to veatchine and which to garryfoline. As already stated the structure assigned to garryfoline must
involve an explanation for the extreme ease of acid catalysed ketonisation
of this compound as compared to veatchine. It has been suggested by
Professor R. B. Woodward (cf. 58, 12) that protonation of the exo-

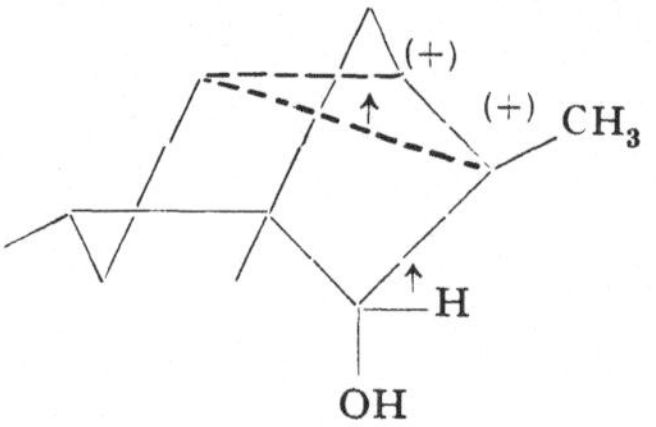

(XLVI.) 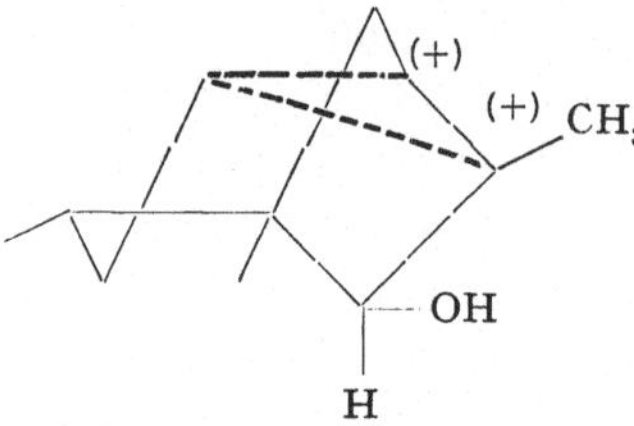   (XLVII.)

cyclic double bond in (XLIV) and (XLV) will give the two non-classical
ions (XLVI) and (XLVII). Of these clearly (XLVI) has the steric
requirements for an easy elimination of a proton, according to the arrows
in formula (XLVI), which leads to the enol of cuauchichicine.

Thus (XLIV) may be assigned to garryfoline and (XLV) to veatchine.
The bridged cation intermediate is not absolutely necessary for the
ketonisation of exocyclic allylic alcohols since it is known (13) that also
for instance compound (XLVIII) will undergo an easy acid catalysed
isomerisation to (XLIX). However, the dramatic stereospecificity of
the ketonisation in the bicyclo-octane system (XLIV)–(XLV) is hardly

explainable in other terms. The search for other possible causes of stereospecificity in the acid catalysed ketonisation of garryfoline and

(XLVIII.) (XLIX.)

veatchine has revealed that all other factors (by far less compelling than the bridged cation concept) necessitate the assignment of (XLIV) to the easily ketonising compound, i. e. garryfoline.

In conclusion, it seems interesting to discuss briefly the configuration of the methyl group in some of the tetrahydro derivatives.

The ketones as the pyrolysis base (V, p. 29) and F-dihydrocuauchichicine are stable to acid and base and may be assigned the presumably more stable exo configuration (L).

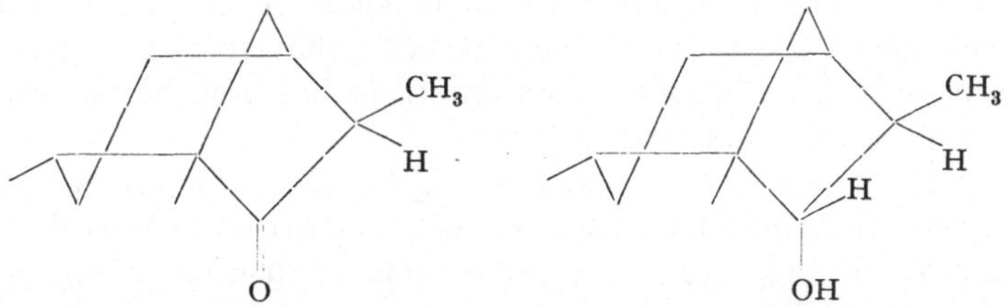

(L.) (LI.)

DJERASSI et al. have found (12) that reduction under equilibrating conditions of F-dihydrocuauchichicine gives epi-tetrahydroveatchine and, possibly, some other compounds. It may be clearly seen on models of the whole molecule that, with the methyl group exo, the hydroxy group has fewer non-bonded interactions when in the endo configuration. Thus, from the two epimeric reduction products of (L), (LI) seems to be more stable and represent epi-tetrahydroveatchine. This is in agreement with the previous discussions which have shown that the configurations of the secondary hydroxyl in garryfoline and epi-tetrahydroveatchine are identical.

The reduction of ketoveatchine (XLIII) with lithium aluminum hydride is known to give F-dihydrogarryfoline represented by the partial formula (XLIV). Since it appears that this epimer is the less stable one, it is necessary to assume that in this case the hydride reduction is kinetically controlled. Tetrahydroveatchine may be represented by the partial structure (LII) or (LIII) (p. 44). There does not appear to be any clear cut argument to decide between these two possibilities.

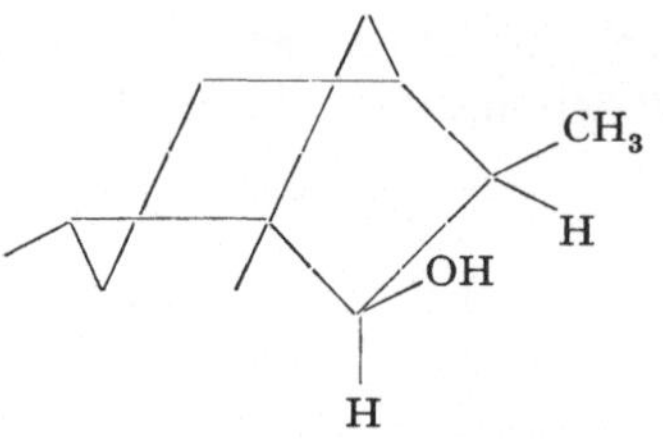

CH₃
H
OH
H
(LII.)

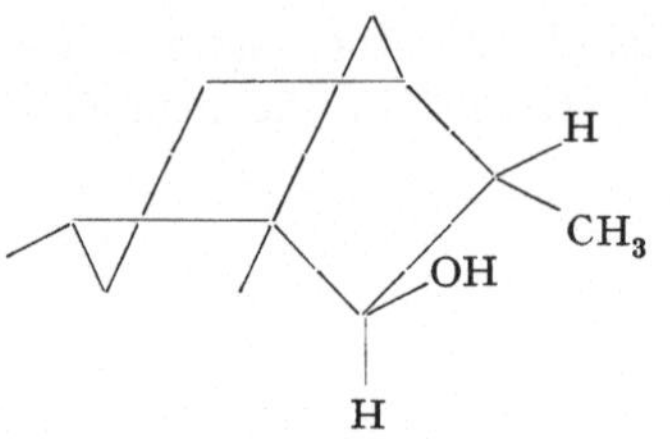

H
CH₃
OH
H
(LIII.)

III. The Structure of Atisine.

The alkaloid atisine has been isolated from the roots of *Aconitum heterophyllum* by Wright (63), Jowett (42) and Lawson and Topps (44).

Original Degradation Work on Atisine. Prior to the structural elucidation of the garrya alkaloids a large amount of data on atisine has been accumulated mainly by the efforts of Jacobs and his coworkers. The similarity between the reactions of veatchine and atisine is so striking that only a brief survey of the atisine transformations will be given in order to avoid repetition. Furthermore, it should be stated that in many instances where deviations in the reported behavior of veatchine and atisine have been found, later work on atisine has brought the disputed point in line with veatchine chemistry.

Atisine is isomeric with veatchine $C_{22}H_{33}O_2N$. It is an exceedingly strong base which by alkali may be easily isomerized to a weaker base termed isoatisine (*33, 34*). The preparation of diacetyl atisine hydrochloride was taken as evidence for the presence of two hydroxyls in atisine (*33*). On vigorous treatment of atisine with alcoholic sodium hydroxide dihydroatisine $C_{22}H_{35}O_2N$ was obtained which was later found to be identical with the product of sodium borohydride reduction of atisine or isoatisine (*24*). Atisine has one N-Alkyl group and the Kuhn-Roth oxidation revealed the presence of one C-methyl group (*33*).

Oxidation of isoatisine with potassium permanganate gave in good yield a neutral compound, oxoisoatisine $C_{22}H_{33}O_3N$. At first atisine and isoatisine were reported to give in the Zerewitinoff determination values corresponding to two active hydrogens. This was later corrected to one, in agreement with the behavior of veatchine (see below, p. 47).

Hydrogenation of atisine gave a mixture of tetrahydroatisines from which one component $C_{22}H_{37}O_2N$ was isolated in pure form.

Further similarities between atisine and veatchine are revealed by dehydrogenation studies which were initiated by Lawson and Topps (*44*) and carried on by Jacobs (*33*). The neutral dehydrogenation products were identified by Jacobs as 1-methylphenanthrene and 1-methyl-6-ethylphenanthrene (*33, 29*). Basic dehydrogenation products were also obtained by both Lawson and Jacobs. Among these was a base $C_{16}H_{15}N$

and another, $C_{20}H_{29}ON$. However, the crucial importance of these two products did not emerge until after the clarification of analogous derivatives in the garrya series.

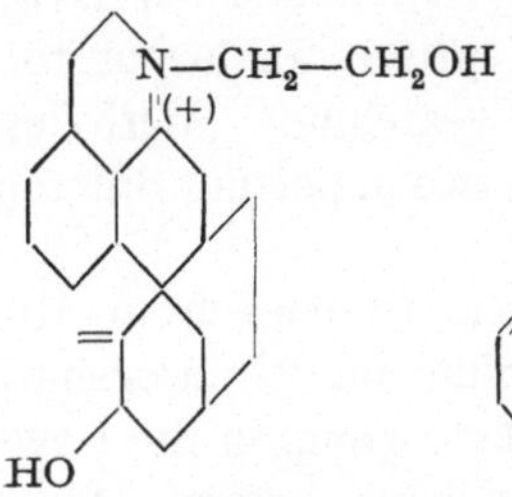

(LIV.) 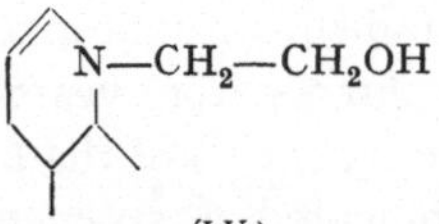 (LV.) 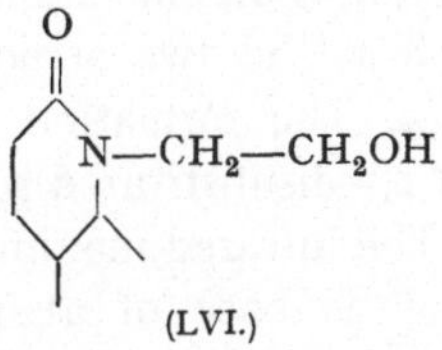 (LVI.)

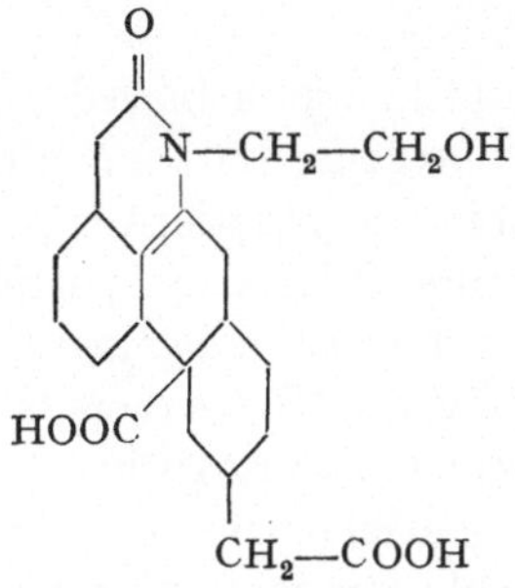

(LVII.) 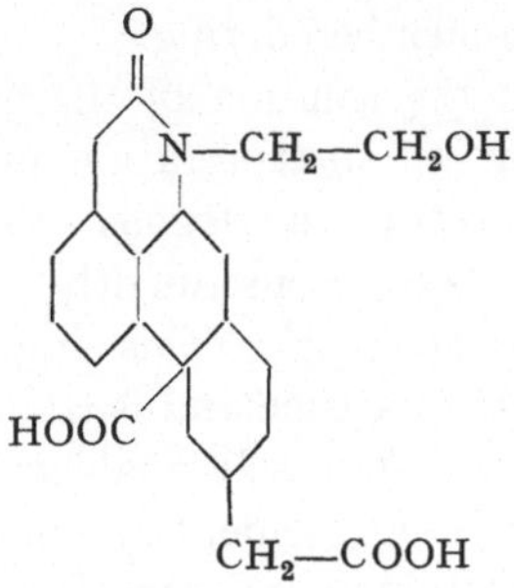 (LVIII.)

Finally, JACOBS (*32*) has performed a series of meticulous oxidation studies and made an attempt to rationalise the large array of experimental data by tentative structural formulae assigned to atisine and its degradation products. Atisine, isoatisine and isooxoatisine were represented by structures (LIV), (LV), and (LVI). Formula (LIV) for atisine incorporated the 1-methyl-6-ethylphenanthrene, identified among the dehydrogenation products. It further accommodated well the oxidative degradations to be described.

Permanganate oxidation of atisine gave a lactam dicarboxylic acid $C_{21}H_{29}O_6N$ formulated as (LVII). This formulation was in agreement with the finding that the dimethylester gave on saponification a monoester and furthermore, with the isolation of 1,6-dimethylphenanthrene upon selenium dehydrogenation of the acid. A similar acid $C_{21}H_{31}O_6N$ was obtained by permanganate oxidation of isoatisine and represented by (LVIII). The relationship of the two acids was corroborated by catalytic hydrogenation of (LVII) which gave (LVIII).

Similarity of Atisine with the Garrya Alkaloids and the Deduction of the Atisine Structure. This short exposition leaves no doubt that the parallelism between the reactions of atisine and veatchine cannot possibly be due to coincidence. This impression was further corroborated by the fact that the $C_{16}H_{15}N$ dehydrogenation product of atisine showed an ultraviolet spectrum practically identical with the 1-methyl-7-ethyl-3-azaphenanthrene from veatchine. Also the $C_{20}H_{29}ON$ base of Lawson was shown (1) to be a Δ_1-piperideine ketone, entirely analogous to the pyrolysis base $C_{20}H_{29}ON$ (V, p. 29) from veatchine. With every reaction of atisine duplicated in the garrya series two important differences between the two structure types emerged:

(a) The formation of 1,6-disubstituted phenanthrenes from atisine and 1,7-disubstituted phenanthrenes from veatchine on dehydrogenation.

(b) The infrared maximum (1705 cm.$^{-1}$) of the keto group in the Lawson pyrolysis base of atisine indicated a six-membered ketone (1); the analogous compound in the veatchine series (see above) has a keto group in a five-membered ring.

When the solution of the garrya problem was achieved (57) it was clear that an interpretation of all the foregoing data on atisine in a manner precisely analogous to the veatchine degradations is exceedingly plausible (55). Consequently, the structure (LIX) was postulated for dihydroatisine (57). This implies the arbitrary assignment of (LX) and (LXI) to atisine and isoatisine or vice versa. Since there is no doubt that atisine is precisely analogous to veatchine, and isoatisine to garryine, the rigorous assignement (58) of the point of attachment of the oxazolidine ring in veatchine and garryine automatically pinpointed atisine as (LX) and isoatisine as (LXI). The differences in the properties of atisine and isoatisine are clearly due to the same factors governing the behavior of veatchine and garryine. These have been already subjected to a thorough discussion. The reinterpretation of the main degradation products is clear and does not require special comment. Lawson's pyrolysis base $C_{20}H_{29}ON$ must be (LXII) and the $C_{16}H_{15}N$ dehydrogenation base of Lawson and Jacobs is best represented as (LXIII). Oxoatisine dicarboxylic acid represented formerly by (LVII) must now be given the structure (LXIV). The conversion of (LXIV) into isooxoatisine dicarboxylic acid may be interpreted as a hydrogenolysis of the oxazolidine ring and the latter compound must be assigned structure (LXV).

The same difficulty, which was already encountered in the garrya problem, occurs at this point. By selenium dehydrogenation (LXIV) and (LXV) give 1,6-dimethylphenanthrene.

Analogously, the corresponding acid from veatchine (XXIV, p. 34) yields 1,7-dimethylphenanthrene. In the case of (XXIV) it has been demonstrated that this is due to a reduction of the secondary carboxyl in the course of the selenium dehydrogenation. While such rigorous evidence is still missing for the atisine case, it is certainly plausible to explain the formation of (LXVI) from (LXIV) and (LXV) in the same manner, since the veatchine case is a good precedent for this unusual reduction.

Some Confirmatory Evidence. With the structure of atisine and its relationship to the garrya alkaloids clarified, the way was open for a more detailed study of atisine chemistry and for the clarification of some isolated pieces of contradictory evidence. This was achieved in a series of careful investigations by PELLETIER and JACOBS (*46–49*) and by O. E. EDWARDS (*16*).

Thus for instance it was shown (*46*) that pure undistilled atisine shows in agreement with the behaviour of veatchine only one active hydrogen in the ZEREWITINOFF test. This was further corroborated by oxidation of the secondary hydroxyl of atisine to a α,β-unsaturated

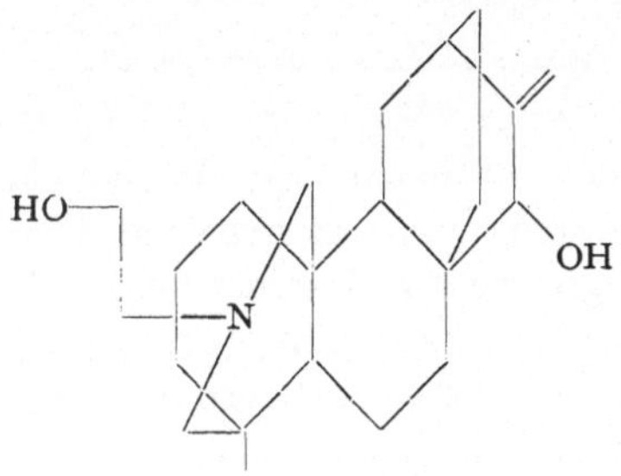

(LIX.) Dihydroatisine.

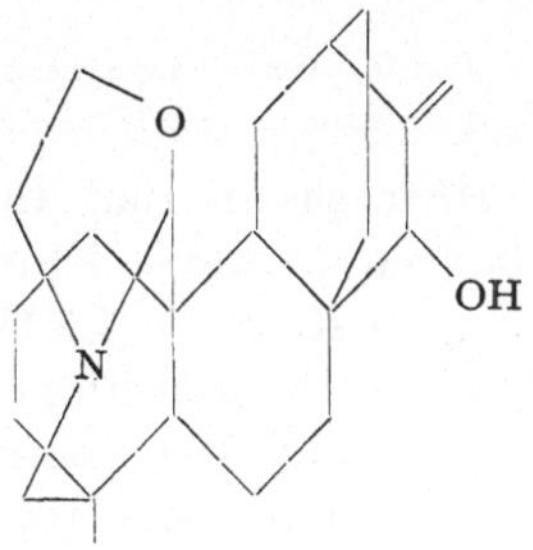

(LX.) Atisine.

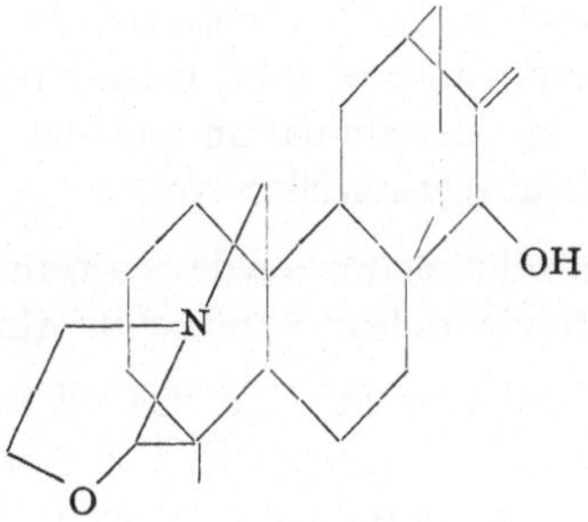

(LXI.) Isoatisine.

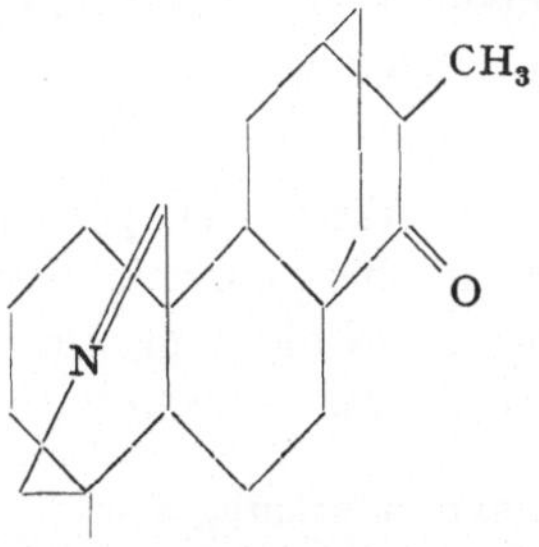

(LXII.)

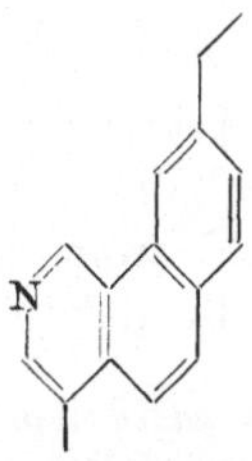

(LXIII.)

ketone. This compound showed the expected ultraviolet spectrum (λ_{max} 228 mμ, $\varepsilon = 9.100$) and no hydroxyl band in the infrared. It

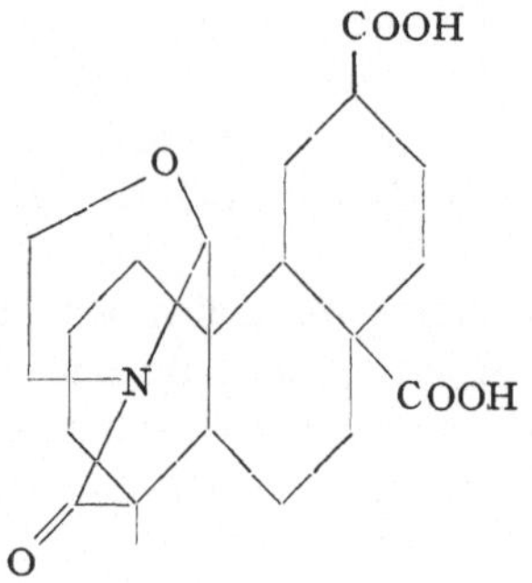

(LXIV.) Oxoatisine dicarboxylic acid. 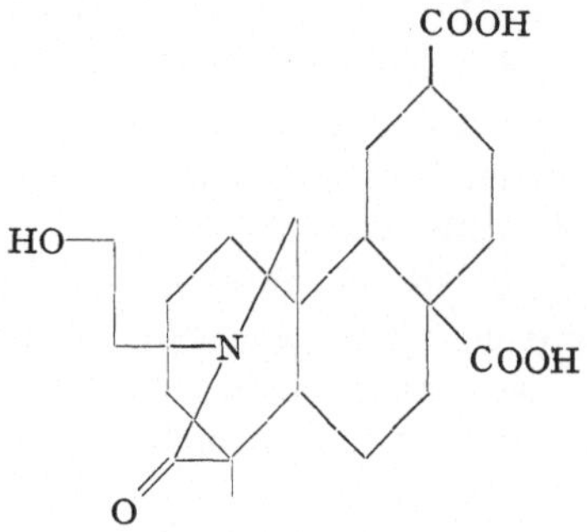   (LXV.) Isooxoatisine dicarboxylic acid.

was further shown that the dimethylester of oxoatisine dicarboxylic acid has no hydroxyl band in the infrared curve as expected from structure (LXIV) but in disagreement with structure (LVII). The basicities of atisine (pK = 12.2), isoatisine (pK = 10.0), and dihydroatisine (pK = 8.2) have been determined (24), and these three values are in the same relationship as the corresponding basicities of veatchine, garryine and dihydroveatchine. The consistently higher values of the atisine compounds may be at least partly due to a different method of determination (atisine in dilute methanol, veatchine in methylcellosolve).

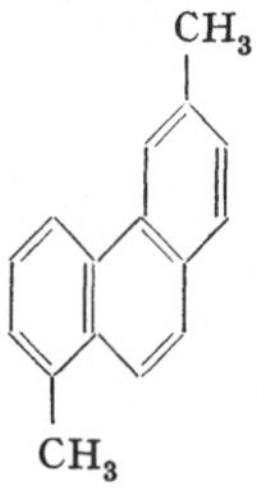

(LXVI.)

The presence of an ethanolamine side-chain in tetrahydroatisine was corroborated by oxidation with lead tetraacetate which gave glyoxal (24)*.

While the oxidation of veatchine has yielded two lactams $C_{22}H_{31}O_3N$, one of which has the lactam carbonyl in a five- and the other in a six-membered ring, such compounds have previously not been obtained from atisine. This discrepancy has been remedied when PELLETIER and JACOBS (47) isolated two comparable compounds by oxidation of atisine with permanganate. These compounds must clearly be formulated as (LXVII) and (LXVIII).

A third product $C_{20}H_{29}ON$ comparable to the second pyrolysis base of veatchine (VI, p. 29) was also isolated in this experiment and must have resulted from an oxidative loss of the N-alkyl group. It was represented by (LXIX). This compound was now reduced with sodium

---

* *Footnote added in proof:* This evidence has to be discounted, since the reaction is unspecific (cf. p. 61).

borohydride to the secondary base (LXX) which on treatment with ethylene chlorohydrine gave dihydroatisine (48). Thus, rigorous evidence

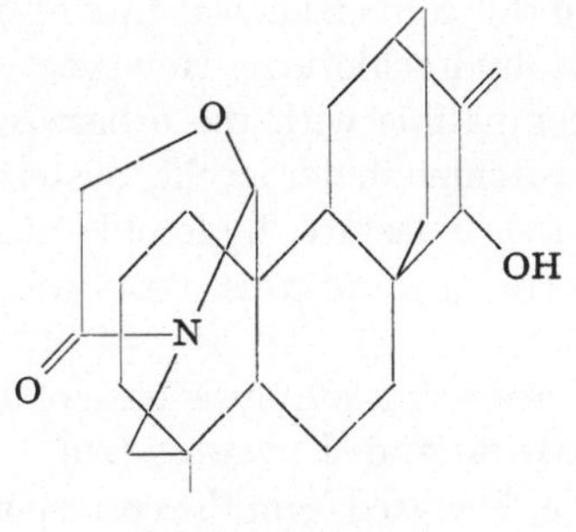

(LXVII.) Oxoatisine A. 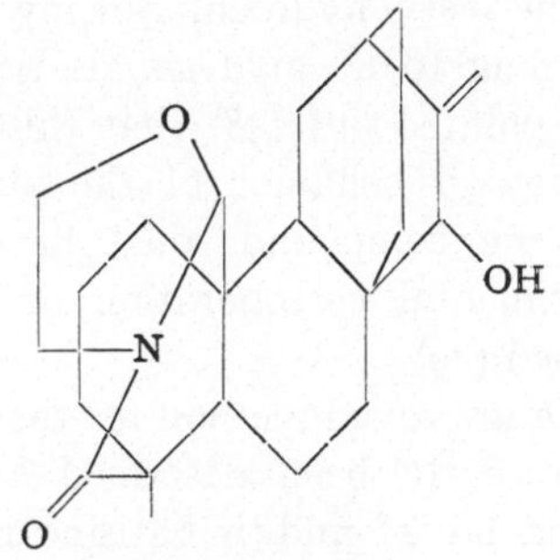   (LXVIII.) Oxoatisine B.

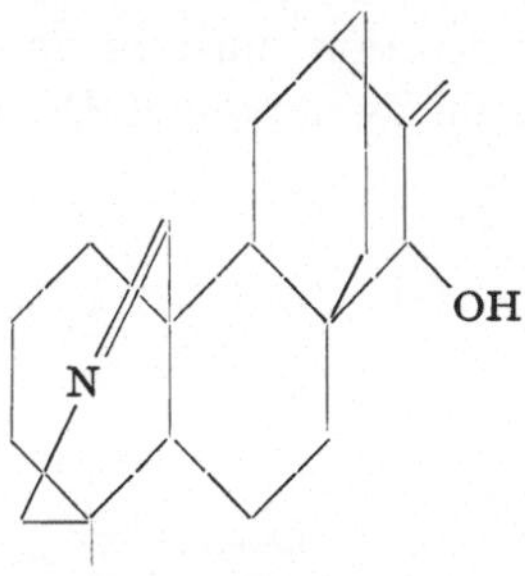

(LXIX.) 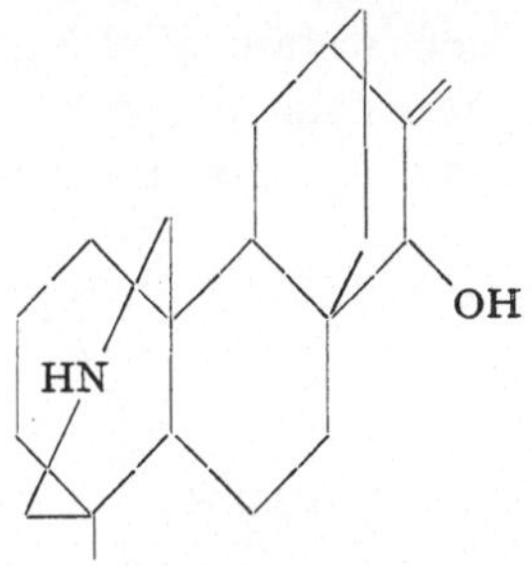   (LXX.)

of the presence of an ethanolamine side-chain in (LIX) has been accomplished by a sequence of reactions analogous to those used in the garrya series.

O. E. EDWARDS has performed an interesting series of reactions which effected a transformation of isoatisine into atisine (23). This seems to be at first startling since we know that the reverse change is effected very easily by alkali. However, it may be easily rationalised if we consider that the nitrogen environments in atisine and veatchine are identical and consequently all our previous deductions on the behavior of this system in veatchine are directly applicable to atisine. Thus, one must realise that the greater stability of isoatisine in strongly alkaline solution is due to the greater stability of the carbinolamine and oxazolidine forms of isoatisine as compared to the same forms of atisine. For the salts existing exclusively in the quaternary SCHIFF forms the reverse is true. Atisine salts should be more stable than isoatisine salts, since the former compounds have a less bulky trigonal carbon in the more hindered position flanking the nitrogen (58).

The actual transformations were as follows. By the action of acetic anhydride on isoatisine hydrochloride an acetyl hydrochloride was obtained which gave atisine by mild hydrolysis with sodium carbonate.

The acetyl hydrochloride involved in this conversion was first regarded, according to its analysis, as triacetate hydrochloride. However, it has been pointed out (58) that this is incompatible with the otherwise well understood behavior of the atisine-veatchine heterocyclic system and that the compound must have been a diacetate hydrochloride. An experimental corroboration of the correctness of this view was soon reached (49).

An improved method for the preparation of a pyrolysis base of atisine has recently been described by Dvornik and Edwards (16). These authors have found that atisine diacetate, liberated from the corresponding hydrochloride (which has been discussed in the preceding paragraph), decomposes easily by heating in nonpolar solvents into acetaldehyde and the monoacetate of (LXIX). This reaction must be regarded (as has been also pointed out by the authors) as a normal Hofmann degradation and is represented by (LXXI).

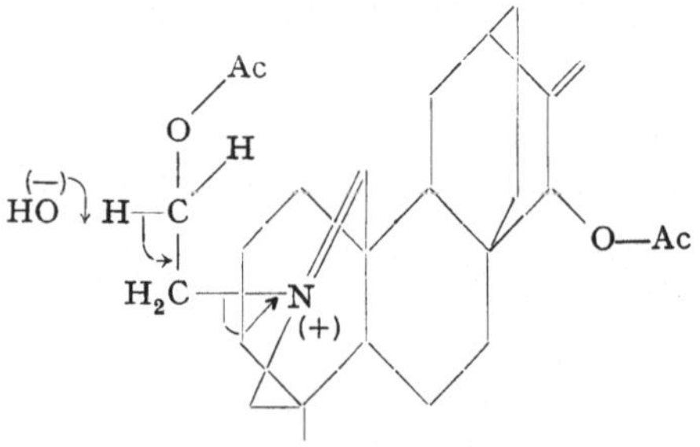

(LXXI.)

A direct comparison of (LXIX) as obtained by Pelletier and Jacobs (47) and by Dvornik and Edwards (16) has to the knowledge of the reviewers not yet been made, but the physical constants reported by both groups are in fair agreement.

Treatment of compound (LXIX) with ethylene chlorohydrine in dimethylformamide gave atisine (16).

IV. Biogenesis of Veatchine and Atisine and Consideration of the Stereochemistry of Atisine.

The biogenetic origin and correlation of veatchine and atisine were postulated in an ingenious hypothesis by Wenkert (54). This author assumes that both compounds originate from a precursor which (disregarding the nitrogen and oxygen functions) is essentially a primaradiene

(LXXII), in which the $C_{(7)}$ vinyl group is quasi-axial and the $C_{(7)}$ methyl quasi-equatorial. The biogenesis of the veatchine skeleton (LXXIV)

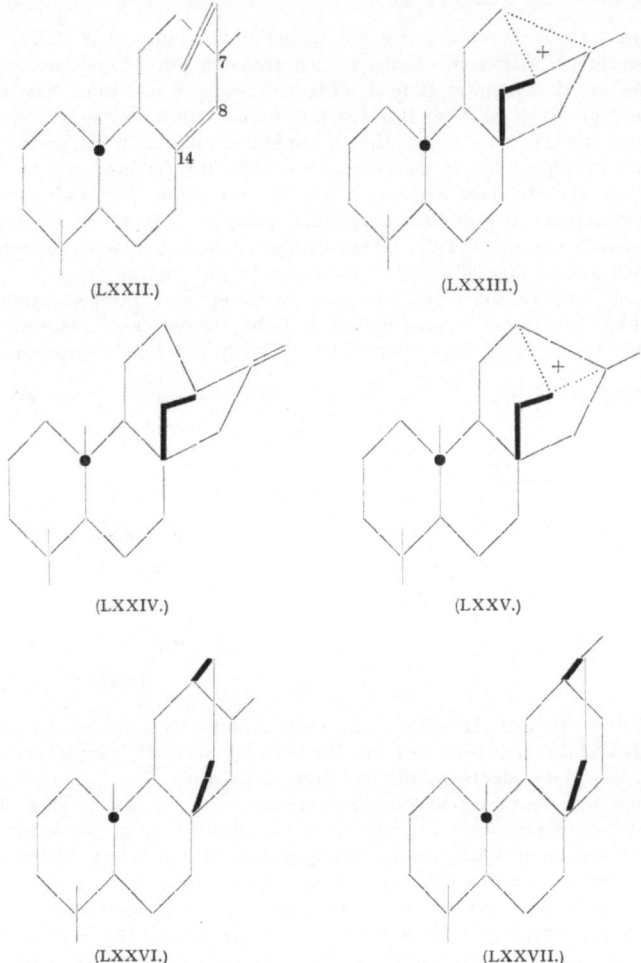

(LXXII.)   (LXXIII.)

(LXXIV.)   (LXXV.)

(LXXVI.)   (LXXVII.)

would then proceed by protonation of the 8,14-double bond *via* the bridged cation (LXXIII). The latter may be transformed by hydride shift* to the bridged cation (LXXV) which can give rise to the atisine skeleton (LXXVI).

We have already discussed the chemical evidence which makes it probable that for the garrya alkaloids the biogenetic steric arrangement

---

* In this abbreviated discussion the question whether the hydride shift transforms (LXXIII) to (LXXV) or gives rise to a more complex non-classical ion in the sense of J. D. ROBERTS is neglected.

(LXXIV) is indeed correct. However, in the case of atisine it is still undecided whether the skeleton is represented by (LXXVI) or by the second alternative (LXXVII).

Recently, O. E. Edwards (*14*) mentioned in a lecture that (LXXVII) is in better agreement with some basicity measurements on atisine derivatives. In this connection it is of interest to describe the recent work which has been going on at the University of New Brunswick for some time and is recorded in two dissertations (*17, 43*). It concerns the attempted correlation of the garrya alkaloids and atisine by vigorous acid isomerisation of dihydroveatchine.

We have already discussed on p. 42, Woodward's proposal according to which protonation of *F*-dihydrogarryfoline gives a bridged cation represented by the partial formula (XLVI). This bridged cation is then transformed to a ketone with a veatchine skeleton as indicated by the arrows.

No such simple ketonisation is possible for the epimeric bridged cation (XLVII, p. 42) which can arise by protonation of *F*-dihydroveatchine. However, it was anticipated in our laboratory that (XLVII) might be slowly transformed by a

(LXXVIII.)   (LXXIX.)

hydride shift to (LXXVIII) which may then ketonise as indicated by the arrows to give (LXXIX). If atisine has the Wenkert biogenetic stereochemistry (*54*), (LXXIX) should be identical with keto-tetrahydroatisine.

By vigorous treatment of dihydroveatchine with hydrochloric acid it was indeed possible to prepare a ketone which was purified by a multistage counter-current distribution until it appeared homogeneous and showed an infrared carbonyl frequency typical of a six-membered ring. It was characterised by its sodium borohydride reduction product which was crystalline and homogeneous.

Since a probable explanation of this rearrangement is the anticipated transformation (LXXVIII) → (LXXIX), it was tentatively concluded that the new ketone, "tetrahydro-ketoisoveatchine" is to be represented by (LXXIX). Subsequently, pure tetrahydro-ketoatisine was prepared by mild hydrogenation of ketoatisine or by isomerisation of *F*-dihydroatisine at room temperature with palladium.

It was also shown that borohydride reduction of keto-tetrahydroatisine gives a mixture of two well-defined diastereoisomeric tetrahydroatisines. Thus, comparable compounds were made available both in the isoveatchine and atisine series and it was found that they are clearly not identical.

It seems at this point, that the simplest interpretation of all this work would be the assignment of structure (LXXX) to tetrahydro-ketoatisine and (LXXIX) to tetrahydro-ketoisoveatchine. However, the structure of tetrahydro-keto-isoveatchine is by no means certain, and also O. E. Edwards (private communication.

in press) has very recently reversed his opinion about the stereochemistry of atisine which will require further experimental studies.

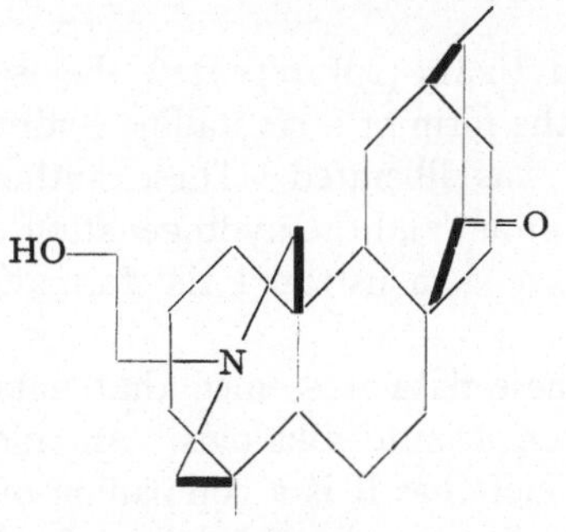

(LXXX.)

V. The Structure of Ajaconine.

From the many known aconite alkaloids only one has been now definitely proven to possess the atisine skeleton. It is the alkaloid ajaconine isolated by GOODSON (28) from the seeds of *Delphinium ajacis*. O. E. EDWARDS (15) in a careful experimental study has shown that ajaconine is a hydroxyatisine of the structure (LXXXI).

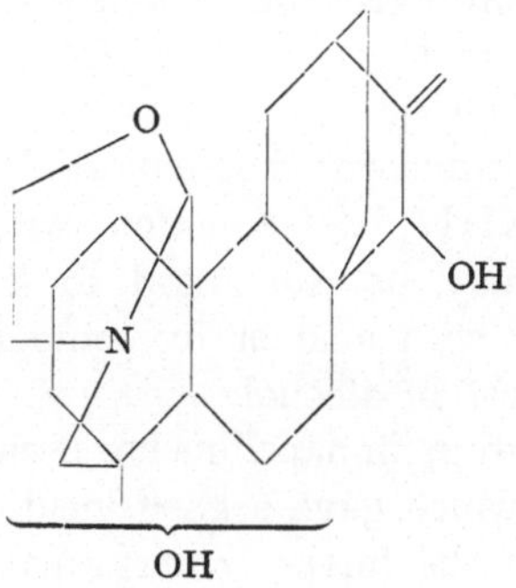

(LXXXI.) Ajaconine. 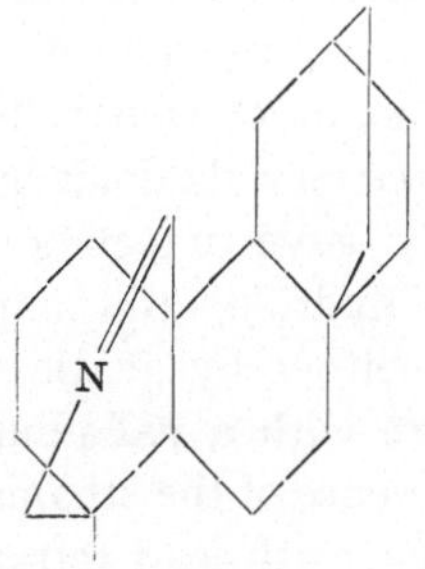   (LXXXII.)

The evidence includes a conversion of both atisine and ajaconine into the same azomethine (LXXXII) by a series of oxydations and WOLFF-KISHNER reductions.

Clearly, ajaconine with its new oxygen function provides an additional opening for testing of the atisine skeleton and undoubtedly this will be utilised by the Ottawa workers in the near future.

VI. The Structure of Napelline.

In 1937 FREUDENBERG and ROGERS (25) reported the isolation of a new alkaloid napelline from commercial amorphous aconitine, which is being manufactured by E. Merck from *Aconitum napellus*. The compound was characterised in form of various salts and given the formula $C_{22}H_{33}O_3N$.

Later Jacobs and Craig (*10*) repeated the isolation and obtained a similar material in the form of a crystalline hydrobromide, from which a resinous free base was liberated. These authors also reported the presence of a N-alkyl group and the hydrogenation to a dihydronapelline. Furthermore, they have demonstrated the formation of phenanthrenes on dehydrogenation.

On the basis of these data it seemed that napelline may be one of the simplest hexacyclic aconite alkaloids. An additional attraction of this alkaloid was the fact that it is a companion of aconitine and might conceivably be structurally related to this extremely challenging compound. As a result of these considerations, the problem was taken up in the laboratory of the writers.

It was first shown that "napelline" obtained by the published procedures is a mixture of at least three compounds. They are represented by the partial structures (LXXXIII), (LXXXIV), and (LXXXV) in which an additional bond between $C_{(17)}$ and another carbon atom in the molecule has to be closed (*61*). Napellonine (LXXXIV) may be separated from the mixture by the use of Girard reagent T, and the residue is resolved by countercurrent distribution into napelline (LXXXIII) and a small amount of isonapelline (LXXXV).

The three new alkaloids have been interrelated as follows: Napellonine (LXXXIV) gave napelline (LXXXIII) by reduction with lithium aluminum hydride, and napelline may be isomerised to isonapelline (LXXXV) either by gentle warming with acid or by contact at room temperature with a palladium catalyst in alcohol.

The presence of the ethanolamine group in napelline has been indicated by oxidation with lead tetraacetate which gave a good yield of glyoxal (cf. *20*) and is confirmed by some of the further transformations. The presence of an exocyclic methylene group follows from both ozonolysis and Kuhn-Roth oxidation, and from infrared spectroscopy.

Finally, the functions of all three oxygens are confirmed by acetylation and by oxydation to various ketonic derivatives.

Napellonine (LXXXIV) shows a carbonyl peak in the infrared, characteristic of a six-membered ring, and an abnormal ultraviolet spectrum (*62, 9*) ($\lambda_{max} = 290$ mμ, log $\varepsilon = 2.6$). These spectral properties find their expression in the formula (LXXXIV) in which an interaction between the carbonyl and the exocyclic methylenic group is possible.

It may be shown very simply that no skeletal rearrangement has taken place in the ketonisation (LXXXIII) → (LXXXV). Isonapelline (LXXXV) gives by reduction with lithium aluminum hydride dihydro-

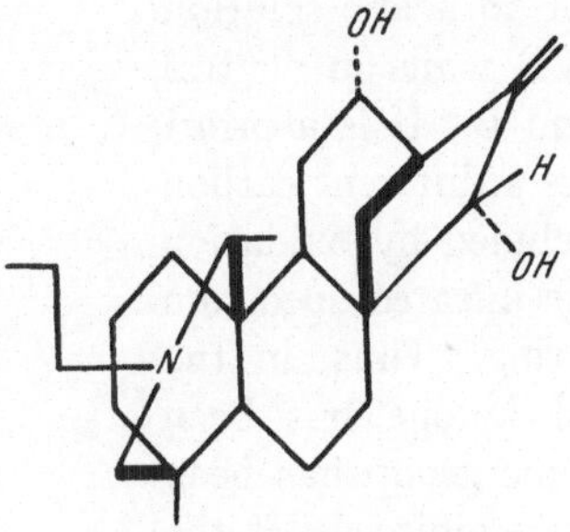

(LXXXIII.) Napelline.

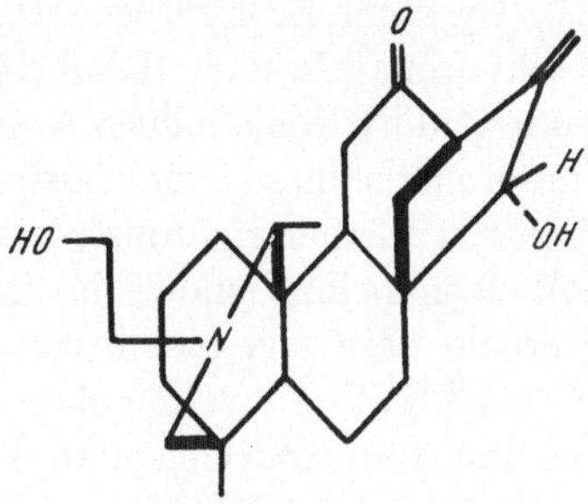

(LXXXIV.) Napellonine.

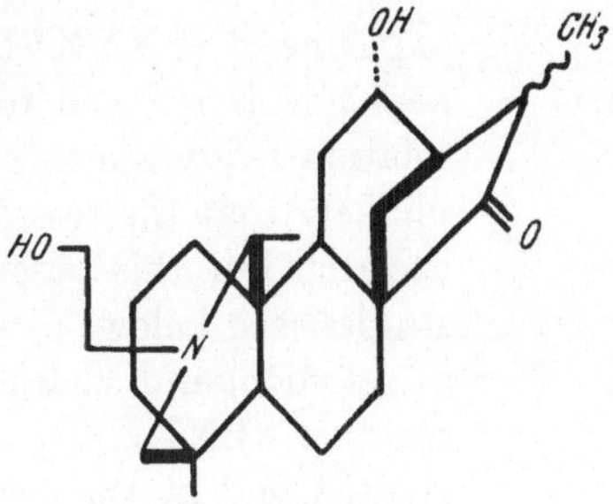

(LXXXV.) Isonapelline.

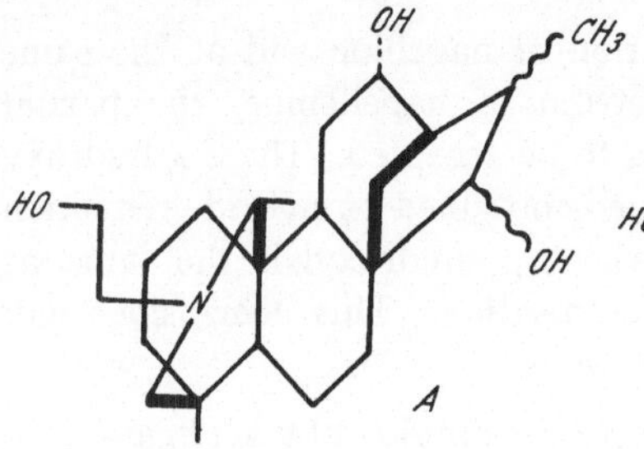

(LXXXVI.) Dihydronapelline A.

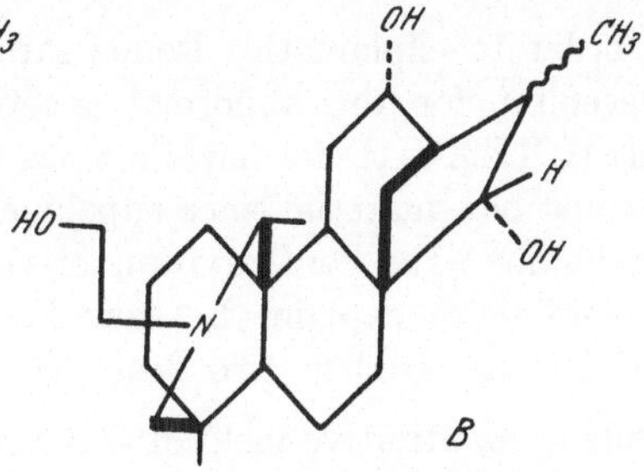

(LXXXVII.) Dihydronapelline B.

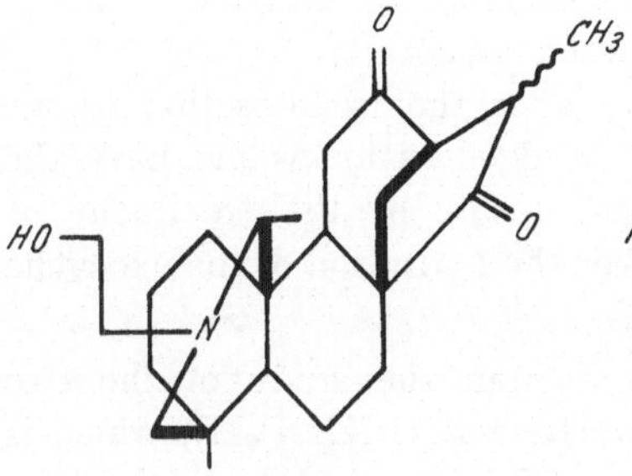

(LXXXIX.)

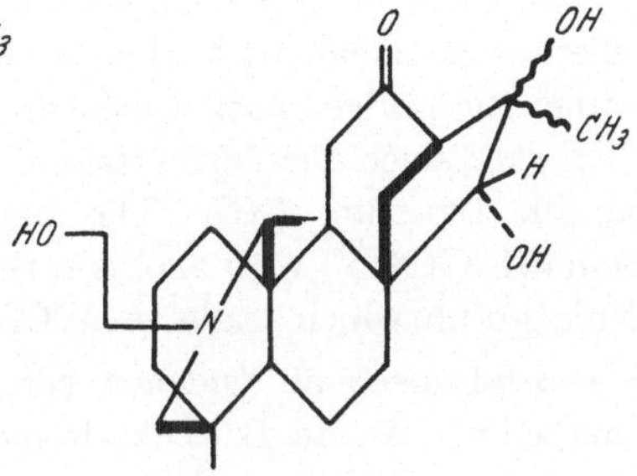

(LCI.)

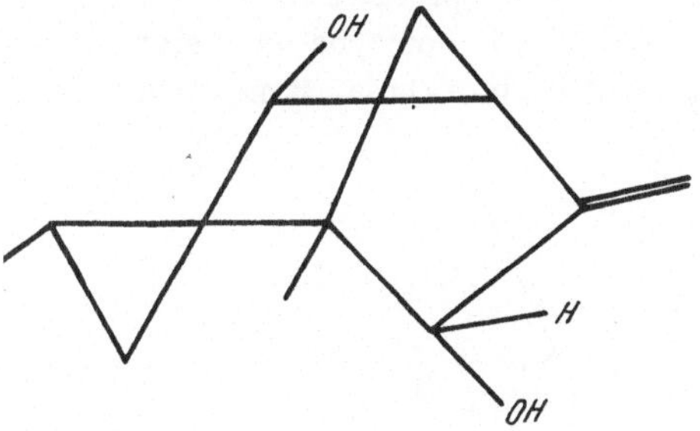

(LXXXVIII.)

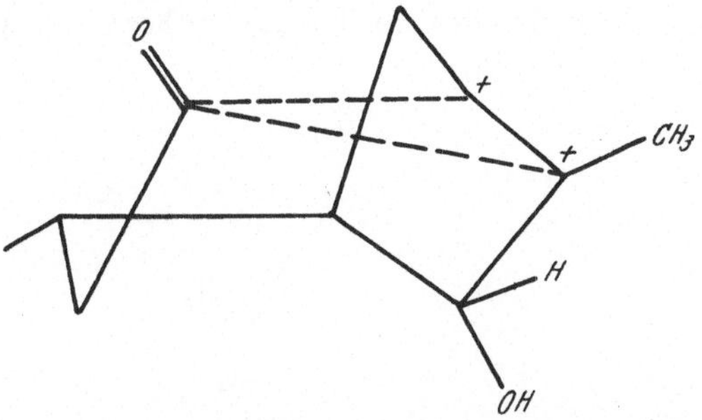

(XC.)

napelline B (LXXXVII). By catalytic hydrogenation in acidic solution napelline is reduced to dihydronapelline A (LXXXVI) while in neutral solution it gives a mixture of dihydronapellines A and B. It is a priori unlikely that the two dihydronapellines should possess a different carbon skeleton but this possibility is further rigorously excluded by oxidation studies (see below, p. 59). Isonapelline shows in the infrared spectrum a ketonic band characteristic of a five-membered ring. Thus, in the reaction (LXXXIII) → (LXXXV) a five-membered ketone has been created and at the same time the exocyclic methylenic group has been transformed into a C-methyl group. This change is reminiscent of the conversion of garryfoline into cuauchichicine (*12*) (p. 42).

In order to explain this isomerisation of napelline and at the same time account for the abnormal spectrum of napellonine, the partial formula (LXXXVIII) for napelline has to be accepted. The $C_{(6)}$ hydroxy group must be equatorial since napelline is obtained by hydride reduction of napellonine. The configuration of the $C_{(19)}$ hydroxyl is the same as in garryfoline to explain the easy ketonisation. This conclusion may now be corroborated in two different ways.

While a quantitative yield of (LXXXV) is obtained by acid catalysed isomerisation of (LXXXIII), the same treatment of (LXXXIV) gives only a poor yield of the corresponding diketone (LXXXIX). This is in excellent agreement with the proposed structure. In the case of napellonine, the bridged cation mechanism for the isomerisation (p. 42) cannot operate since the corresponding bridged cation would have the unfavourable structure (XC). The main result of the treatment of napellonine (LXXXIV) with acid is rather the hydration of the exocyclic methylenic group which leads to (XCI).

The second piece of evidence comes from the study of the two dihydronapellines A and B. Dihydronapelline B (LXXXVII), which is also obtained by hydride reduction of (LXXXV), must have the $C_{(19)}$ hydroxyl in the same configuration as tetrahydroepiveatchine (tetra-hydrogarryfoline). Dihydronapelline A (LXXXVI) may be epimeric to dihydronapelline B at either $C_{(18)}$ or $C_{(19)}$ or at both of these asymmetric centers.

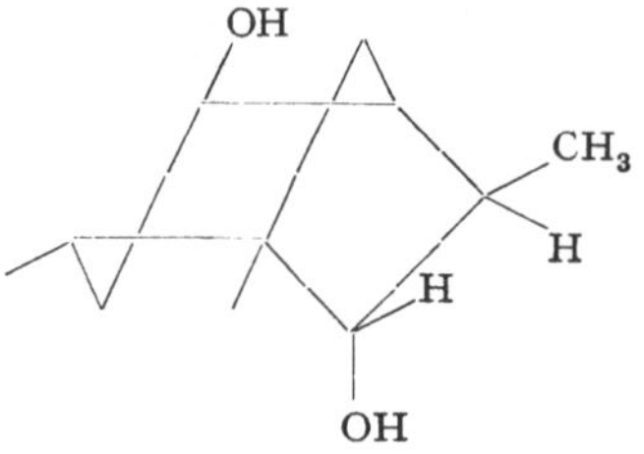

(XCII.) 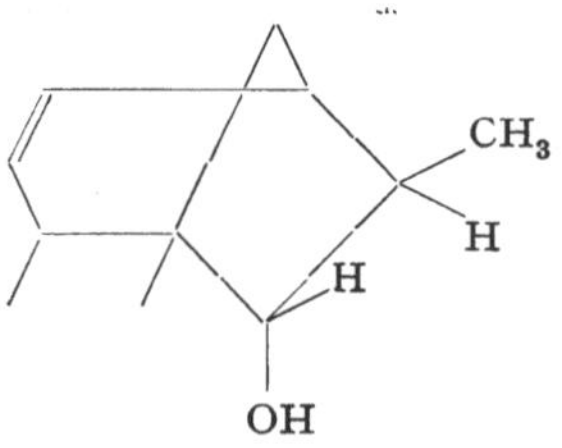　　　　　　　　　　　　(XCIII.)

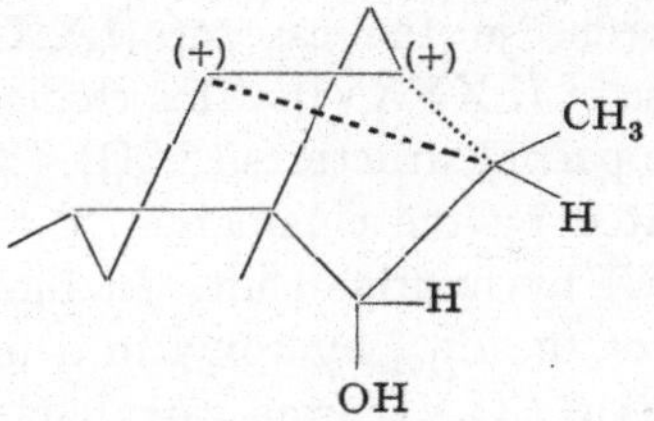

(XCIV.)

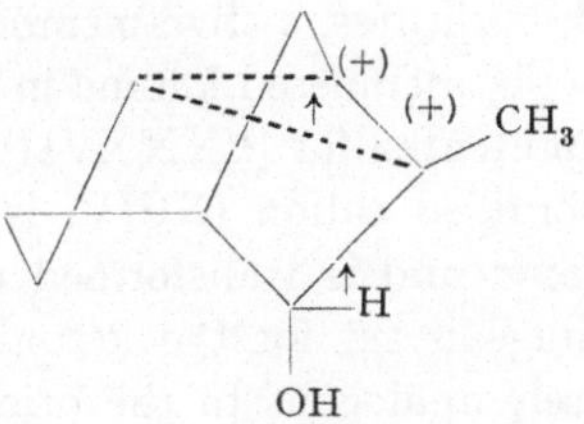

(XCV.)

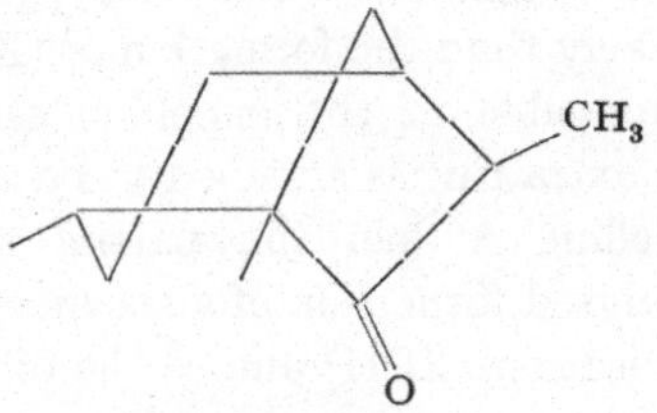

(XCVI.)

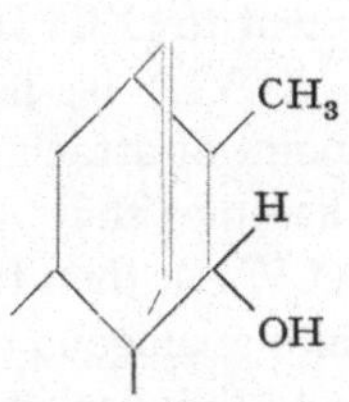

(XCVII.)

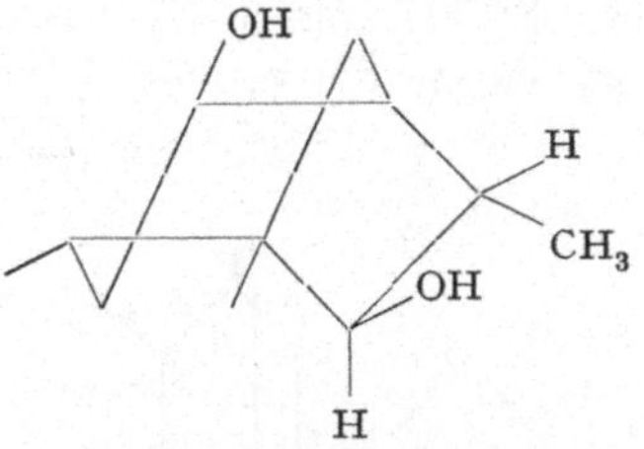

(XCVIII.)

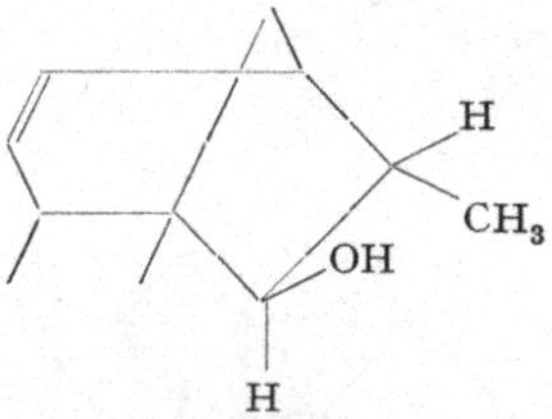

(XCIX.)

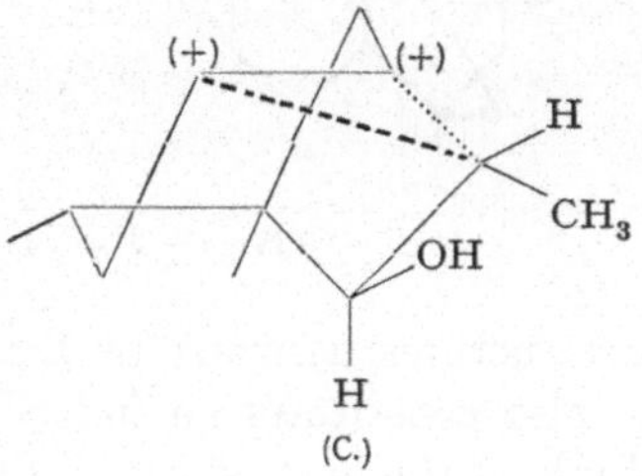

(C.)

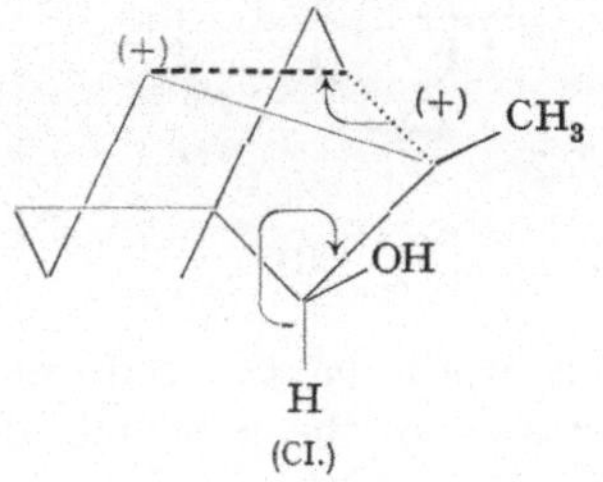

(CI.)

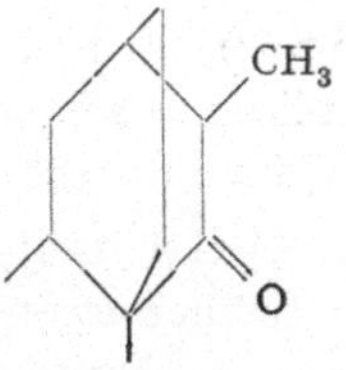

(CII.)

If the two dihydronapellines are warmed with hydrochloric acid, the reaction mixtures develop a strong carbonyl band in the infrared which indicates a five-membered ketone in the case of (LXXXVII) and a six-membered ketone in the case of (LXXXVI). This change may be represented for (LXXXVII) by the partial structures (XCII)–(XCVI). The bridged cation (XCIV) is generated by the elimination of the $C_{(6)}$ hydroxyl and is transformed to (XCV) by hydride shift. In the latter the stage is set for the ketonisation of the $C_{(19)}$ hydroxyl in a manner precisely analogous to the bridged cation (XLVI) from garryfoline. One point in the above reaction sequence appears pertinent. It might seem that in a simple system portrayed by the partial structures (XCII)-(XCVI) a rearrangement to (XCVII) is more likely than the formation of (XCVI). Compound (XCVII) is, however, impossible in the napelline series if the second point of attachment of the extra ring is assumed to be at $C_{(8)}$.

If we assume that dihydronapelline A has the partial stereo-structure (XCVIII), then the acid catalysed formation of a six-membered ketone is quite analogous to the ketonisation (XCII) and to the dihydro-veatchine → tetrahydroketo-isoveatchine rearrangement. The reaction sequence may be represented by the partial structures (XCVIII)–(CII) and is self-explanatory. The ketones (XCVI) and (CII) have been characterised in the form of their oxazolidine derivatives (CIII) and (CIV) (*31*).

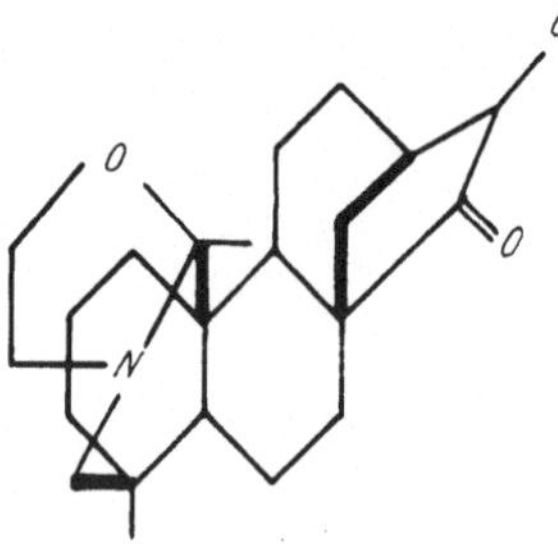

(CIII.) 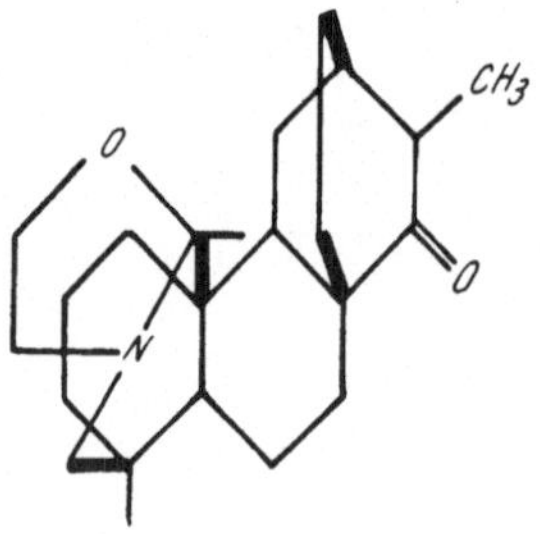   (CIV.)

This work, besides confirming the structures assigned to the *C–D* ring system of the napelline alkaloids, also constitutes an independent confirmation of Woodward's proposed role of the bridged cation (XCVI) in the ketonisation of garryfoline (cf. p. 42).

We shall now turn our attention to experiments designed to develop the partial formula (LXXXVIII) into the complete structure.

The partial formula (LXXXVIII) itself is reminiscent of the garrya skeleton. This impression is further strengthened by the isolation of dehydrogenation products which are characteristic of the garrya alkaloids.

Selenium dehydrogenation of isonapelline gave a good yield of a mixture of phenanthrenes. Although it was not yet possible to resolve this mixture, the phenanthrenes were purified by crystallisation of the trinitrobenzene complexes and characterised spectroscopically. As the basic dehydrogenation product the azaphenanthrene (XIX, p. 32) was unambiguously identified.

On the basis of these results, one may extend the napelline structure from (LXXXVIII) to (CV). Since it was possible to show clearly by spectroscopy in the near ultraviolet that the two dihydronapellines are

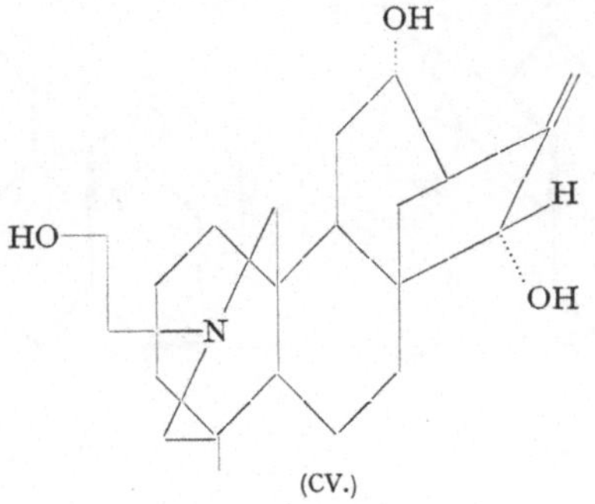

(CV.)

saturated compounds, the napelline skeleton must be hexacyclic and the partial structure (CV) still requires the formation of one carbon-carbon bond.

It is now possible to show that one point of attachment of this sixth ring is a carbon adjacent to the nitrogen atom.

On oxidation with silver oxide in methanol dihydronapelline B gave two compounds formulated as (CVI) and (CVII). These two compounds are analogous to garryine hydrate and veatchine, with one important difference. While (CVI) corresponds in basicity exactly to garryine hydrate, (CVII) is not only a much weaker base than veatchine, but is more weakly basic than dihydronapelline B (pK = 7.8). Consequently, one is led to believe that $C_{(17)}$ is the site of a new bridgehead not present in veatchine. If (CVII) is oxidised by chromium trioxide in pyridine, a diketo-oxazolidine lactam (CVIII) is obtained. This compound shows infrared frequencies characteristic of a five- and six-membered ketone and a six-membered lactam but does not possess a hydroxyl group.

An analogous series of oxidations performed on dihydronapelline A (LXXXVI) yields the same compound (CVIII). This finding shows rigorously that the two dihydronapellines are diastereoisomers.

Up till now there is only one piece of evidence which permits the deduction that the second point of attachement of the new ring is carbon 8. It is a consideration of the remarkable properties of the oxazolidine (CVII). In this compound as already stated, the oxazolidine oxygen must be attached to a bridgehead, since if this was not the case, the compound like veatchine would be an extremely strong base. On the other hand, the salts of (CVII) appear to have quaternary SCHIFF salt character, since they show a strong band at 1670 cm.$^{-1}$ in the infrared spectrum. Thus, the $C_{(17)}$ bridgehead clearly must be small enough to make the quaternary SCHIFF form energetically unfavorable, but large enough to make it capable of existence.

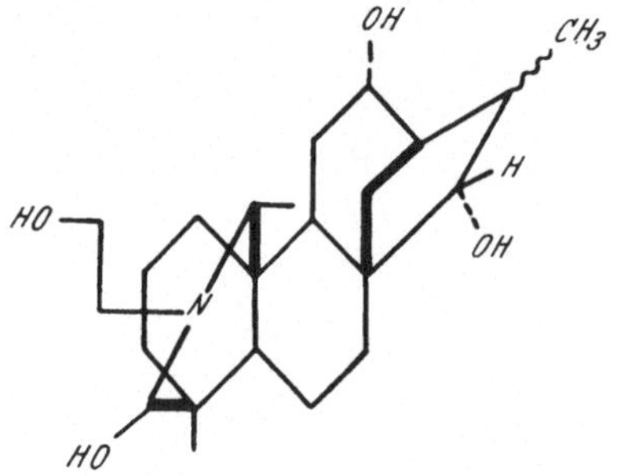

(CVI.) (pK = 8.8.) 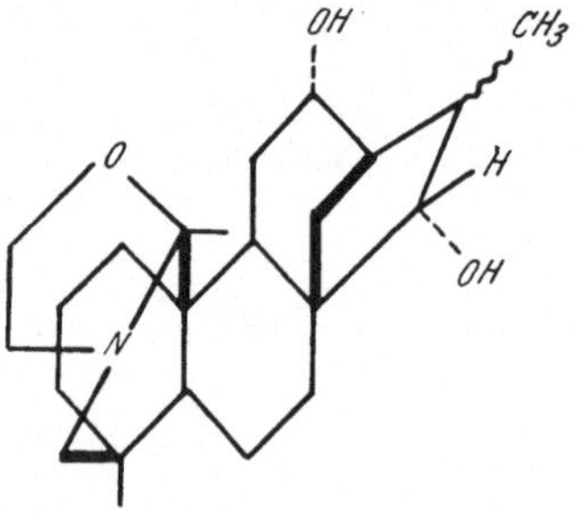   (CVII.) (pK = 6.8.)

These requirements seem to be fulfilled in the formula (CIX a), whereas the attractive alternative structure (CIX b) is impossibly strained and would be contrary to all experience in organic chemistry.

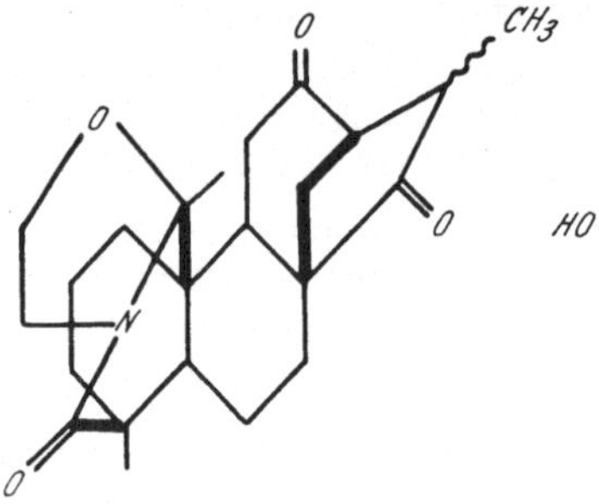

(CVIII.) 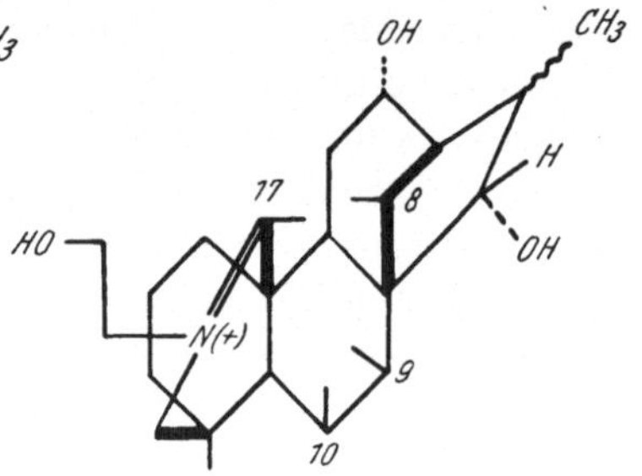   (CIX a.) $C_{(17)}$—$C_{(8)}$ bond. (CIX b.) $C_{(17)}$—$C_{(9)}$ bond.

Finally, it is possible to show that no rearrangement has taken place in the silver oxide oxidation of *dihydronapelline*, since this later compound is recovered by lithium aluminum hydride reduction of the oxazolidine (CVII).

Some time ago we had been asked for a sample of napellonine by Dr. A. KUZOVKOV (Moscow). Dr. KUZOVKOV now kindly informed us that napellonine is identical with the alkaloid songorine, the selenium dehydrogenation of which he has been studying. He has been able to isolate a $C_{18}H_{18}$ trisubstituted phenanthrene and identify it as 1,10-dimethyl-7-ethyl-phenanthrene, by synthesis. This shows obviously (as has also been pointed out by Dr. KUZOVKOV) that if the dehydrogenation is taken at its face value, the second point of attachment of the sixth ring in napelline is $C_{(10)}$. An alternative to this is the already considered $C_{(17)}-C_{(9)}$ bond. As KUZOVKOV's synthesis involved in the last step a dehydrogenation, migration of the 10-methyl group might have occurred and the synthetic compound might actually be 1,9-dimethyl-7-ethyl-phenanthrene.

Whichever of these two possibilities is correct, the chemistry of napelline required re-interpretation since (CIXb) and (CIX) with a $C_{(17)}-C_{(10)}$ bond are clearly impossible. This situation has been resolved quite recently as follows (59a). It has been shown that the oxidation with lead tetra-acetate (24) is not a specific reaction for the ethanolamine group and that napelline in fact contains an N-ethyl group.

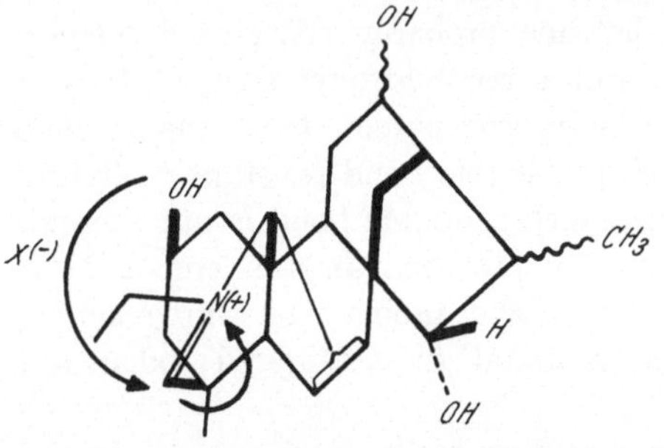

(CIXc.) 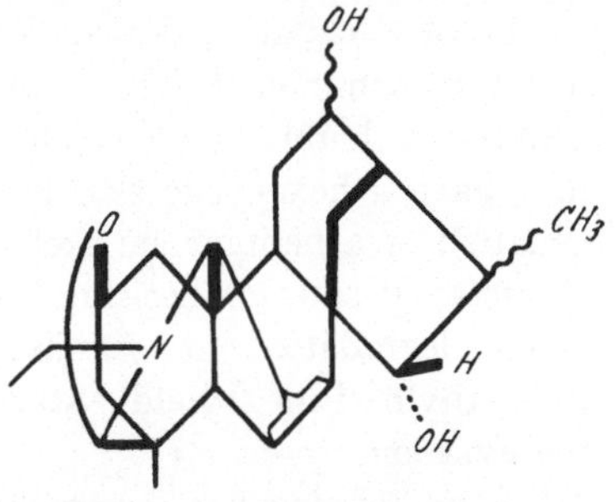   (CX.)

The carbinolamine ether (CVII) may be formulated, consequently, as (CX), its salts as (CIXc), and the "diketooxazolidine" (CVIII) must be reformulated as the triketone (CXI). That the latter is indeed a triketone has been rigorously shown by infrared spectroscopy of the mono- and bis-2,4-dinitrophenylhydrazones. The weak basicity of (CX) may now be

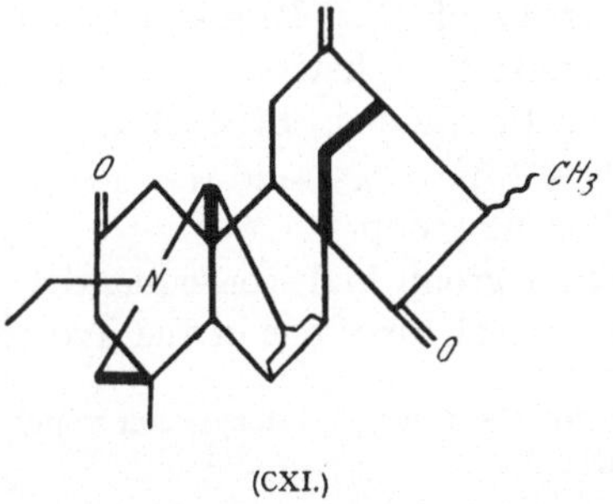

(CXI.) 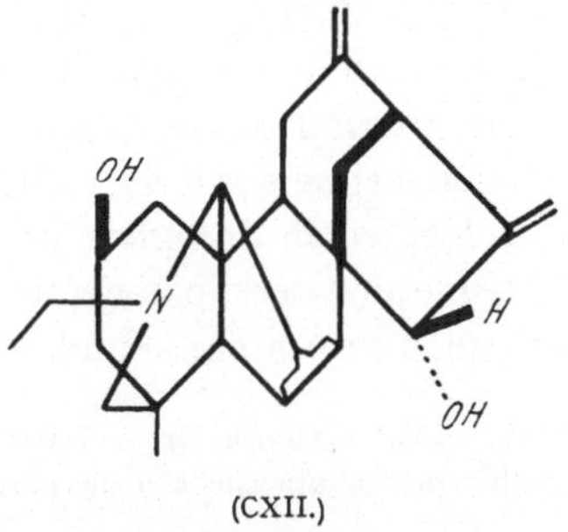   (CXII.)

understood not in terms of a bridgehead position of the carbinolamine ether group but in terms of a rigid 1–3 diaxial interaction between the trigonal carbon and $C_{(3)}$ hydroxyl in (CIXc). This interaction is relieved by ether formation as indicated in (CIXc). It has been further shown that in the strongly basic carbinolamine (CVI) the carbinolamine hydroxyl is located in the α-position of the N-ethyl group.

The structure of songorine (napellonine) may thus be represented by the formula (CXII, p. 61); and napelline is the corresponding ring C, equatorial alcohol*.

VII. The Chemistry of Hetisine.

The alkaloid hetisine $C_{20}H_{27}O_3N$ has been isolated from *Aconitum heterophyllum* and preliminarily investigated by Jacobs and Huebner (*36*). While it is possible on the basis of this work to present only speculations, it is nevertheless felt that hetisine does represent a remarkable structural type and should be briefly mentioned.

Jacobs has first shown that hetisine probably contains a tertiary nitrogen common to two rings and three hydroxyl groups. It also contains a double bond which can be hydrogenated. Thus, the alkaloid may have a hexacyclic skeleton with a double bond resistant to hydrogenation or a heptacyclic skeleton if such a double bond is not present. Quaternary salts of hetisine are readily prepared and undergo a Hofmann degradation in the first stage. The product of this reaction, desmethylhetisine, yields starting material in attempted exhaustive methylation.

Since hetisine has originally one double bond which can be hydrogenated, it would be expected that desmethylhetisine will take up two moles of hydrogen. However, only a crystalline dihydroderivative was obtained by Jacobs.

Dehydrogenation of hetisine gave 1,7-dimethylphenanthrene and some basic materials characterised as picrates. Although no spectra of these bases are recorded, the analyses of the picrates indicate that these compounds in all probability are azaphenanthrenes similar to the typical basic dehydrogenation product of veatchine (XIX, p. 32).

L. G. Humber (*30*) at the University of New Brunswick had the opportunity to investigate a small amount of hetisine and thus add some information to the data gathered by Jacobs. First of all, he has shown that hetisine is a very strong base (pK = 9.5) and that its double bond is located in an exocyclic methylenic group.

Hetisine does not possess a C-methyl group and, consequently, if it is analogous to veatchine and atisine it must carry one of the hydroxyls

---

* We wish to thank Dr. A. Kuzovkov for the valuable information which has led to this rather unexpected development.

at $C_{(15)}$. HUMBER was able to prepare a normal desmethylhetisine which gave a tetrahydro derivative on hydrogenation. This compound (pK = 7.8) is a much weaker base than hetisine and resembles in basicity the corresponding derivatives in the veatchine or atisine series. Finally, spectroscopy in the near ultraviolet in acidic solution revealed that dihydrohetisine shows extinctions lower than tetrahydroveatchine and hence is in all probability a heptacyclic compound with no double bond.

All these data taken together seem to indicate an alkaloid of the veatchine-atisine type with an extra carbon-carbon bond and a carbon-nitrogen bond instead of the N-alkyl group. If this conclusion should be correct, there would not be many structures possible which could fill these requirements. One of them is (CXIII). The desmethylhetisine according to JACOBS which gives only a dihydroderivative may then be formulated as (CXIV).

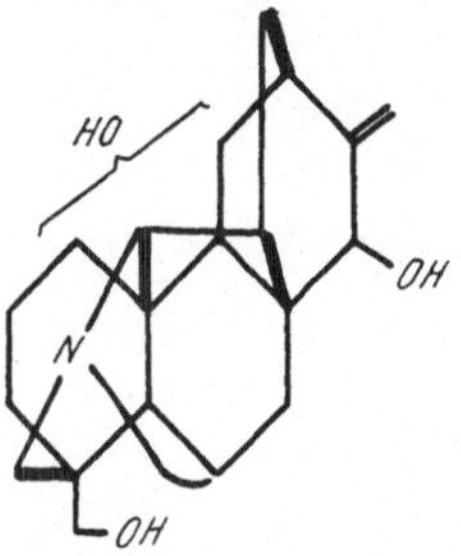

(CXIII.) Hetisine 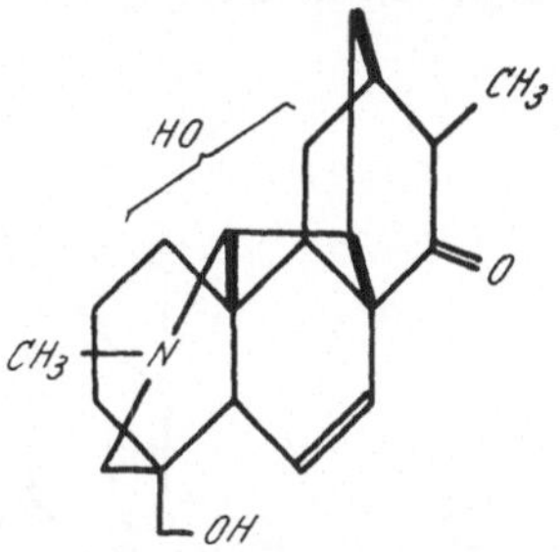 (?).   (CXIV.)

The structures (CXIII) and (CXIV) also explain satisfactorily the difference in basicity between hetisine and desmethylhetisine.

VIII. Lycoctonine and Delpheline.

The excellent investigations of lycoctonine by EDWARDS, MARION and their collaborators (18–22) and of delpheline by COOKSON and TREVETT (3–8) represent a significant contribution to our knowledge of the more highly hydroxylated *Aconitum* and *Delphinium* alkaloids. Both research teams have been able to determine the functional groups and their environment and have described numerous transformations which limited considerably the number of structural possibilities. However, no degradation or conversion product of known structure has been obtained as yet from the two alkaloids, and the chemical evidence alone is not extensive enough for an unambiguous structural proposal.

A great progress was achieved by PRZYBYLSKA's brilliant X-ray analysis (50) which resulted in structure (CXV) (R = H) for des(oxymethylene)-lycoctonine. This formulation can be extended without ambiguity to (CXV) ($R = CH_2OH$) as a representation of lycoctonine (22).

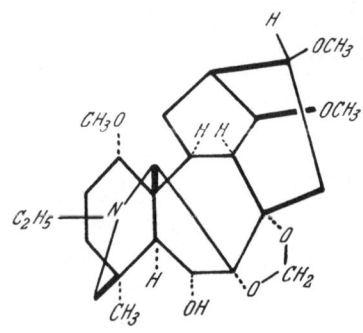

(CXV.)
$R = H$. Des(oxymethylene)-lycoctonine.
$R = CH_2OH$. Lycoctonine.

(CXVI.) Demethylene-delpheline.

(CXVII.) Delpheline.

Because of the remarkable similarity of many degradation products of lycoctonine and delpheline and the fact that all the known chemistry of demethylene-delpheline (p. 75) and its degradation products can be reconciled with the same carbon skeleton, COOKSON and TREVETT (4) considered structure (CXVI) for demethylene-delpheline and (with some reservations) structure (CXVII) for delpheline.

For the sake of simplicity and clarity, the proposed structures (CXV) ($R = CH_2OH$) and (CXVII) will be used in the discussion of the chemistry of the two alkaloids, although these structures are not rigorously proven at the present time.

1. The Chemistry of Lycoctonine.

The distribution in plants of various monoester alkaloids containing lycoctonine as the aminoalcohol moiety has been reviewed by STERN (52). These alkaloids are present in both the *Aconitum* and the *Delphinium* species and hence lycoctonine provides an interesting link between the two genera.

Lycoctonine is a monoacidic tertiary base (pKa 8.8 in 50% aqueous methanol) containing four methoxyl and three hydroxyl groups. The empirical formula $C_{25}H_{41}O_7N$ was finally chosen for lycoctonine (*19*), after several formulae had been considered by different authors. The low end-absorption in the ultraviolet spectrum of lycoctonine perchlorate, the absence of peaks due to unsaturation in the infrared spectrum, and the fact that no hydrogen was taken up on catalytic hydrogenation of lycoctonine, indicate that the base is probably saturated and therefore hexacyclic. Analytical studies indicate the presence of one C-methyl group and three active hydrogens (*19*).

Oxidation Studies. The environment of the nitrogen atom and the presence and properties of the hydroxyl groups were established by extensive oxidation studies. Permanganate oxidation of lycoctonine, $C_{25}H_{41}O_7N$, in acetone yielded two neutral products, lycoctonam* (CXVIII) ($R = CH_2OH$), $C_{25}H_{39}O_8N$, and des(oxymethylene)-lycoctonam (CXVIII) ($R = H$), $C_{24}H_{37}O_7N$, and one acidic product, lycoctonamic acid (CXVIII) ($R = COOH$), $C_{25}H_{37}O_9N$ (*19*). Both lycoctonam and the methylester of lycoctonamic acid could be reduced back to lycoctonine with lithium aluminum hydride, indicating that there were no skeletal changes during the oxidation. The lactam maxima in the infrared spectrum of lycoctonam (1613 cm.$^{-1}$), des(oxymethylene)-

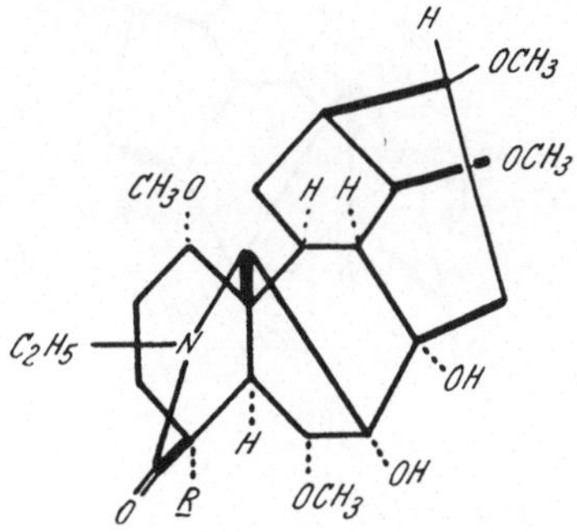

(CXVIII.)   (CXIX.)
R = H. Des(oxymethylene)-lycoctonam.
R = CH₂OH. Lycoctonam.

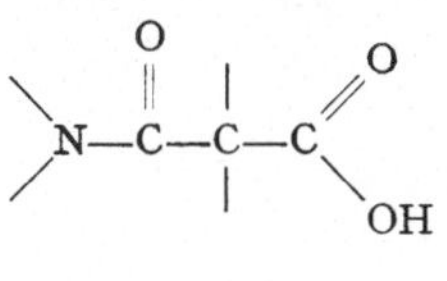

lycoctonam (1630 cm.$^{-1}$) and lycoctonamic acid (1597 cm.$^{-1}$) show that the nitrogen atom is located in a six-membered or a larger ring. The acid can be readily decarboxylated by heat to des(oxymethylene)-

---

* Chemical names chosen by the original authors will be used throughout this Section.

lycoctonam. In the absence of other unsaturation in the molecule, this fact can be best explained by the presence of the grouping (CXIX) in lycoctonamic acid. The shift to lower wave numbers of the lactam maximum in the infrared spectrum of the acid and other degradation products containing the lactam carboxylic acid grouping is in good agreement with the partial structure (CXIX) (*19*). Although the tertiary nature of the carboxyl group in lycoctonamic acid has not been rigorously proven, the described reactions are in satisfactory agreement with structure (CXVIII) ($R = COOH$).

The acid does not react appreciably with bromine in methanol (*19*), and the relative ease of esterification and hydrolysis of the carboxyl group can be compared with the similar behavior of the bridgehead carboxyl group in the bicyclo (2,2,2)-octane series described by Roberts and Mooreland (*51*, cf. *22*).

The presence of a primary hydroxyl group in lycoctonine was further confirmed by chromium trioxide oxidation of lycoctonine which yielded lycoctonal (CXV) ($R = CHO$), $C_{25}H_{39}O_7N$ (aldehyde peaks in the infra-red spectrum at 2706 and 1725 cm.$^{-1}$) (*19*).

Oxidation of lycoctonine with silver oxide in aqueous methanol gave hydroxylycoctonine, $C_{25}H_{41}O_8N$ (*19*). The low pKa of the new base

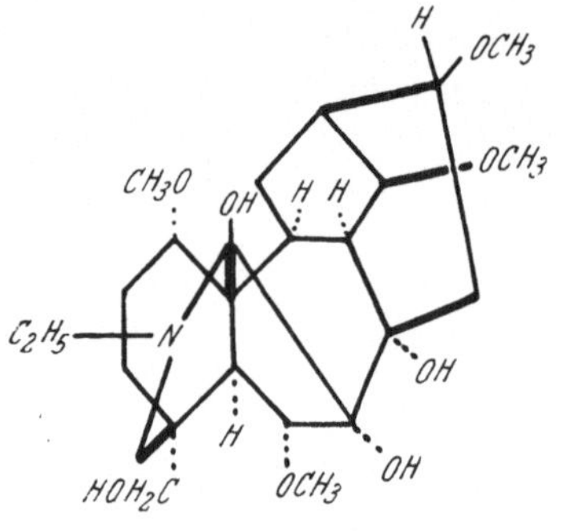

(CXX.) 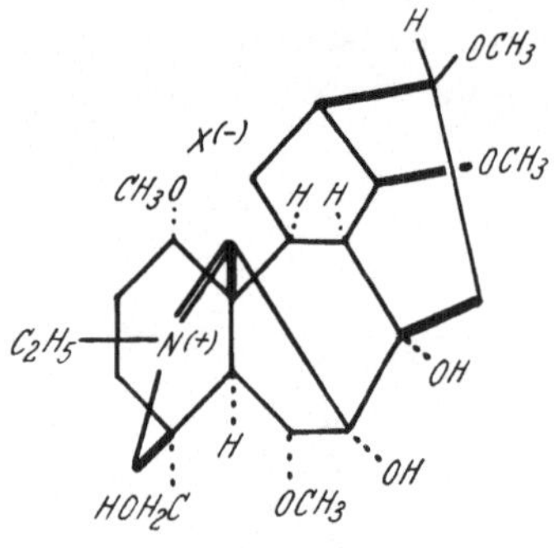   (CXXI.)

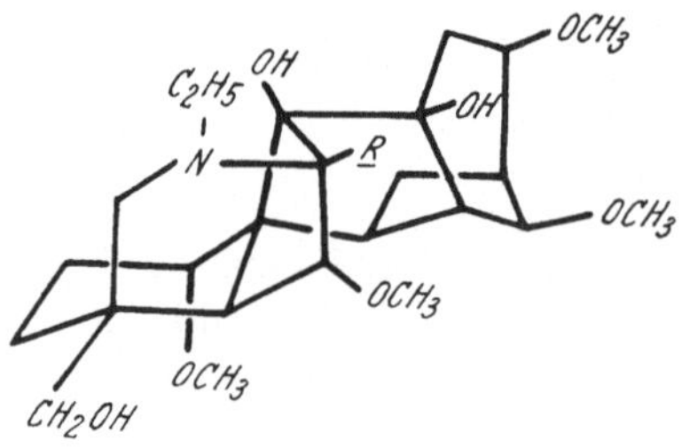

(CXXII.)

(5.6 in 50% aqueous methanol) and the negative outcome of most diagnostic tests for carbinolamines can be reconciled with structure (CXX) for hydroxylycoctonine. The proximity of the hydroxyl group to the nitrogen atom may account for the lowering of basicity, while its bridgehead position makes the participation of the more strongly basic anhydronium form in the equilibrium between the possible structures of the base unimportant. However, it was found by COOKSON and TREVETT (7) that hydroxylycoctonine does in fact form a crystalline anhydro-hydriodide, $C_{25}H_{40}O_7NI$, and a perchlorate, $C_{25}H_{40}O_{11}NCl$, from which hydroxylycoctonine can be regenerated on alkaline treatment. The infrared spectrum of the anhydro-perchlorate is somewhat unusual, showing an intense band at 1710 cm.$^{-1}$ and a weaker one at 1666 cm.$^{-1}$. COOKSON and TREVETT (7) tentatively assign the band at 1710 cm.$^{-1}$ to a strained $\diagdown \underset{\diagup}{C} = \overset{(+)}{\underset{\diagdown}{N}}$ stretching vibration and suggest an influence of this strain upon the basicity of hydroxylycoctonine. They also propose that the separation of the relatively unstable anhydronium salts from the solution may be governed by their relative insolubility or speed of crystallization.

If (CXX) is the correct expression for hydroxylycoctonine, then the anhydronium salts should have the structure (CXXI) and the unprecedented position of the double bond at the bridgehead of an azabicyclo (1,2,3)-octane system in this structure represents a serious objection against structure (CXV) ($R = CH_2OH$) for lycoctonine. However, it is not experimentally proven that hydroxylycoctonine and lycoctonine possess the same skeleton; in fact, there are some indications that hydroxylycoctonine might have a rearranged structure. Hydroxylycoctonam (lactam of hydroxylycoctonine) is oxidised much more slowly than lycoctonam with periodic acid, and methyl hydroxylycoctonamate (see below) shows greater resistance to alkaline hydrolysis than methyl lycoctonamate (18). A rearranged structure such as (CXXII) ($R = OH$) as an expression for hydroxylycoctonine could possibly explain these two rate differences, since the steric situation of the two groups involved, namely the vicinal glycol and the hydroxymethyl group (i. e. the carbomethoxyl group in the lycoctonamic esters) changes in the right direction during the transformation (CXV) ($R = CH_2OH) \rightarrow$ (CXXII) ($R = OH$). However, the formulation of the anhydronium salt formation based on (CXXII) ($R = OH$) clearly encounters difficulties similar to those described for structure (CXX). Unless the structural formulae considered above are incorrect, an unusual structure (possibly of a non-classical type) might have to be considered for the anhydronium salts.

Reduction of hydroxylycoctonine over ADAMS' catalyst in ethanolic hydrochloric acid yielded isolycoctonine, $C_{25}H_{41}O_7N$, (pKa 6.7 in 50% aqueous methanol) (19). The rearranged structure (CXXII) ($R = H$) was proposed for this base (22).

Oxidation of hydroxylycoctonine with permanganate in acetone yielded hydroxylycoctonam, des(oxymethylene)-hydroxylycoctonam, and hydroxylycoctanamic acid in complete analogy to the oxidation of lycoctonine itself (18). The formation of these products in good yield confirms the tertiary nature of the new hydroxyl group in hydroxy-lycoctonine. Vigorous acid hydrolysis of hydroxylycoctonam gave ethylamine, proving the presence of an N-ethyl group in lycoctonine (18).

Oxidation of lycoctonam (CXVIII) ($R = CH_2OH$) with periodic acid gave secolycoctonam diketone (CXXIII) ($R = CH_2OH$), $C_{25}H_{37}O_8N$; and oxidation of des(oxymethylene)-lycoctonam (CXVIII) ($R = H$) yielded des(oxymethylene)-secolycoctonam diketone (CXXIII) ($R = H$), $C_{24}H_{35}O_7N$. Since the oxidation products are stable to further oxidation with silver oxide, potassium permanganate or chromium trioxide, both carbonyls are ketonic and the vicinal hydroxyl groups in lycoctonine are, therefore, both tertiary (20). The infrared maxima in the carbonyl region of (CXXIII) ($R = CH_2OH$) (1766, 1707, and 1631 cm.$^{-1}$) and of (CXXIII) ($R = H$) (1765, 1707, and 1644 cm.$^{-1}$) indicate one five-membered (or smaller) and one six-membered (or larger) ketone. The difference in the position of the lactam maxima in the two compounds can be ascribed to the proximity of the primary hydroxyl group to the lactam carbonyl in (CXXIII) ($R = CH_2OH$) (20). Since all the compounds in this series containing the primary hydroxyl group show such a hydrogen bonding effect, the spectroscopic evidence can be used as a further confirmation of the relative position of the hydroxymethyl group and the nitrogen atom. Both diketones have an unusual ultraviolet spectrum, showing a high intensity end-absorption in addition to a maximum at 318 mμ (log $\varepsilon = 2.45$) in the spectrum of (CXXIII) ($R = CH_2OH$) and at 322 mμ (log $\varepsilon = 2.42$) in that of (CXXIII) ($R = H$). The two compounds can be reduced catalytically or with sodium borohydride to give seco-lycoctonam ketol, $C_{25}H_{39}O_8N$, and des(oxymethylene)-secolycoctonam ketol, $C_{24}H_{37}O_7N$. The five-membered ketone is reduced in both cases, whereas the larger-membered ketone is resistant to reduction under these conditions (single ketonic maximum at 1710 cm.$^{-1}$ in the carbonyl region). The ultraviolet spectrum of the ketols is normal (λ_{max} 270 mμ; log $\varepsilon = 1.60$).

The diketones reduce TOLLENS' reagent and FEHLING's solution whereas the corresponding ketols are inert to these reagents. This led EDWARDS and MARION (20) to postulate a methoxyl group in α-position to the five-membered ring ketone. However, attempts to hydrolyse this α-ketol methylether with hot acid were unsuccessful.

Treatment of the diketones with acid or alkali resulted in the loss of the elements of methyl alcohol and yielded the α,β-unsaturated ketones (CXXIV) ($R = CH_2OH$) and (CXXIV) ($R = H$).

The fact that the corresponding ketols also underwent this elimination and that the ketone maximum at 1707 cm.$^{-1}$ in the diketone spectra shifted to 1679 cm.$^{-1}$ in (CXXIV) ($R = CH_2OH$) and to 1678 cm.$^{-1}$

in (CXXIV) ($R = H$), clearly places a methoxyl group in a position β to the six-membered (or larger) ring ketone in (CXXIII) (*20*). The

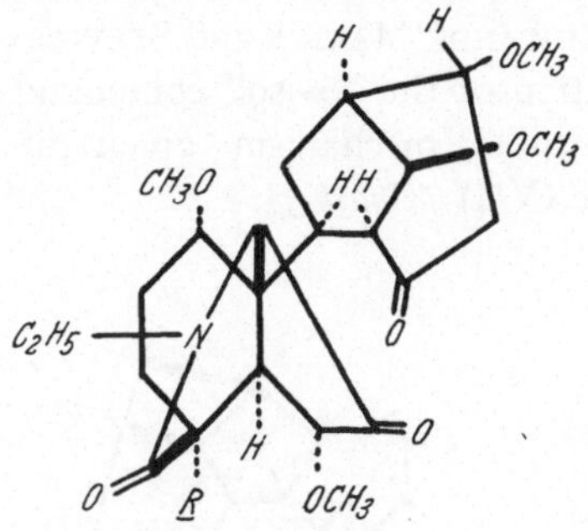

(CXXIII.) 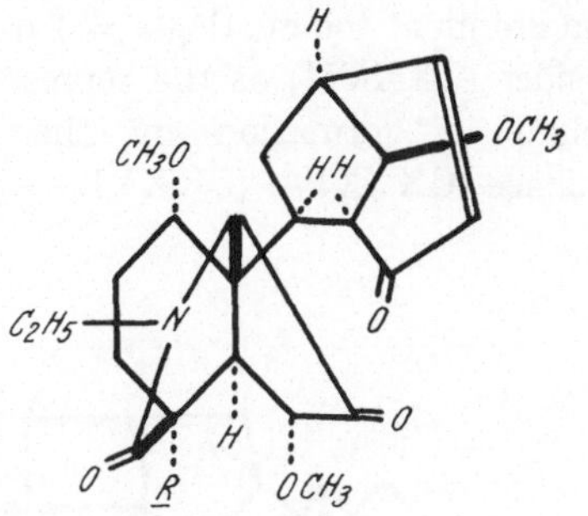   (CXXIV.)

position of the maximum in the ultraviolet spectrum of the unsaturated ketones ($\lambda_{max} = 223$ mμ) indicates a low degree of substitution and is in agreement with the proposed structures.

Skeletal Changes. Activated alumina or bicarbonate convert *seco*-lycoctonam diketone (CXXIII) ($R = CH_2OH$) to a so-called "iso" compound, $C_{25}H_{37}O_8N$ (*20*). The infrared spectrum of the latter (maxima at 1743 and 1632 cm.$^{-1}$ in the carbonyl region) indicates the disappearance of the six-membered (or larger) ring ketone during this isomerisation. Since the five-membered carbonyl stretching frequency is normal in the "iso" compound (1743 cm.$^{-1}$), its unusual position in (CXXIII) (1766 cm.$^{-1}$) can probably be ascribed to an interaction of the two carbonyl groups. The ultraviolet spectrum of the "iso" compound is unusual, showing maxima at 218 mμ (log $\varepsilon = 3.82$) and 335 mμ (log $\varepsilon = 2.26$). A new hindered hydroxyl group was formed in the isomerisation, since the infrared spectrum of "iso" compound mono-acetate still contains an OH-band (3500 cm.$^{-1}$). The reaction is reversible as shown by the formation of *seco*lycoctonam ketol on catalytic reduction or reduction with sodium borohydride and by the formation of (CXXIV) ($R = CH_2OH$) on treatment with strong alkali (*20*). Structure (CXXV) was proposed for the "iso" compound by EDWARDS, MARION and STEWART (*22*), the product being formed by a simple aldol condensation. An analogous isomerisation was achieved with des(oxymethylene)-*seco*lycoctonam diketone (CXXIII) ($R = H$).

Treatment of the "iso" compound (CXXV) with hot 6N acid yielded a mixture, consisting mainly of the "α-iso" compound, $C_{25}H_{37}O_8N$, and the "anhydro-iso" compound, $C_{25}H_{35}O_7N$ (*20*). The "α-iso" compound contains no ketonic function (infrared) and cannot be readily acetylated; the primary hydroxyl group has therefore taken part in the reaction.

However, the compound still contains two active hydrogens. The "anhydro-iso" compound has no hydroxyl groups (infrared) and contains a ketonic function (band at 1736 cm.$^{-1}$). Its ultraviolet spectrum has a maximum at 300 mμ (log $\varepsilon = 1.9$). Edwards, Marion and Stewart (22) consider (CXXVII) as the representation of the "α-iso" compound and explain its formation by the plausible mechanism given in the sequence (CXXV) → (CXXVI) → (CXXVII) (Chart 3).

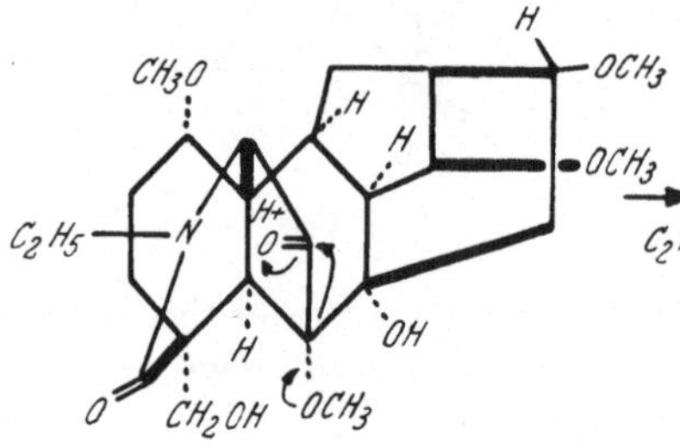

(CXXV.) "Iso" compound. 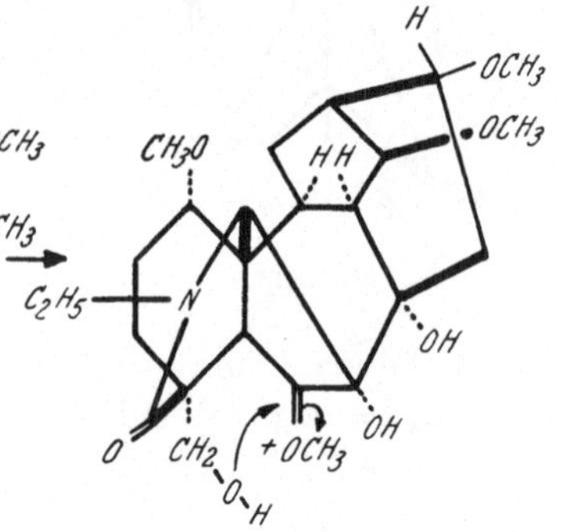   (CXXVI.)

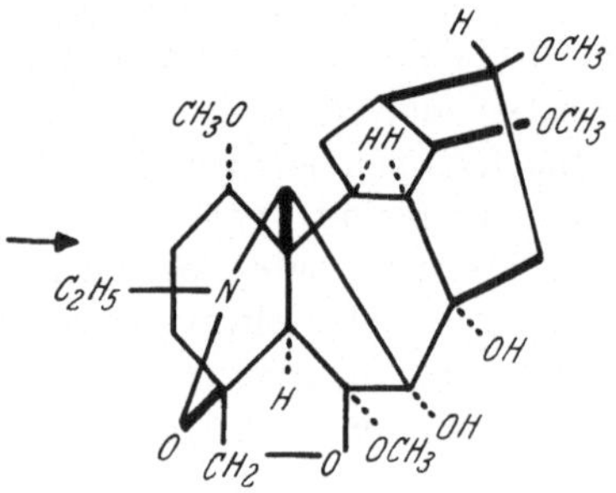

(CXXVII.) "α-Iso" compound.

Chart 3.

These authors point out that the original lycoctonine skeleton has been reformed in this structure. Formula (CXXVIII) was originally considered for the "anhydro-iso" compound (22), the reaction involving an acyl migration (CXXV) → (CXXVIII) (probably a two-step process in analogy to the formation of the "α-iso" compound). It is now assumed, however (private communication from Dr. O. E. Edwards), that the "α-iso" compound is an intermediate in the formation of the "anhydro-iso" compound, the reaction involving the dehydration (CXXVII) → → (CXXIX) in complete analogy to the formation of anhydrolycoctonam

(CXXX a) $(R = CH_2OH)$ (see below). The compounds of the anhydro-lycoctonam type are described to have an unusually high six-membered

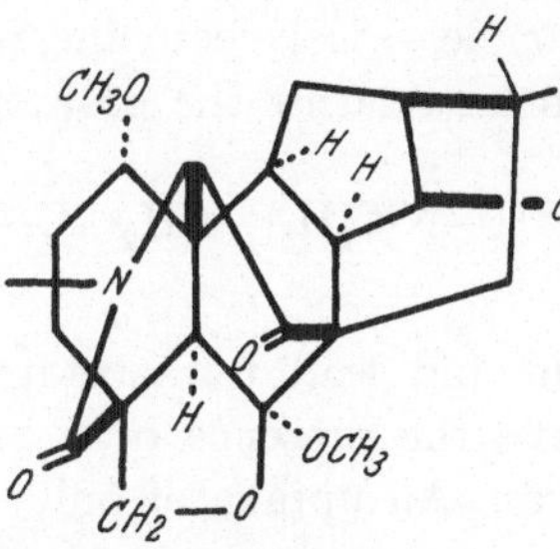

(CXXVIII.) 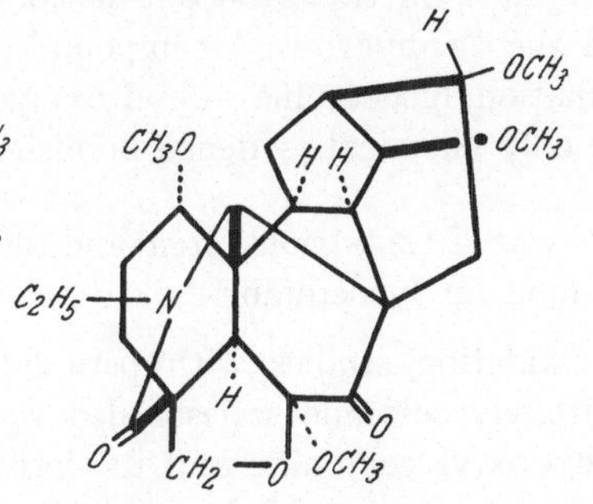   (CXXIX.) "Anhydroiso"compound.

ketone maximum (around 1730 cm.$^{-1}$) (*19*); and formula (CXXIX) agrees with the properties and reactions of the "anhydro-iso" compound.

Treatment of lycoctonam (CXVIII) $(R = CH_2OH)$, $C_{25}H_{39}O_8N$, with acetyl chloride yielded anhydrolycoctonam acetate, $C_{27}H_{39}O_8N$, which was hydrolysed with base to give anhydrolycoctonam, $C_{25}H_{37}O_7N$ (*19*). The infrared spectrum of anhydrolycoctonam (maximum at 1728 cm.$^{-1}$) indicated that a keto group was formed during dehydration. Stereo-electronic considerations and the fact that the anhydro compound did not eliminate methanol on treatment with acid or base, led EDWARDS, MARION and STEWART (*22*) to postulate structure (CXXX a) $(R = CH_2OH)$ for anhydrolycoctonam.

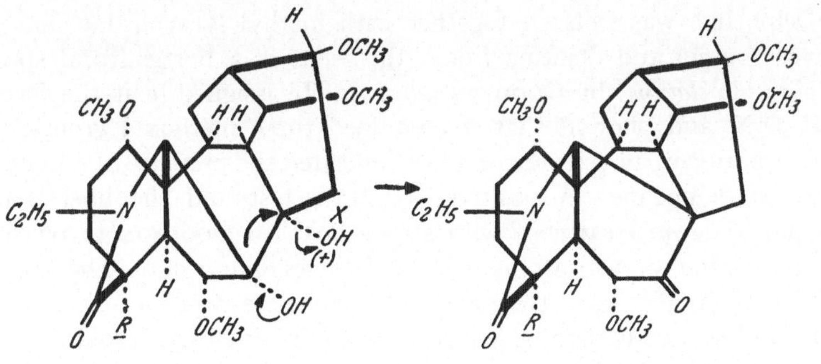

(CXXX.)   (CXXX a.)

R = H. Anhydro-des(oxymethylene)-lycoctonam.
R = CH$_2$OH. Anhydrolycoctonam.

A similar reaction in the desoxymethylene series led to anhydro-des(oxymethylene)-lycoctonam (CXXXa) $(R = H)$.

While (CXXXa) $(R = CH_2OH)$ is stable to hot dilute alkali, (CXXXa) $(R = H)$ undergoes isomerisation. This was explained as an epimerisation of the methoxyl group α to the ketone carbonyl (*22*). The reversed stability in the two compounds was ascribed to the steric influence of the hydroxymethyl group. This correlation, the formation of the "α-iso" and the "anhydro-iso" compound, and the somewhat tentative transformation, lycoctonine → hydroxylycoctonine → isolycoctonine, represent the only chemical evidence published so far relating the position of the

methylated 1,2,3-triol system and the $\underset{\diagup}{\overset{\diagdown}{C}}H—N(C_2H_5)—CH_2—\overset{|}{\underset{|}{C}}—CH_2OH$ grouping in lycoctonine.

Oxidation studies with periodic acid and lead tetraacetate in the hydroxylycoctonine series failed to prove the presence of a 1,2,3-triol in hydroxylycoctonine and its derivatives. An uptake of only one mol of the reagent or, in the case of oxidation of hydroxylycoctonine with lead tetraacetate, an unspecific uptake of over three moles was observed.

In conclusion, it can be said that formula (CXV, p. 64) ($R = CH_2OH$) for lycoctonine is eminently sound on biogenetic grounds (*53*, *4*) and accommodates most of the chemistry of this compound. The two major objections, namely the results of the oxidation of hydroxylycoctonine and its derivatives with periodic acid and lead tetraacetate and the unprecedented highly strained structure required for the anhydro salts of hydroxylycoctonine, can probably be overcome by a re-interpretation of the chemistry of hydroxylycoctonine, independently from the structure of lycoctonine itself.

2. The Chemistry of Delpheline.

Delpheline was isolated together with methyl lycaconitine (an ester of lycoctonine) and delatine from "the seeds of a horticultural species, *Delphinium elatum*" by GOODSON (*26*, *27*). He assigned to it the formula $C_{25}H_{39}O_6N$ and showed that it contained three methoxyl groups (*27*). The formation of a basic monoacetate indicated the presence of a hydroxyl group in delpheline. A positive GAEBEL's test and the liberation of formaldehyde on treatment with strong acid led GOODSON to place the two remaining oxygen atoms in a methylenedioxy group. The HERZIG-MEYER hydriodic acid treatment of delpheline yielded ethyl iodide, indicating a two-carbon side-chain (possibly an ethyl group) on the tertiary nitrogen.

Environment of the Nitrogen. COOKSON and TREVETT (*3*) confirmed the formula, $C_{25}H_{39}O_6N$, for delpheline and proved the presence of an N-ethyl group in two independent ways: Treatment of delpheline (CXXXI) ($R = C_2H_5$) with nitrous acid yielded the neutral de-ethyl-N-nitroso-delpheline (CXXXI) ($R = NO$) which was reduced to N-aminodeethyldelpheline (CXXXI) $(R = NH_2)$. Hydrogenolysis of the latter gave de-ethyldelpheline (CXXXI) $(R = H)$ which could be acetylated to the neutral N-acetyl-deethyldelpheline (CXXXI) $(R = CH_3CO)$. Delpheline was regenerated on reduction of (CXXXI) $(R = CH_3CO)$ with lithium aluminum hydride.

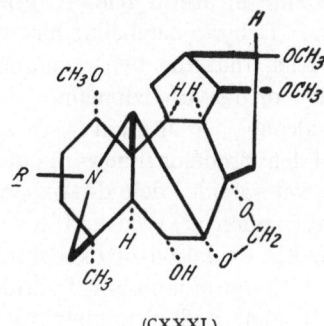

(CXXXI.)

In the second experimental series, delpheline was oxidised with mercuric acetate to give, among other products, the water-soluble α-hydroxydelpheline (CXXXI) $(R = CH_3CHOH)$. This could be oxidised with potassium permanganate to the N-acetyl compound (CXXXI) $(R = CH_3CO)$ and hydrolysed with dilute acid to de-ethyldelpheline (CXXXI) $(R = H)$.

Since delpheline in acid solution has a very low extinction coefficient at 210 mμ and its suitable neutral derivatives consume no perbenzoic acid, the base is saturated and contains seven rings $(3, 6)$.

Oxidation of delpheline with permanganate in acetone or with chromium trioxide in pyridine yielded the neutral oxodelpheline (CXXXII),

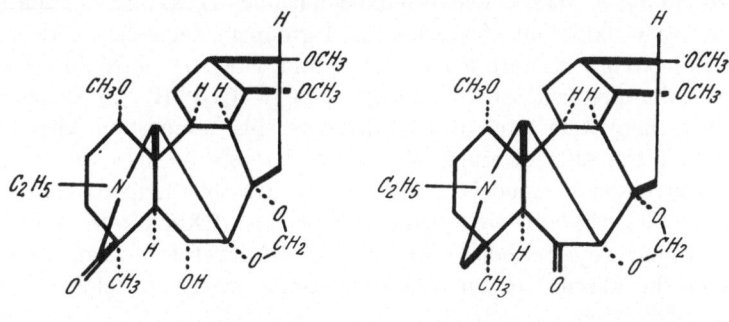

(CXXXII.) Oxodelpheline.   (CXXXIII.) Dehydrodelpheline.

$C_{25}H_{37}O_7N$ (3). Its lactam stretching frequency at 1648 cm.$^{-1}$ shows that the nitrogen atom is situated in a six-membered (or larger) ring.

The formation of a basic monoacetate on acetylation of delpheline with acetyl chloride and the oxidation of delpheline to dehydro-delpheline (CXXXIII), $C_{25}H_{37}O_6N$, (ketone stretching frequency at 1748 cm.$^{-1}$) shows the presence in delpheline of a secondary hydroxyl group situated in a five-membered ring (3). The oxidation can be carried out with chromium trioxide in acetic acid, N-bromosuccinimide, silver oxide or mercuric acetate. Dehydrodelpheline has no bands in the O—H stretching region, indicating that the two remaining oxygen atoms are both ethereal. Reduction of dehydrodelpheline with lithium aluminum hydride, sodium and alcohol, or lithium and ammonia regenerated delpheline. Oxidation of dehydrodelpheline with potassium permanganate in acetone yielded the keto-amide, dehydro-oxodelpheline, $C_{25}H_{35}O_7N$, (maxima in the carbonyl region at 1760 and 1656 cm.$^{-1}$), which could also be obtained by further oxidation of oxodelpheline with chromium trioxide in pyridine. Lithium aluminum hydride reduced dehydro-oxodelpheline to oxodelpheline and ultimately to delpheline.

The variation of physical properties of delpheline and its oxidation products clearly indicates some interaction between the nitrogen function and the five-membered alcohol group (3). For instance, the basicity falls in the series: delpheline (pKa 7.6 in 50% aqueous ethanol), acetyl-delpheline (pKa 6.9), dehydrodelpheline (pKa 5.5). This sequence is best interpreted by electron withdrawal from the nitrogen atom. The high stretching frequencies of the lactam (1656 cm.$^{-1}$) and the ketone carbonyl (1760 cm.$^{-1}$) in the spectrum of dehydro-oxodelpheline, compared with the normal frequencies in oxodelpheline (1649 cm.$^{-1}$) and dehydro-delpheline (1748 cm.$^{-1}$), are also indicative of the mutual interaction of the two carbonyls in dehydro-oxodelpheline. Furthermore, the ultra-violet maximum due to the ketone carbonyl group changes from 269 mμ ($\varepsilon = 160$) in dehydrodelpheline to 259 mμ ($\varepsilon = 200$) in its ammonium salt, to 313 mμ ($\varepsilon = 44$) in dehydro-oxodelpheline. These interrelationships and the easy oxidation of the hydroxyl group in delpheline with silver oxide or mercuric acetate led Cookson and Trevett (6, 8) to consider at first the presence of the group N—CH—CH(OH) in delpheline. However, they were later able to disprove this assumption when they found that the salts of aminocamphor and its derivatives which could be used as reasonable models did not absorb ultraviolet light at abnormally short wavelengths and that de-ethyldelpheline (CXXXI) ($R = H$) did not consume any periodic acid or lead tetraacetate. Their rigorous proof of the absence of an α-hydroxy-amine group in delpheline will be described later.

Transformations in the Demethylene Series. Delpheline and its oxidation products, dehydrodelpheline, oxodelpheline and dehydro-oxodelpheline, liberated one mol of formaldehyde on treatment with strong acid and yielded the crystalline "demethylene" compounds (*4, 6*). Analytically, these compounds differed from their precursors only in the loss of one carbon atom, and like their precursors were saturated. On the basis of the chemistry of the demethylene compounds, the hydrolysis is best explained by the simple change: —O—CH_2—O— → —OH HO—.

While delpheline and its oxidation products are stable to lead tetra-acetate, demethylene-delpheline (CXXXIV) ($R' = H_2$; $R'' = H$) and demethylene-oxodelpheline (CXXXIV) ($R' = O$; $R'' = H$) consumed two mols of the reagent. Demethylene-oxodelpheline acetate (CXXXIV) ($R' = O$; $R'' = COCH_3$) and dehydrodemethylene-oxodelpheline (CXXXV) consumed one mol of lead tetraacetate (*4, 6*).

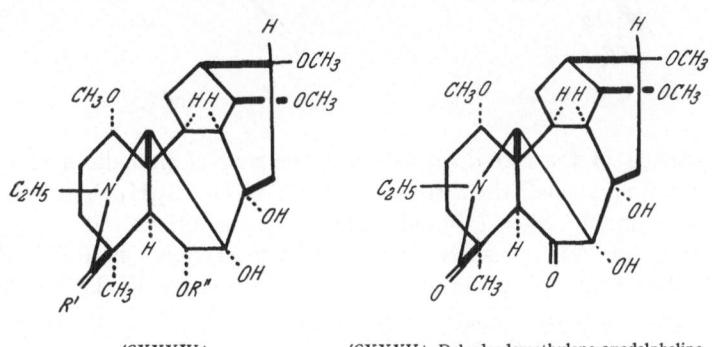

(CXXXIV.)   (CXXXV.) Dehydrodemethylene-oxodelpheline.

Thus, COOKSON and TREVETT (*6, 8*) have been able to prove that demethylene-delpheline is a 1,2,3-triol. Furthermore, since dehydro-demethylene-oxodelpheline (CXXXV) could be reduced by lithium aluminum hydride to demethylene-oxodelpheline, no ketol rearrangement took place during the acid hydrolysis of dehydro-oxodelpheline, and the secondary hydroxyl group produced in the reduction was probably the one originally present in delpheline. Since (CXXXV) is inert to bismuth acetate, this secondary hydroxyl group cannot be adjacent to another secondary hydroxyl (*4*). In fact, subsequent work indicated that the two remaining hydroxyl groups in demethylene-delpheline are both tertiary.

Oxidation of demethylene-oxodelpheline with one mol of periodic acid yielded the crystalline *seco*-diketone (CXXXVI) ($R = H$), $C_{24}H_{35}O_7N$, with infrared bands at 1754 and 1710 cm.$^{-1}$ (*4, 6*); no bands characteristic of an aldehyde grouping were present. The reported maxima suggest the presence of one ketonic group in a five-membered ring and one in a six-membered (or larger) ring. The two hydroxyls involved in this oxidation were therefore both tertiary and the originally present secondary hydroxyl group was uneffected. In agreement, the *seco*-diketone could be acetylated to the monoacetate (CXXXVI) $(R = COCH_3)$, also obtainable by oxidation of demethylene-oxodelpheline acetate (CXXXIV) $(R' = O; R'' = COCH_3)$ with lead tetraacetate (4).

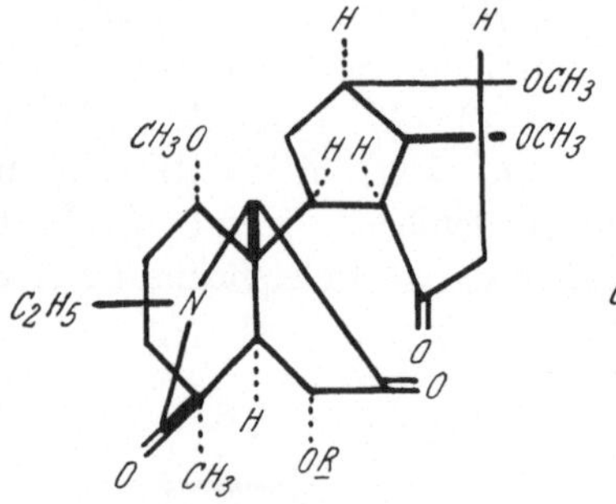

(CXXXVI.) 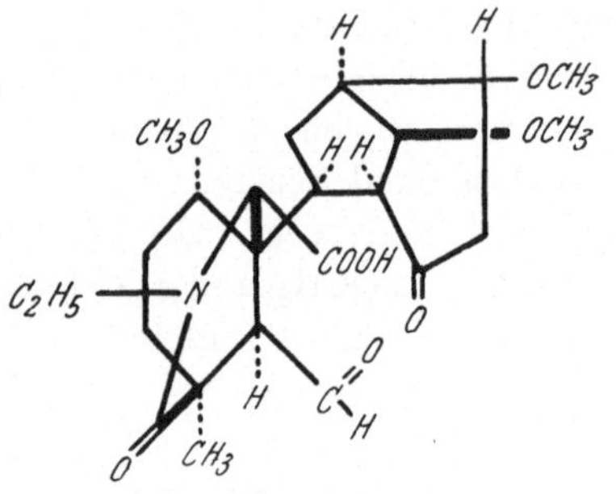   (CXXXVII.)

Oxidation of the *seco*-diketone with one mol of periodic acid in an alkaline buffer yielded the *diseco*-acid (CXXXVII), $C_{24}H_{35}O_8N$, (pK' 4.1 in 50% ethanol) with infrared bands, among others, at 1710 and 2800 cm.$^{-1}$ (4). The presence of an aldehyde grouping in (CXXXVII) and the fact that no carbon atom was lost during the oxidation clearly defines the 1,2,3-triol function. Formula (CXXXVII) is in agreement with the strength of the *diseco*-acid.

Oxidation of dehydrodemethylene-oxodelpheline (CXXXV) with lead tetraacetate gave the red α-diketone (CXXXVIII) with infrared bands at 1775 and 1755 cm.$^{-1}$ (cyclopentanedione) and at 1712 cm.$^{-1}$ (cyclohexanone) (4).

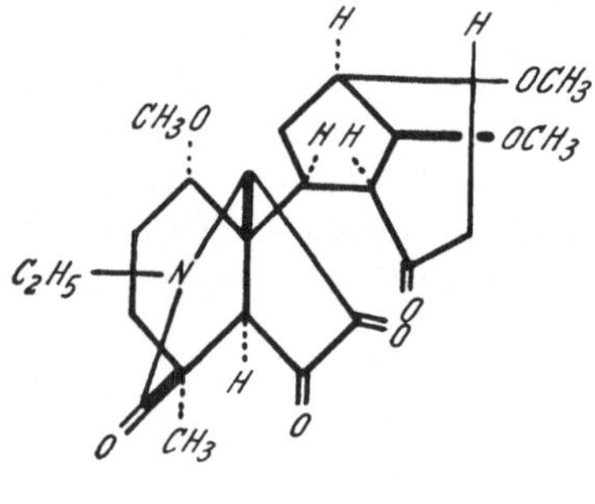

(CXXXVIII.) 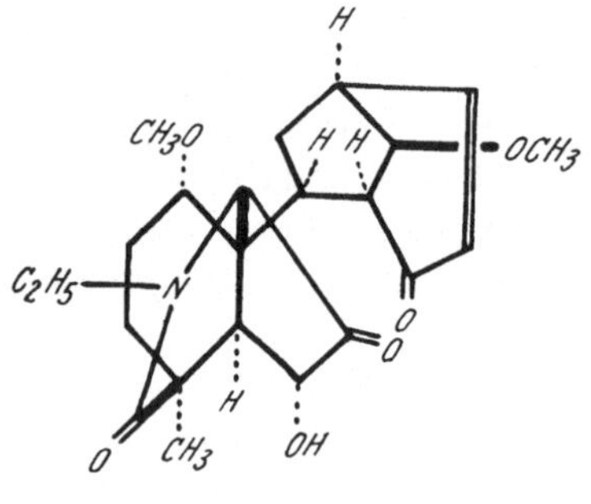   (CXXXIX.)

In complete analogy to the transformations in the lycoctonine series, the *seco* compounds eliminated the elements of methanol on treatment with dilute acid (4, 6). The *seco*-diketone (CXXXVI) $(R = H)$ gave the unsaturated *seco*-diketone (CXXXIX), $C_{23}H_{31}O_6N$, with ultraviolet maximum at 225 mμ ($\varepsilon = 11800$) and infrared bands at 1768 and 1680 cm.$^{-1}$. These spectroscopic properties indicate that the six-membered ring ketone is involved in the conjugated system and that the newly formed α,β-unsaturated ketone is only slightly substituted.

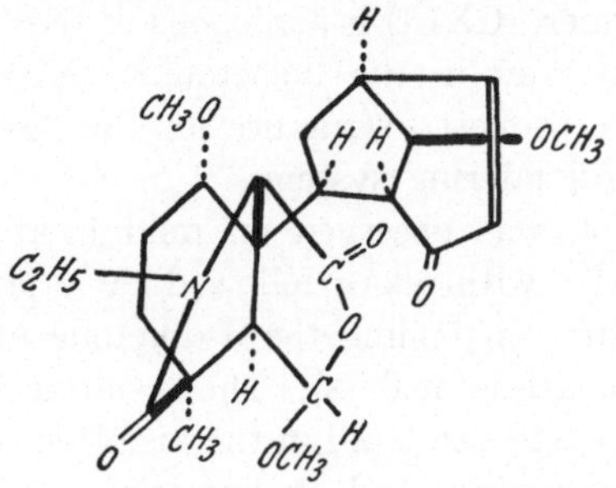

(CXL.)

Similarly, the red α-diketone (CXXXVIII) gave a red α,β-unsaturated ketone with an ultraviolet maximum at 223 mμ ($\varepsilon = 10000$). Elimination of methanol from the *diseco*-acid (CXXXVII) yielded the corresponding α,β-unsaturated keto-acid which was converted on treatment with hydrochloric acid in methanol to the *pseudo*-ester (CXL), the infrared band of which at 1755 cm.$^{-1}$ suggests that of a δ-lactone, raised by about 20 cm.$^{-1}$ by the adjacent methoxyl group (4).

Treatment of the *seco*-diketone (CXXXVI) ($R = H$) with alkaline alumina or potassium carbonate converted it to an isomer of substance (CXXXIX) which could also be prepared from (CXXXIX) itself in the same way. The ultraviolet spectrum of this isomer did no longer show

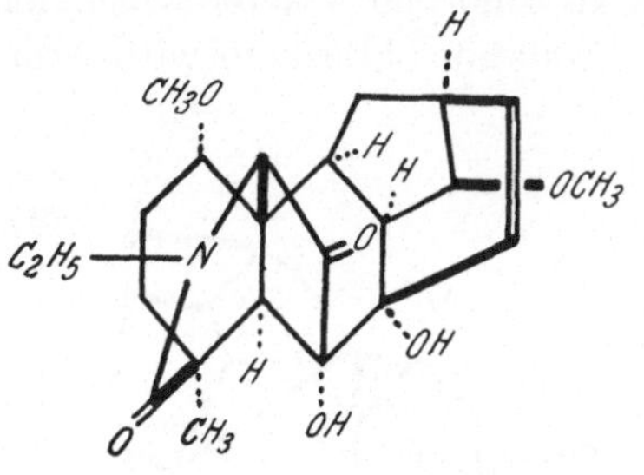

(CXLI.) 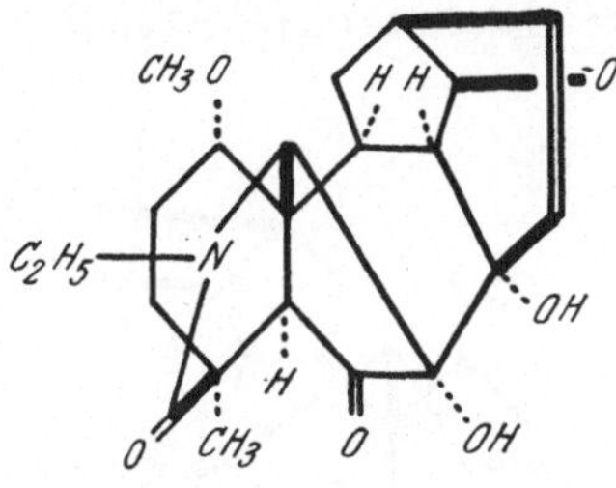   (CXLII.)

the maximum characteristic of an α,β-unsaturated ketone system, and its infrared spectrum lacked bands corresponding to a saturated or unsaturated cyclohexanone grouping while the cyclopentanone band was retained (4). The compound was stable to boiling dilute acid and to chromium trioxide in pyridine, but was oxidised with chromic acid in acetic acid. Furthermore, lead tetraacetate oxidation converted this isomer into the red α,β-unsaturated ketone-α-diketone. The isomerisation is clearly an aldol addition of one of the two possible enolates of the ketol to the six-membered ring carbonyl in (CXXXIX) (4). Of the two possible structures of the isomer, (CXLI) is analogous to the "iso" compound (22) formed in a similar way in the lycoctonine series, while (CXLII) is comparable with the proposed structure for the "α-iso" compound (22) and contains the original ring system.

A similar reaction was observed on mild treatment of the red α-diketone (CXXXVIII) with alkali followed by rapid extraction at low temperature or, better, on boiling the α-diketone in ethyl acetate that contained a trace of acetic acid (4). The resulting colorless isomer did not show the long wavelength band in the visible spectrum characteristic of the α-diketone grouping, and its infrared spectrum indicated the presence of only one five-membered (1755 cm.$^{-1}$) and one six-membered ring ketone (1720 cm.$^{-1}$). Further treatment with alkali converted the colorless isomer into the red α,β-unsaturated ketone-α-diketone. Cookson and Trevett (4) explain this change by another aldol condensation, this time involving a carbon atom α to the cyclohexanone carbonyl and one of the carbonyl carbons of the cyclopentanedione ring in (CXXXVIII).

While the treatment of demethylene-oxodelpheline with acetic anhydride in pyridine yielded the secondary monoacetate, treatment with acetyl chloride caused acetylation and dehydration, with the formation of a substance that was recognised as a ketone by its characteristic absorption spectra, λ_{max} 293–294 mμ ($\varepsilon = 114$) and ν_{max} 1738 cm.$^{-1}$ (ketone and acetate band superimposed) (5, 8). Hydrolysis of the pinacone acetate gave an amorphous ketol which formed an isomeric acetate on reacetylation. Oxidation of the ketol with chromium

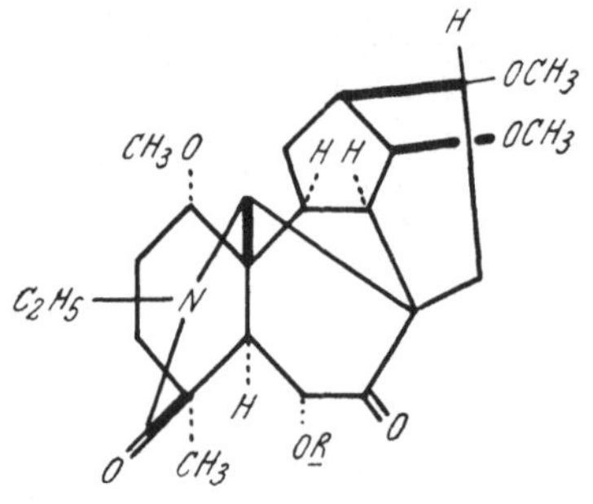

(CXLIII.) 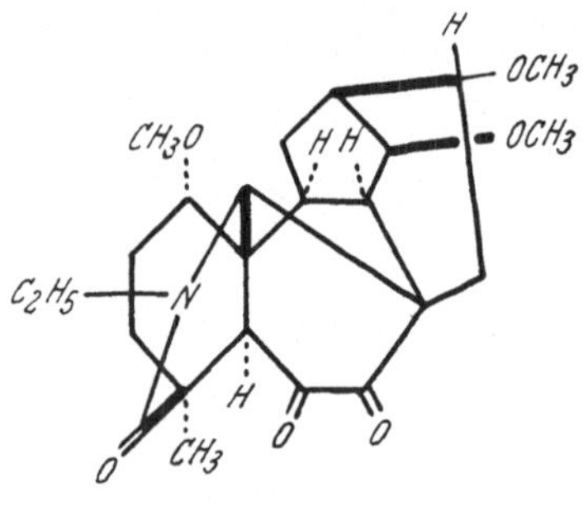   (CXLIV.)

trioxide in pyridine yielded a pale yellow α-diketone, $C_{24}H_{33}O_6N$, with infrared maxima at 1758 and 1720 cm.$^{-1}$ and an ultraviolet maximum at 402 mμ ($\varepsilon = 53$). On the basis of the structures discussed, the pinacone acetate, the ketol and the yellow α-diketone can therefore be represented, respectively, as (CXLIII) ($R = COCH_3$), (CXLIII) ($R = H$), and (CXLIV).

The rather high infrared frequencies of the pale yellow α-diketone are in reasonable agreement with the frequencies of some comparable six-membered models (5).

Oxidation of the α-diketone with alkaline hydrogen peroxide yielded the dicarboxylic acid (CXLV), $C_{24}H_{35}O_8N$, which could be converted to the dimethyl ester with diazomethane, and to a cyclic anhydride (ν_{max} 1800 and 1758 cm.$^{-1}$) with acetic anhydride (5). The acid titrated as a very strong monobasic acid (pK' 3.75 in 50% ethanol); the acid ester, obtained by methoxide opening of the cyclic anhydride titrated as a weaker acid (pK' 5.85). COOKSON and TREVETT explain these titration results by the stabilisation of the anion-acid by internal hydrogen bonding and destabilisation of the dianion by electrostatic repulsion and by hindrance to external hydrogen bonding and solvation (5).

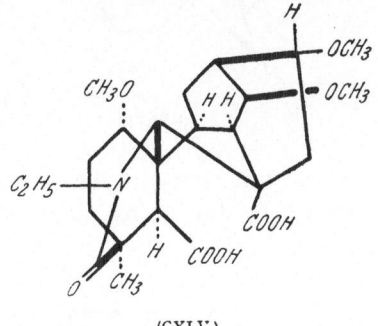

(CXLV.)

Opening of the cyclic anhydride with ammonia yielded an amide acid, which was converted to the methyl ester with diazomethane. Treatment of the amide ester with hypobromite gave an amino ester that was stable to boiling concentrated hydrochloric acid (5). If delpheline, as suspected earlier, were a 1,2-aminoalcohol, the aminoester would contain the grouping —CO—N(C_2H_5)—C—NH$_2$, since the carboxyl group involved in the HOFMANN degradation originates from the carbon atom bearing the secondary hydroxyl in delpheline. The stability of the aminoester to acid therefore rigorously excludes the presence of a 1,2-aminoalcohol grouping in delpheline.

The Carbon Skeleton of Delpheline. The brilliant experiments performed by COOKSON and TREVETT in the delpheline series can clearly all be accommodated by the carbon skeleton suggested by the X-ray analysis of des(oxymethylene)-lycoctonine. The remarkable similarity of many properties and reactions in the two series led to experiments by which the two alkaloids could be converted to an identical product. This has now been achieved (private communication from Dr. O. E. EDWARDS) by transforming lycoctonine to compound (CXLVI) which on reduction with sodium amalgam yielded the ketone (CXLVII). Compound (CXLVII) was oxidised with selenium dioxide to the α-diketone (CXLIV) obtained by COOKSON and TREVETT in the delpheline series.

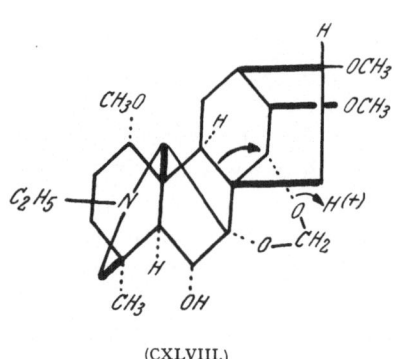

(CXLVI.)   (CXLVII.)

While COOKSON and TREVETT (4) assume that the carbon skeleton indicated by the X-ray analysis is probably present in demethylene-delpheline, they suggest that there is a possibility (even if quite remote) that delpheline itself might have, for example, the structure (CXLVIII), which would rearrange during the acid hydrolysis as indicated by arrows. However, the fact that a small amount of demethylene-oxodelpheline is formed in the oxidation of delpheline with permanganate (i. e. in the absence of strong acid) makes a skeletal rearrangement during the acid hydrolysis unlikely (4).

(CXLVIII.)

3. Biogenesis of Lycoctonine and Delpheline.

The proposed skeleton of lycoctonine and delpheline can be related biogenetically to the atisine skeleton (53, 4). For instance, the atisine type precursor (CXLIX) can lose a methyl group in a biological demethylation and undergo a WAGNER-MEERWEIN rearrangement and a MANNICH condensation to give (CL) which contains the complete lycoctonine and delpheline skeleton. The exact nature of the precursor and the reaction steps, the stage at which the oxygen functions are introduced and removed, and the order of the steps can of course not be specified at the present time.

(CXLIX.)　　　　　　　　　　(CL.)

IX. The Chemistry of Delphinine.

The alkaloid contained in the seeds of *Delphinium staphysagria*, delphinine, is together with aconitine possibly the most important representative of the polysubstituted class of aconite alkaloids. Although it has been studied most extensively, its structure is still unknown and only a speculative interpretation of delphinine chemistry can be attempted.

In the present article consideration will be given mainly to some beautiful experimental studies of JACOBS and his collaborators and to various possible interpretations of these results in the light of recent progress in the chemistry and biogenesis of diterpenoid alkaloids. Much of the older work or studies which do not contain structural information obvious to the writers will not be considered.

Recently, JACOBS (*39*) summarised with PELLETIER the results of his extensive studies on delphinine and attempted to rationalise them by means of a complete structure.

This structure does not fit the biogenetic ideas developed in the foregoing discussions and also presents great intrinsic difficulties in the interpretation of experimental data. Nevertheless, the paper mentioned is an excellent summary of the main attainments in JACOBS' laboratory and can be conveniently used as starting point for a discussion of delphinine.

In the structure (CLI) delphinine $C_{33}H_{45}O_9N$ is portrayed by JACOBS as a hexacyclic compound containing an N-methyl, four methoxyls, one hydroxyl, one acetoxyl, and one benzoxyl group.

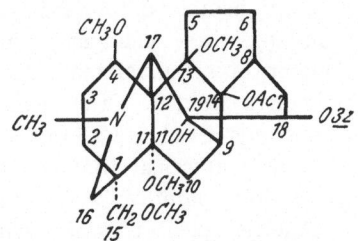

(CLI.) Delphinine? (*Bz* = benzoyl.)

This disposition of rings and functional groups has been extensively documented (*38, 40*, and references therein).

Oxidation of delphinine gives α-oxodelphinine which is a N-formyl compound and β-oxodelphinine, probably containing a δ-lactam group (*40*).

An interesting series of reactions studied by JACOBS is as follows. Pyrolytic loss of acetic acid from delphinine derivatives gives some unsaturated compounds which may be easily isomerised by acid. Thus, pyro-α-oxodelphinine represented by the partial structure (CLII) is isomerised to isopyro-α-oxodelphinine (CLIII).

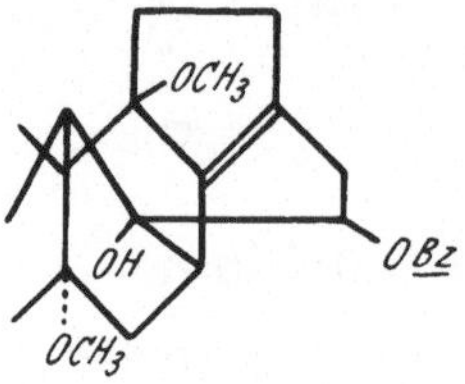

(CLII.)

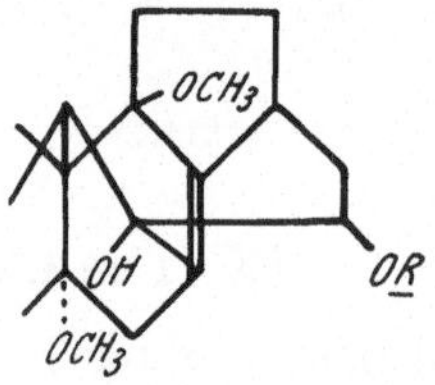

(CLIII.) (*R* = benzoyl.) (CLIV.) (*R* = H.)

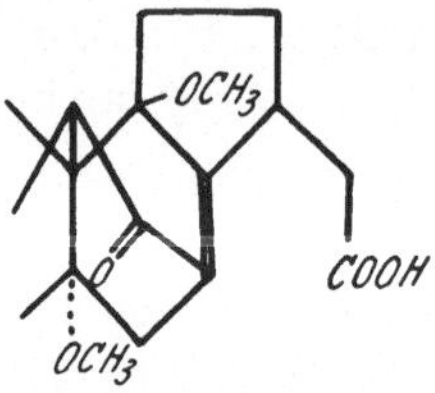

(CLV.)

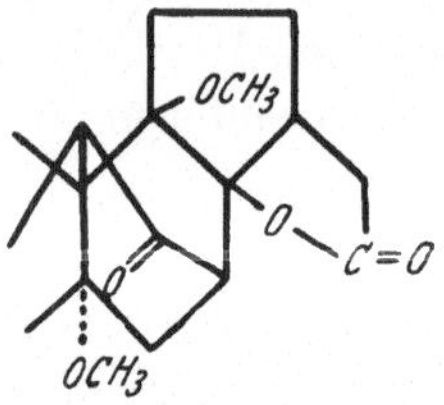

(CLVI.)

Saponification of the benzoyl group gives isopyro-α-oxodelphonine (CLIV). This compound must contain a vicinal secondary-tertiary glycol system, since oxidation with chromic acid converts it into an α,β-unsaturated ketoacid represented by JACOBS as (CLV). Compound (CLV) can lactonise by addition of the carboxyl across the conjugated double bond and form a saturated γ-lactone (CLVI) (*41*). Infrared evidence indicates that the keto group in (CLV) and (CLVI) is present in a six-membered (or larger) ring. According to JACOBS and PELLE-TIER (*39*), these changes cannot be accommodated in a lycoctonine skeleton or some related structure. On the other hand, the proposed structures in the series just discussed are not acceptable because compounds (CLIII), (CLIV) and (CLV) are impossibly strained. However, it seems quite clear that this degradation sequence is well defined and has a deep significance for the elucidation of the delphinine skeleton.

Another piece of work which sheds much light on the delphinine structure is a meticulous study on the demethylation of delphinine derivatives (*38*). It has been shown that nitric acid at 25° demethylates one methoxyl of isopyrooxo-delphinine. The liberated hydroxyl group has been considered by JACOBS to be tertiary, because no well-defined ketone was produced by chromic acid oxidation of the demethylated compound, and it has been placed at $C_{(11)}$ of his formula. A slightly more vigorous action of nitric acid demethylates two methoxyls and simultaneously oxidises one of the liberated new hydroxyl groups to an aldehyde. The aldehyde group may be oxidised further to a hindered carboxyl. JACOBS assumed that the two methoxyls attacked in this case were the ones at $C_{(11)}^-$ and $C_{(15)}$. These methoxy groups indeed seem to possess a considerable ability for being displaced. Thus, when isopyrooxo-delphinine is treated with hydrochloric acid in methanol, they are replaced by chlorine and the resulting dichloro compound may be reconverted into isopyrooxo-delphinine by heating with methanol.

The chlorine atoms in the dichloro compound may also be replaced by hydroxyl groups. The product of this reaction may then be oxidised to the same aldehyde which was obtained by interaction of isopyrooxo-delphinine with nitric acid. This shows rigorously that the same two methoxyls were involved also in this reaction sequence.

Complete demethylation of isopyrooxo-delphinine with zinc chloride results in a desmethyl anhydro compound formulated as (CLVII).

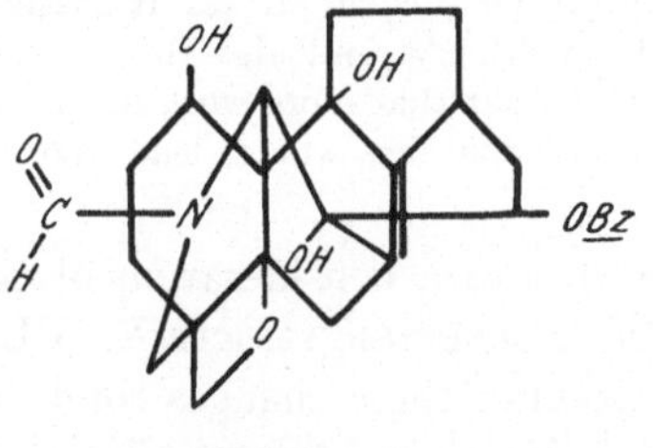

(CLVII.) 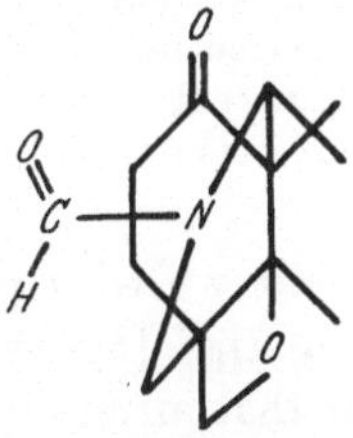   (CLVIII.)

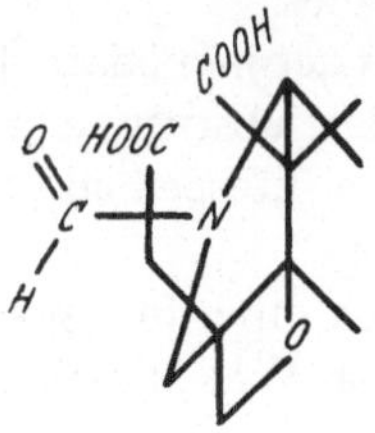

(CLIX.) 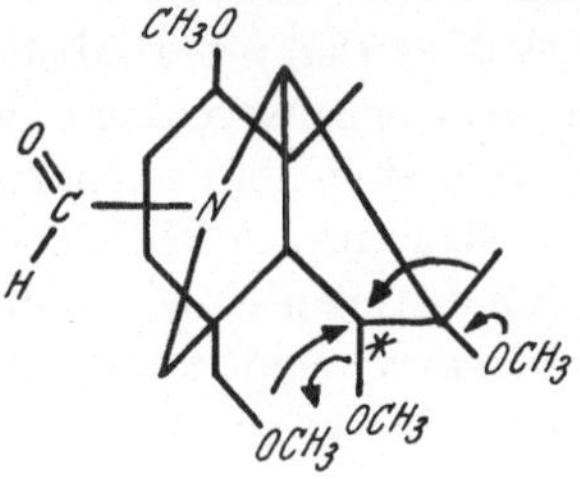   (CLX.)

Compound (CLVII) on oxidation gives first the six-membered ketone (CLVIII) (I. R. max 1712 cm.$^{-1}$) and by further, more vigorous treatment the dicarboxylic acid (CLIX). The diester of (CLIX) may be saponified to a monoester which indicates the tertiary character of one of the carboxyls.

This sequence indicates not only the secondary character of one of the last liberated methoxyls, but also the fact that the primary hydroxyl is not free in the anhydro compound and, consequently, it may be possibly involved in an ether bridge as shown in (CLVII).

These degradations may indeed, as postulated by JACOBS, be correctly represented as taking place in ring A of a diterpenoid alkaloid skeleton. If this is true, then the nature of the methoxyl which is most easily displaced and assumed to be tertiary, is puzzling. The four-membered ether (CLVII) is badly strained and moreover a methoxyl at $C_{(11)}$ in a diterpenoid alkaloid skeleton is in a bridgehead position and would be difficult to displace.

There seem to be two alternatives if the ring A hypothesis is to be maintained. The first one is that the easily displaced methoxyl is secondary and that the partial structure (CLX) represents the area in question of α-isopyrooxo-delphinine. The displacement of the first methoxyl and ether formation would then be easily understood as indicated by the arrows in (CLX). Such a reaction not only appears plausible but is even to some extent precedented in the lycoctonine series (p. 70). However, one should question the correctness of this explanation, since a secondary alcohol at the position marked with an asterisk in (CLX) is known in the chemistry of delpheline to oxidise extremely well by a variety of reagents to a ketone (see above). The second alternative is the possibility that in delphinine the nitrogen containing ring is modified. For instance, modifications by a biogenetic WAGNER-MEERWEIN rearrangement can be envisaged which makes it possible to explain the smooth displacement of the first methoxyl and also the ether formation by a simple rearrangement. However, it is clear that more work in this area is needed and that a discussion of all these possibilities would lead too far into pure speculation.

At this point we wish to return to a consideration of the reactions represented by the JACOBS and PELLETIER structures (CLII)–(CLVI). It seems clear that an interpretation of these changes combined with the work discussed above would provide at least approximate information about practically the whole molecule of delphinine.

It is striking that a partial structure which can accommodate JACOBS' reaction sequence exceedingly well is a suitably substituted 1,2,3-bicyclooctane system that is characteristic of all the diterpenoid alkaloids of known structure or their biogenetic precursors.

One can distinguish two main possibilities which differ in representing the pyro-isopyro acid catalysed change as (a) a WAGNER-MEERWEIN rearrangement, and (b) the shift of a double bond.

(a) The partial structures (CLXI)–(CLXIV) represent the first possibility. Structure (CLXI) shows two rings and the disposition of three substituents in a delphinine derivative. In this scheme there are two possibilities for the location of the benzoxy group and two possibilities for the configuration of the acetoxy group. The pyrolytic elimination of acetic acid which generally takes place without rearrangement is represented by (CLXI)–(CLXII), and the acid-catalysed pyro-isopyro rearrangement by (CLXII)–(CLXIII). Finally, the α,β-unsaturated ketoacid which can form a γ-lactone is portrayed by the partial structure (CLXIV).

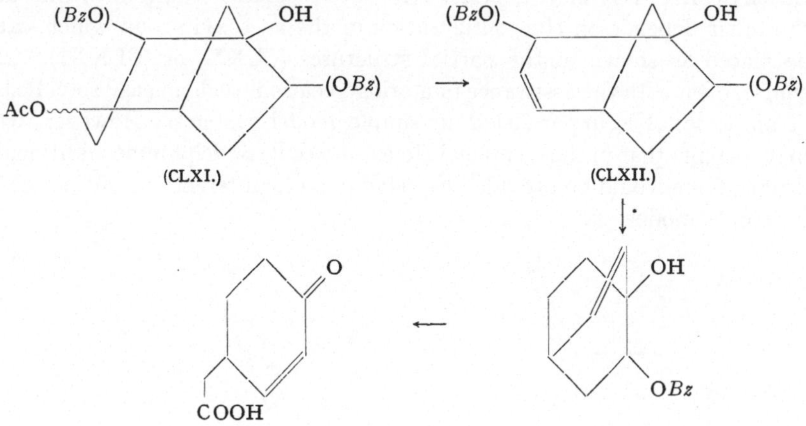

(CLXI.)   (CLXII.)   (CLXIV.)   (CLXIII.)

(b) The second possibility is given by the partial structures (CLXV) to (CLXVII) which are self-explanatory*.

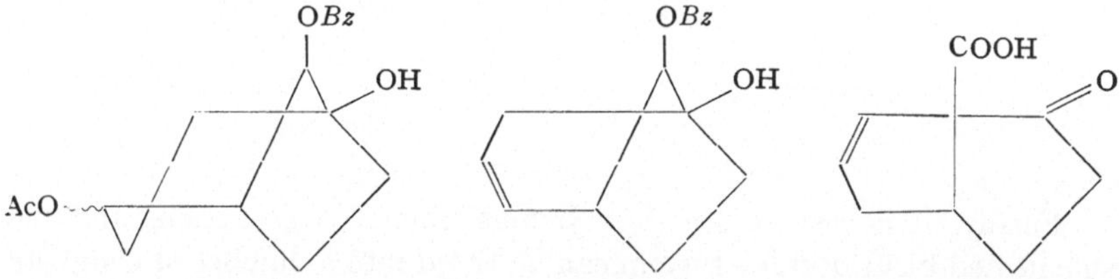

(CLXV.)   (CLXVI.)   (CLXVII.)

If this interpretation of JACOBS' data is correct, then it must account for the easy replacement of the acetoxyl group by a methoxyl which is

---

* *Note added in proof:* A further communication on delphinine chemistry (*40 a*) appeared after the submission of this manuscript. The experimental evidence cited there can be interpreted as a proof of possibility (b), i. e. the transformations (CLXV) → (CLXVI) → (CLXVII). Furthermore, the base-catalysed isomerisation of the α-hydroxyketones in the dihydropyro and dihydroisopyro series which JACOBS aud PELLETIER tentatively explain as an epimerisation α to the keto group, is undoubtedly an α-ketol rearrangement. The change of pyro- into isopyro-derivatives may however be not a simple double bond shift, but for instance an allylic rearrangement, if there is a methoxyl α to the tertiary hydroxyl group. This might be a simple interpretation of the difference between the dihydropyro and dihydroisopyro compounds. In the same paper, JACOBS has definitely confirmed the first stage of the HOFMANN degradation of delphinine. This may be explained by a simple biogenetic rearrangement of ring B. This kind of rearrangement performed, for instance, on the lycoctonine skeleton leads to a skeleton as in (CXXV, p. 70) which may undergo a HOFMANN reaction.

known to occur in some delphinine derivatives (*35*). One can visualise this solvolysis as being aided by the stability of the non-classical bridged cations (CLXVIII) or (CLXIX). The intermediate formation of one or the other depends on the configuration of the acetoxyl group which can be placed as shown in the partial structures (CLXX) or (CLXXI). Of course, even with the assistance of a bridged cation, such an easy solvolysis of an acetoxyl is unparalleled in simple model systems. However, we may imagine that in the complex bridged molecule of delphinine additional driving force would be provided by relief of steric interactions not present in simple models.

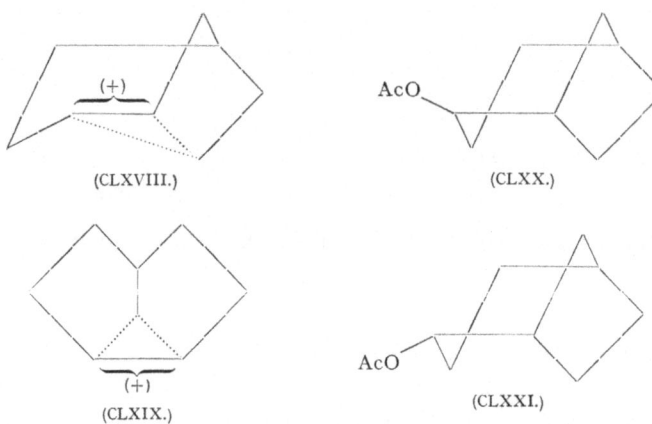

(CLXVIII.)   (CLXX.)

(CLXIX.)   (CLXXI.)

Finally, it is easy to see that Jacobs' ring *A* degradations and the substituted bicyclooctane system can be fitted into a number of complete structures based on the veatchine or lycoctonine skeleton or simple biogenetically plausible modifications of these structures. Again, a complete discussion of all possibilities would lead too far but one can conclude that the data known at present on delphinine support rather than refute a biogenetically anticipated constitution.

That this hypothesis may be correct is further in agreement with the fact that Jacobs (*37*) has isolated by dehydrogenation of an iso-pyrodelphinine derivative a hydrocarbon which, judging from the analysis of the trinitrobenzene complex and the ultraviolet spectrum of the crude product, may be a phenanthrene.

Aconitine chemistry shows several similarities to the chemistry of delphinine, and frequently the assumption has been made (*39*) that aconitine and delphinine possess the same skeletal structure. However, there is not yet available a sufficient amount of experimental data on aconitine to warrant even a tentative discussion of this alkaloid within the scope of the present article.

References.

1. BARTLETT, M. F., J. A. EDWARDS, W. I. TAYLOR and K. WIESNER: Evidence of an Ethanolamine Structure in the Garrya Alkaloids and Atisine; and a Partial Synthesis of Garryine. Chem. and Ind. **1953**, 323.

2. BARTLETT, M. F. and K. WIESNER: Synthesis of 3-Aza-1-methyl-7-ethyl-phenanthrene. Chem. and Ind. **1954**, 542.

3. COOKSON, R. C. and M. E. TREVETT: *Aconitum* and *Delphinium* Alkaloids. Part I. The Environment of the Nitrogen Atom in Delpheline. J. Chem. Soc. (London) **1956**, 2689.

4. — — *Aconitum* and *Delphinium* Alkaloids. Part II. Interrelation of the Functional Groups of Delpheline. J. Chem. Soc. (London) **1956**, 3121.

5. — — *Aconitum* and *Delphinium* Alkaloids. Part III. Some Rearrangements in the Delpheline and Lycoctonine Series. A Monobasic Dicarboxylic Acid. J. Chem. Soc. (London) **1956**, 3864.

6. — — The Functional Groups of Delpheline. Chem. and Ind. **1954**, 1324.

7. — — Hydroxylycoctonine.. Chem. and Ind. **1956**, 276.

8. — — The Pinacolic Dehydration of Demethylene-Oxodelpheline. Chem. and Ind. **1954**, 1391.

9. COOKSON, R. C. and N. S. WARIYAR: Absorption Spectra of Ketones. Part IV. The Steric Requirements for Spectroscopic Interaction between a Carbonyl Group and a $\beta\gamma$-Double Bond. J. Chem. Soc. (London) **1956**, 2302.

10. CRAIG, L. C. and W. A. JACOBS: The Aconite Alkaloids. X. On Napelline. J. Biol. Chem. **143**, 611 (1942).

11. DJERASSI, C., C. R. SMITH, S. K. FIGDOR, J. HERRAN and J. ROMO: Alkaloid Studies. VI. Cuauchichicine, a New Diterpenoid Alkaloid. J. Amer. Chem. Soc. **76**, 5889 (1954).

12. DJERASSI, C., C. R. SMITH, A. E. LIPPMAN, S. K. FIGDOR and J. HERRAN: Alkaloid Studies. VIII. The Structures of the Diterpenoid Alkaloids Laurifoline and Cuauchichicine. J. Amer. Chem. Soc. **77**, 4801 (1955).

13. DREIDING, A. S. and J. A. HARTMAN: The Rearrangement of Some Substituted Allyl Alcohols to their Isomeric Ketones. J. Amer. Chem. Soc. **78**, 1216 (1956).

14. DVORNIK, D. and O. E. EDWARDS: Atisine. Structure Studies. Abstract of a paper, 40th Annu. Conf. Chem. Inst. of Canada; Chem. Canada **9**, No. 4, p. 63 (April 1957).

15. — — Ajaconine. Interrelation with Atisine. Chem. and Ind. **1957**, 952.

16. — — Atisine. A Novel Degradation. Chem. and Ind. **1956**, 248.

17. EDWARDS, J. A.: The Structure and Stereochemistry of Diterpene Alkaloids. Thesis, Univ. New Brunswick, 1955.

18. EDWARDS, O. E. and L. MARION: Lycoctonine: The Environment of the Nitrogen. Canad. J. Chem. **32**, 1146 (1954).

19. — — Lycoctonine and Its Oxidation Products. Canad. J. Chem. **30**, 627 (1952).

20. — — Lycoctonine. Periodate Oxidation Studies. Canad. J. Chem. **32**, 195 (1954).

21. EDWARDS, O. E., L. MARION and R. A. McIVOR: Lycoctonine. Acid Catalyzed Rearrangements. Canad. J. Chem. **32**, 708 (1954).

22. EDWARDS, O. E., L. MARION and D. K. R. STEWART: The Chemistry of Lycoctonine. Canad. J. Chem. **34**, 1315 (1956).

23. EDWARDS, O. E. and T. SINGH: Atisine. The Functional Groups. Canad. J. Chem. **33**, 448 (1955).

24. — — Atisine. The Heterocyclic Ring and Functional Groups. Canad. J. Chem. **32**, 465 (1954).

25. FREUDENBERG, W. and E. F. ROGERS: New Alkaloids in *Aconitum napellus*. J. Amer. Chem. Soc. **59**, 2572 (1937).
26. GOODSON, J. A.: The Alkaloids of the Seeds of *Delphinium elatum* L. J. Chem. Soc. (London) **1943**, 139.
27. — *Delphinium* Alkaloids. Part III. Delpheline. J. Chem. Soc. (London) **1944**, 665.
28. — *Delphinium* Alkaloids. Part IV. The Alkaloids of the Seeds of *Delphinium ajacis*. J. Chem. Soc. (London) **1945**, 245.
29. HUEBNER, C. F. and W. A. JACOBS: The Aconite Alkaloids. XVIII. The Synthesis of the Hydrocarbon Obtained on Dehydrogenation of Atisine. J. Biol. Chem. **170**, 203 (1947).
29a. HUGHES, E. W. and R. NATHAN: Private communication.
30. HUMBER, L. G.: Unpublished experiments, University of New Brunswick.
31. ITÔ, S.: Unpublished experiments, University of New Brunswick.
32. JACOBS, W. A.: The Aconite Alkaloids. XXIV. The Degradation of Atisine and Isoatisine. J. Organ. Chem. (USA) **16**, 1593 (1951).
33. JACOBS, W. A. and L. C. CRAIG: The Aconite Alkaloids. VIII. On Atisine. J. Biol. Chem. **143**, 589 (1942).
34. — — The Aconite Alkaloids. XI. The Action of Methyl Alcoholic Sodium Hydroxide on Atisine. Isoatisine and Dihydroatisine. J. Biol. Chem. **147**, 567 (1943).
35. — — Delphinine. III. The Action of Hydrochloric, Nitric and Nitrous Acids on Delphinine and its Derivatives. J. Biol. Chem. **136**, 303 (1940).
36. JACOBS, W. A. and C. F. HUEBNER: The Aconite Alkaloids. XVII. Further Studies with Hetisine. J. Biol. Chem. **170**, 189 (1947).
37. — — The Aconite Alkaloids. XIX. Further Studies with Delphinine Derivatives. J. Biol. Chem. **170**, 209 (1947).
38. JACOBS, W. A. and S. W. PELLETIER: The Aconite Alkaloids. XXV. The Oxygen-containing Groups of Delphinine. J. Amer. Chem. Soc. **76**, 161 (1954).
39. — — The Aconite Alkaloids. XXXII. The Structure of Delphinine. J. Amer. Chem. Soc. **78**, 3542 (1956).
40. — — The Nature of α-Oxodelphinine and β-Oxodelphinine. Chem. and Ind. **1955**, 948.
40a. — — The Aconite Alkaloids. XXXV. Structural Studies with Delphinine Derivatives. J. Organ. Chem. (USA) **22**, 1428 (1957).
41. JACOBS, W. A. and Y. SATO: The Aconite Alkaloids. XXIII. Oxidation of Isopyrooxodelphonine, Dihydroisopyrooxodelphonine, and their Desmethylanhydro Derivatives. J. Biol. Chem. **180**, 479 (1949).
42. JOWETT, H. A. D.: Contributions to our Knowledge of the Aconite Alkaloids. Part XIII. On Atisine, the Alkaloid of *Aconitum heterophyllum*. J. Chem. Soc. (London) **69**, 1518 (1896).
43. KING, J. F.: Degradation Studies in the Aconite Alkaloid Series. Ph. D. Thesis, Univ. New Brunswick, 1957.
44. LAWSON, A. and J. E. C. TOPPS: Aconitine. Part II. The Relationship between Aconitine and Atisine and Some Degradation Products of the Latter. J. Chem. Soc. (London) **1937**, 1640.
45. ONETO, J. F.: Alkaloids of Species of Garrya. I. Isolation of Alkaloids. J. Amer. Pharmaceut. Assoc. **35**, 204 (1946).
46. PELLETIER, S. W. and W. A. JACOBS: The Aconite Alkaloids. XXVII. The Structure of Atisine. J. Amer. Chem. Soc. **76**, 4496 (1954).
47. — — The Aconite Alkaloids. XXX. Products from the Mild Permanganate Oxidation of Atisine. J. Amer. Chem. Soc. **78**, 4139 (1956).

48. PELLETIER, S. W. and W. A. JACOBS: The Aconite Alkaloids. XXXI. A Partial Synthesis of Atisine, Isoatisine and Dihydroatisine. J. Amer. Chem. Soc. **78**, 4144 (1956).

49. — — The Quaternary Chlorides and Acetates of Atisine. Chem. and Ind. **1955**, 1385.

50. PRZYBYLSKA, M. and L. MARION: The Crystal Structure of Des-(oxymethylene)-lycoctonine Hydriodide Monohydrate. Canad. J. Chem. **34**, 185 (1956).

51. ROBERTS, J. D. and W. T. MOORELAND, Jr.: Electrical Effects of Substituent Groups in Saturated Systems. Reactivities of 4-Substituted Bicyclo[2,2,2]-octane-1-carboxylic Acids. J. Amer. Chem. Soc. **75**, 2167 (1953).

52. STERN, E. S.: The Aconitum and Delphinium Alkaloids. In: R. H. F. MANSKE and H. L. HOLMES, The Alkaloids, Chemistry and Physiology, Vol. IV, p. 275. New York: Academic Press. 1954.

53. VALENTA, Z. and K. WIESNER: Biogenetic Interrelationships of Diterpene Alkaloids. Chem. and Ind. **1956**, 354.

54. WENKERT, E.: Structural and Biogenetic Relationships in the Diterpene Series. Chem. and Ind. **1955**, 282.

55. WIESNER, K.: Lecture before the Chem. Inst. of Canada, Organic Division, Montreal, March 8, 1954.

56. WIESNER, K., J. R. ARMSTRONG, M. F. BARTLETT and J. A. EDWARDS: *Garrya* Alkaloids. III. The Skeletal Structure of the Garrya Alkaloids. J. Amer. Chem. Soc. **76**, 6068 (1954).

57. — — — — The Skeleton of the *Garrya* Alkaloids and Atisine. Chem. and Ind. **1954**, 132.

58. WIESNER, K. and J. A. EDWARDS: The Basicity and Steric Configuration of the Diterpene Alkaloids Veatchine and Atisine. Experientia **11**, 255 (1955).

59. WIESNER, K., S. K. FIGDOR, M. F. BARTLETT and D. R. HENDERSON: *Garrya* Alkaloids. I. The Structure of Garryine and Veatchine. Canad. J. Chem. **30**, 608 (1952).

59 a. WIESNER, K., S. ITÔ and Z. VALENTA: The Structure of Napelline and Songorine. Experientia **14**, 167 (1958).

60. WIESNER, K., W. I. TAYLOR, S. K. FIGDOR, M. F. BARTLETT, J. R. ARMSTRONG und J. A. EDWARDS: *Garrya* Alkaloide. II. Mitt. Weitere Versuche über den Abbau von Garryin und Veatchin. Chem. Ber. **86**, 800 (1953).

61. WIESNER, K., Z. VALENTA, J. F. KING, R. K. MAUDGAL, L. G. HUMBER and S. ITÔ: The Isolation and Structure of Napellonine and Napelline. Chem. and Ind. **1957**, 173.

62. WOODWARD, R. B. and E. G. KOVACH: The Structure of the Santonides. J. Amer. Chem. Soc. **72**, 1009 (1950).

63. WRIGHT, A.: Report on the Aconite Alkaloids. Yearbook of Pharmacy, 1879, p. 422.

(Received, November 26, 1957.)

Structural Chemistry of Actinomycetes Antibiotics.

By **E. E. VAN TAMELEN**, Madison, Wisconsin.

Contents.

Manuscript prepared at Cambridge University, Cambridge, England, during the tenure of a Haight Travelling Fellowship (Research Committee, the Graduate School of the University of Wisconsin, Madison, Wisconsin).

Introduction.

Antibiotics, i. e., metabolic products of microorganisms which possess the capacity, in low concentration, of inhibiting the growth of, or destroying, other microorganisms, are elaborated mainly by fungi, bacteria and actinomycetes, although materials exhibiting antibiotic properties are also produced by lichens, algae, insects and higher plants. Certain substances derived from the first three categories have attained during the past decade considerable success as therapeutic agents: penicillin, streptomycin, bacitracin, terramycin, aureomycin, chloromycetin, and erythromycin are well-known examples. Antibiotics, in addition to being important for reasons of public health, are of great interest to the organic chemist—in no other area of the natural product field has he confronted such novelty, variety and complexity of structure.

The developments in the antibiotic story are ordinarily traced back to 1929, when FLEMING noted that a principle produced by a mold, *Penicillium notatum*, possessed potent bacteriostatic properties. This fortunate observation was capitalized upon during the early part of 1940, when commercial production by fermentation methods of the active agent, penicillin, commenced. Chemical investigations on the constitution of this remarkable substance culminated during the same period. Although the chemistry and utility of the penicillin molecule are, even today, most noteworthy, attention during more recent years has shifted to antibiotic materials produced by the *Actinomycetes* group of microorganisms. As a consequence of this effort, many new weapons have been added to the battery of antibacterial agents at the disposal of the medical profession, and the knowledge of natural product chemistry has been enriched correspondingly. This review is devoted to the *structural chemistry* of antibiotics derived from the *Actinomycetes* group.

Actinomycetaceae, a large group of hardy microorganisms occurring widely in soils, air, dust or associated with plants, have been studied extensively by WAKSMAN and his school. These investigators reported during the early part of 1940 their efforts in the isolation of useful antibiotic principles, crowned by the development of streptomycin as a therapeutic agent. It has been suggested that the *Actinomycetes* group be divided into four genera: *Actinomyces, Nocardia, Micromonospora,* and *Streptomyces*. The members of the last-named subdivision have provided the more useful antibiotics, and consequently it is also this group which has been accorded more attention by chemical investigators. To date, more than two dozen *Streptomyces* antibiotics have been structurally characterized; the constitutions of only several from the other *Actinomycetes* subdivisions have been determined.

In this survey attention has been restricted to those *Actinomycetes* antibiotics which have been the subjects of chemical work sufficient to establish complete organic, gross structures. In most cases the arguments presented or used by the investigators to derive the structures, are not reproduced in detail—instead the broader aspects of reactivity which were of direct use in structure determination are summarized.

1. Actidione (Cycloheximide).

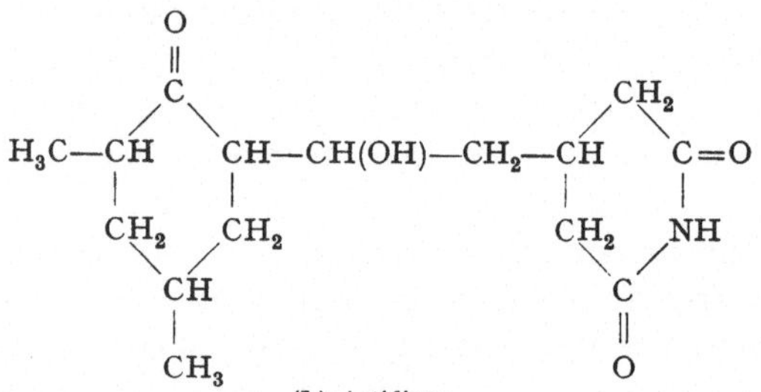

$$
\begin{array}{c}
\text{O} \\
\| \\
\text{C}
\end{array}
$$

H₃C—CH   CH—CH(OH)—CH₂—CH   C=O

(I.) Actidione.

Subsequent to the commercial development of the streptomycin fermentation utilizing *Streptomyces griseus*, it was discovered (*105*) that this same microorganism elaborates a distinctly different antibiotic substance, which was given the trade name Actidione (trivial chemical name, cycloheximide). Whether *S. griseus* produces a preponderence of streptomycin or Actidione depends upon the medium used for its growth.

Under favorable conditions, Actidione of 30–60% purity can be obtained from the beer by extraction with chloroform; adsorption on activated carbon followed by acetone elution and finally chromatography are used for production of the compound in crystalline form (*33*).

Actidione is a market product, active against yeasts and some fungi, but not bacteria.

Recognition of the ketonic nature of Actidione and the belief (*66*) that it possessed the molecular formula $C_{27}H_{42}O_7N_2$ led to the suffix "-dione" in the name; revision of the formula to $C_{15}H_{23}O_4N$ (*33*) makes "Actidione" a misnomer. The key to the structure of the molecule (*58, 59*) lay in the rapid cleavage brought about by dilute alkali at room temperature: there are formed 2,4-dimethylcyclohexanone (II) and the

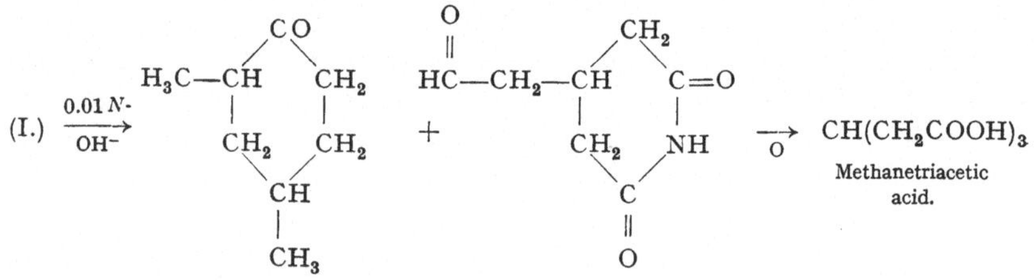

(I.) $\xrightarrow[\text{OH}^-]{0.01\,N\text{-}}$ (II.)   +   (III.) $\xrightarrow[\text{O}]{}$ CH(CH₂COOH)₃

Methanetriacetic acid.

aldehydoglutarimide moiety (III), which was identified by oxidation to methanetriacetic acid. The liberation of ammonia under more drastic alkaline conditions, coupled with spectral and electrometric titration data, confirmed the presence of the glutarimide portion. Interpretation of the basic cleavage as a reverse aldol (IV), supplemented by the observation that the antibiotic could be dehydrated to an α,β-unsaturated ketone (V), led to the expression (I).

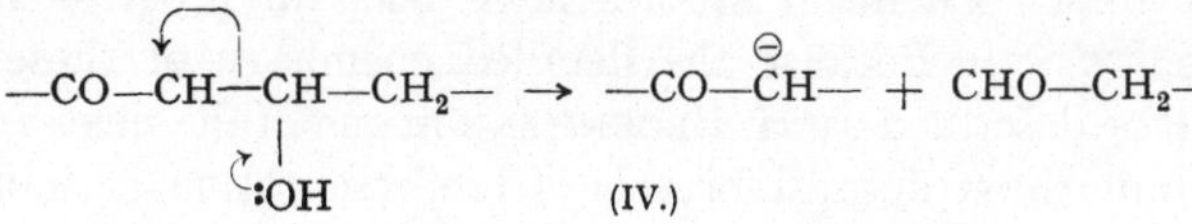

$$-CO-CH-CH-CH_2- \quad \rightarrow \quad -CO-\overset{\ominus}{CH}- \; + \; CHO-CH_2-$$

$$:OH \qquad \text{(IV.)}$$

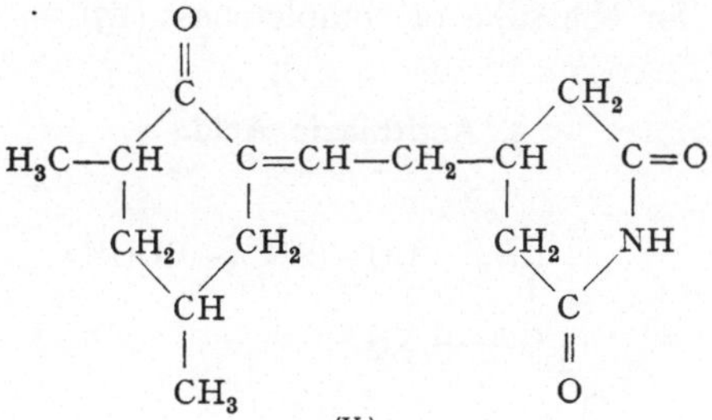

(V.)

2. Actinomycins.

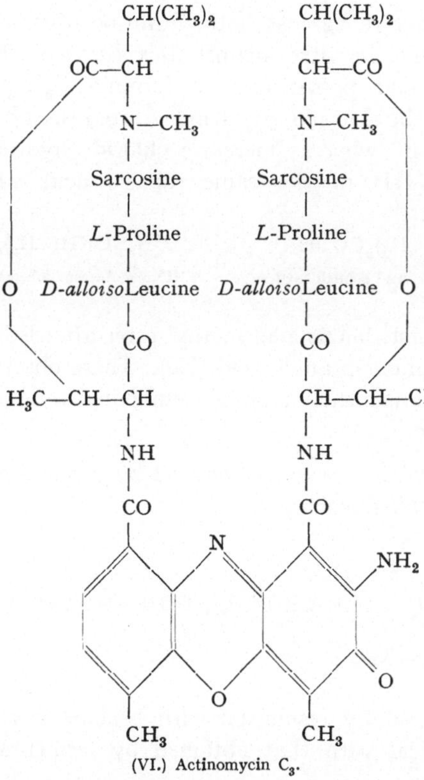

(VI.) Actinomycin C₃.

The actinomycin group of antibiotics, representatives of which were first reported in 1942 (96), have been isolated from cultures of *Streptomyces antibioticus* and other *Streptomyces* species, as well as from a *Micromonospora*. The actinomycins are potent antibiotics, but are also highly toxic. Chemical studies have been in progress in several laboratories (15, 55). Because the detailed chemistry of these chromo-peptides will be described by H. Brockmann in one of the next Volumes of this Series, only the structural formula of a representative, actinomycin C_3, is included here, for the sake of completeness (13).

3. Actithiazic Acid.

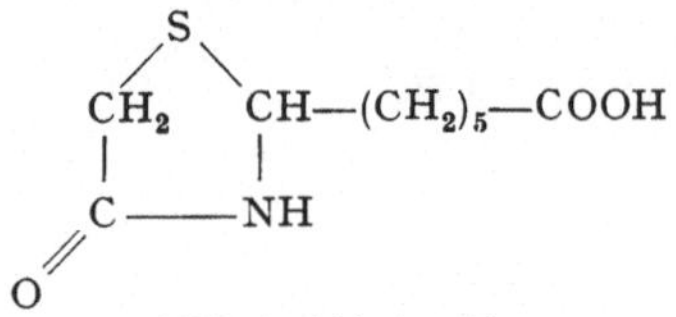

$$
\begin{array}{c}
S \\
\diagup \quad \diagdown \\
CH_2 \quad CH-(CH_2)_5-COOH \\
| \qquad | \\
C-----NH \\
\diagup\diagup \\
O
\end{array}
$$

(VII.) Actithiazic acid.

Streptomyces virginia and *S. cinnamonensis* biosynthesize actithiazic acid (41, 82), inactive against both Gram-positive and Gram-negative bacteria, but active in vitro against *Mycobacteria*. The antibiotic is a monocarboxylic acid possessing the formula $C_9H_{15}O_3NS$ (82, 68, 77). Oxidation to pimelic acid (68, 77) demonstrated the presence of a straight seven-carbon chain, whereas mercuric chloride hydrolysis provided the semi-aldehyde (VIII) of the same dibasic acid. Raney nickel de-

$$CHO(CH_2)_5COOH$$

(VIII.) Pimelic acid semialdehyde.

$$CH_3CONH(CH_2)_6COOH$$

(IX.) ω-Acetamidoheptanoic acid.

sulfurization of actithiazic acid methyl ester afforded the corresponding ester of ω-acetamidoheptanoic acid (IX). These observations, reinforced by spectral (68) and other data, are embodied by the expression (VII) for the antibiotic.

Actithiazic acid has been synthesized by condensing ω-formylcaproic acid ester with thioglycolamide (68, 22).

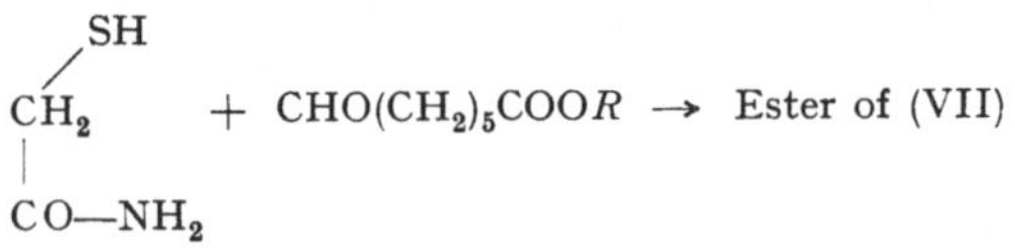

$$
\begin{array}{c}
SH \\
\diagup \\
CH_2 \qquad + \quad CHO(CH_2)_5COOR \quad \rightarrow \quad \text{Ester of (VII)} \\
| \\
CO-NH_2
\end{array}
$$

Hydrolysis followed by resolution with brucine gave rise to synthetic antibiotic, identical with that obtained by fermentation.

4. Amicetin.

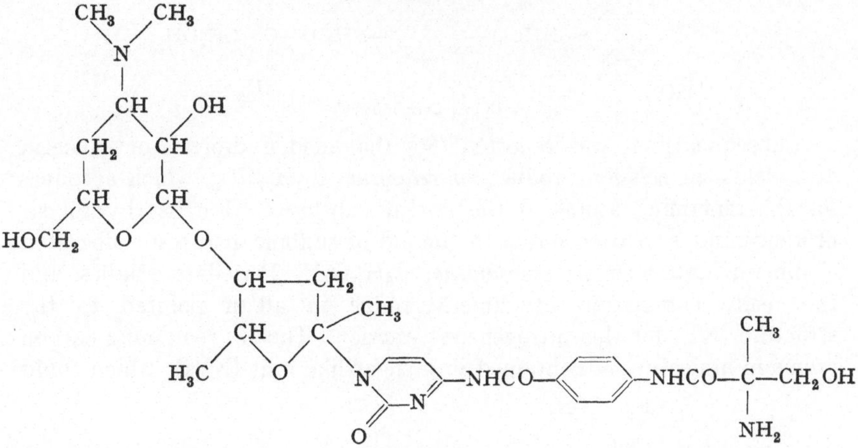

(X.) Amicetin.

HINMAN, CARON and DE BOER (*45*) found that a microorganism *(Streptomyces vinaceus-drappus)* isolated from a soil sample collected near Kalamazoo, Michigan, produced an antibiotic material active against certain acid-fast and Gram-positive bacteria. In another laboratory, this same substance, amicetin, was secured from *S. fasciculatis*. The preparation from *S. vinaceus-drappus* is carried out by deep-culture fermentation followed by extraction with butanol. The antibiotic, $C_{29}H_{42}O_9N_6$, was characterized as the crystalline free base and in the form of various salts.

As in the case of many chemical studies on other antibiotics, important information about the constitution of amicetin was gained by characterization of products resulting from controlled hydrolysis. The action of 6 *N*-HCl at 75–80° brought about the liberation of a $C_{15}H_{17}O_4N_5$ unit named *cytimidine* (*31*). More drastic treatment of this degradation product with the same reagent effected cleavage to three fragments: cytosine (XI), *p*-aminobenzoic acid (XII), and a new amino acid, $C_4H_9NO_3$,

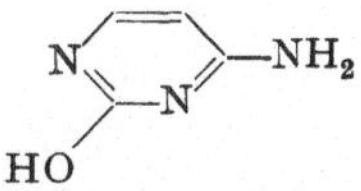

(XI.) Cytosine.    (XII.) 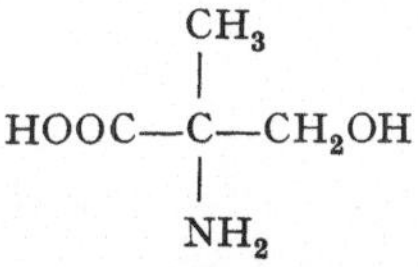   (XIII.)

for which the structure (XIII) was proposed. Potentiometric titration data, color tests and the behavior with periodate of cytimidine allowed the formulation of its structure as (XIV).

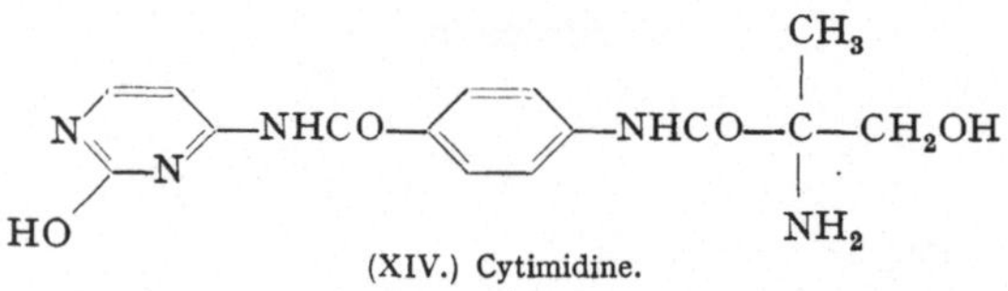

$$N\diagdown\diagup N \text{—NHCO—} \diagup\diagdown \text{—NHCO—} \underset{\underset{NH_2}{|}}{\overset{\overset{CH_3}{|}}{C}} \text{—CH}_2OH$$

HO

(XIV.) Cytimidine.

Subsequently it was reported (*89*) that acid hydrolysis of amicetin also yields the scission product *amicetamine*, $C_{14}H_{27}NO_6$, which accounts for the remaining atoms in the parent substance. Further hydrolysis of amicetamine, carried out with the aid of sulfonic acid resin, liberated a dimethylamino sugar, *amosamine*, $C_8H_{17}O_4N$. Periodate studies and the ready elimination of dimethylamine in alkali pointed to the structure (XV) for this nitrogeneous fraction. The six remaining carbon atoms in amicetin are contained in a sugar-like unit (XVI), which could

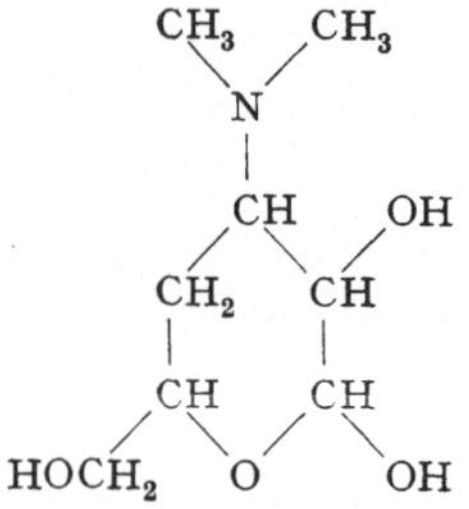

(XV.) Amosamine. 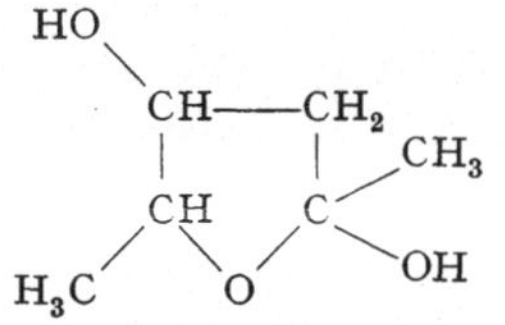   (XVI.)

be isolated by cleavage of amicetin with methanolic hydrogen chloride, followed by resin-induced methanolysis of the resulting methyl glycoside of amicetamine. Periodate determinations on the methyl glycoside of (XVI) as well as on the free ketose, coupled with the observation that (XVI) possesses a potential methyl ketone function, led to the proposed structure. Since amicetamine gives a positive iodoform test, the hemiketal system of the C_6 unit (XVI) must be present per se; and

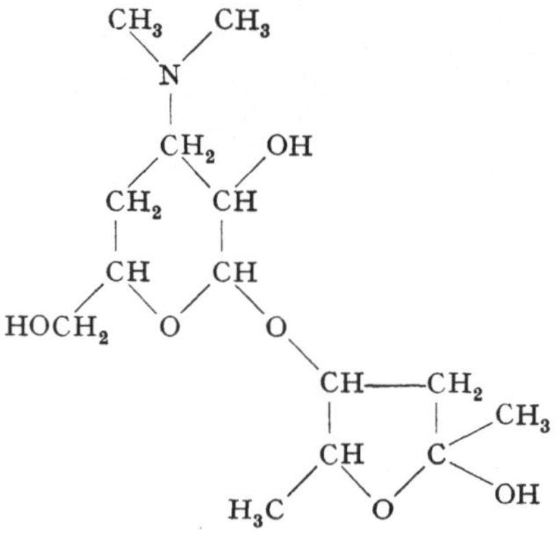

(XVII.) Amicetamine.

consequently the two moieties constituting amicetamine must be joined through the remaining oxygen of the unit (XVI) and the hemiacetal system of amosamine, as depicted in formula (XVII).

Although relatively stable in acid media, amicetin is rapidly inactivated by dilute base at room temperature, which treatment causes precipitation of a still different substance, *cytosamine*, $C_{18}H_{32}O_6N_4$ (*31*). As indicated by the results of further acid hydrolysis, amicetamine (XVII) and the previously detected cytosine (XI) are combined in this cleavage product. The mode of attachment between the two units was apparent from several observations. First, the negative iodoform test observed for cytosamine denoted masking of the hemi-ketal group in (XVII), thereby indicating the point of attachment of the cytosine unit to amicetamine. Of the three possible positions for linkage to cytosine

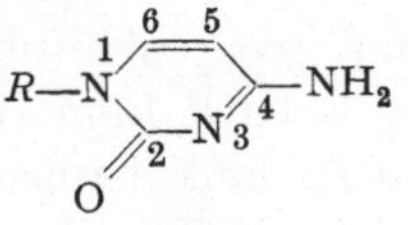

(XVIIIa.) 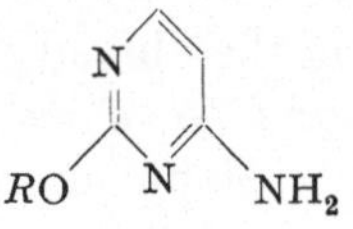 (XVIIIb.) 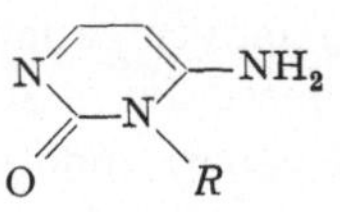 (XVIIIc.)

nucleus (XVIII a–c), (XVIII c) was excluded by conversion of cytosamine with nitrous acid to the 4-hydroxy counterpart; methylation with diazomethane followed by formation of 3-methyluracil (XIX) on hydrolysis, proved that the 3-nitrogen was unsubstituted in cytosamine itself. Possibility (XVIIIb) was considered unlikely by virtue of the fact that, whereas O-glycosides of 2-hydroxypyrimidines are unstable to both base and acid and reduce FEHLING's

(XIX.) 3-Methyluracil.

solution directly, cytosamine gives a positive test with this reagent only after acid hydrolysis. Also, it was considered that the acidic behavior

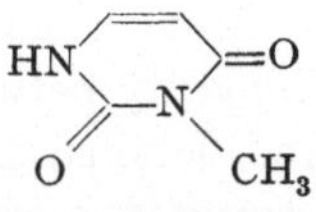

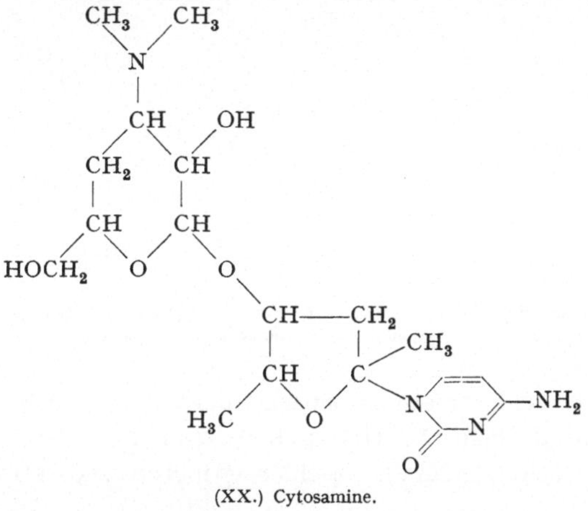

(XX.) Cytosamine.

of amicetin was more in keeping with partial structure (XVIIIa) than (XVIIIb). Cytosamine was therefore formulated as (XX), and amicetin, consequently, was .assigned the structure (X, p. 95).

5. Aureomycin (Chlorotetracycline).

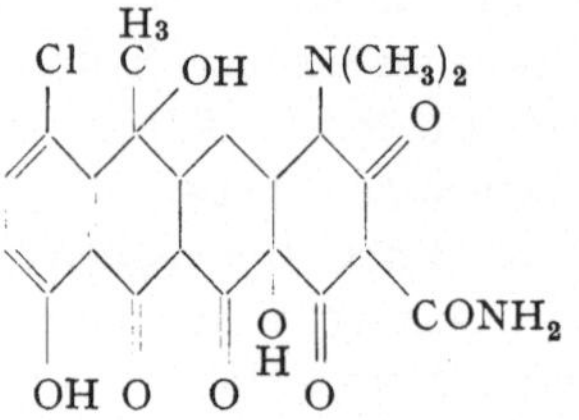

(XXI.) Aureomycin.

Aureomycin, a member of the "broad spectrum" group of antibiotics, is produced by a *Streptomyces (S. aureofaciens)* isolated from the soil of a Missouri timothy field. Manufacture by deep tank fermentation yields the faintly yellow, crystalline material, $C_{22}H_{23}O_8N_2Cl$, readily isolated as the hydrochloride. Aureomycin is a useful and versatile antibiotic, active against numerous Gram-positive and Gram-negative bacteria as well as rickettsiae and certain viruses.

Early X-ray (29) and other physical measurements indicated a close similarity between the aureomycin and terramycin molecules (see below), and detailed chemical investigations, carried out by two research teams, proved the correctness of this lead. Drastic oxidation and alkaline fusion afforded in the earlier stages of the investigations a limited amount of chemical information. Thus, it was shown (54) that alkaline fusion of aureomycin produced ammonia, dimethylamine, and 5-chlorosalicylic acid (XXII). Methylation followed by permanganate oxidation led to

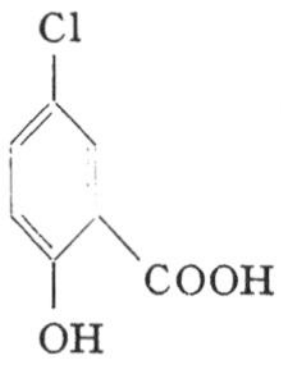

(XXII.) 5-Chlorosalicylic acid.

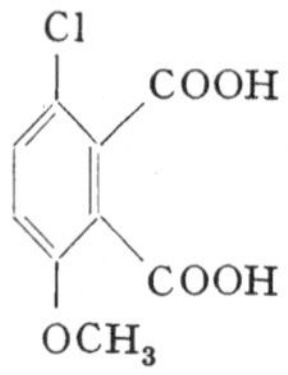

(XXIII.) 6-Chloro-3-methoxyphthalic acid.

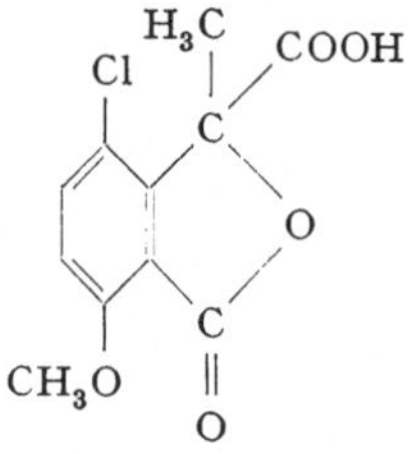

(XXIV.) 4-Chloro-7-methoxy-3-methylphthalide-3-carboxylic acid.

a mixture of *p*-chloromethoxybenzene derivatives, including 6-chloro-3-methoxyphthalic acid (XXIII), 4-chloro-7-methoxy-3-methylphthalide-3-carboxylic acid (XXIV), 4-chloro-7-methoxy-3-methylphthalide-3-

succinic- and -β-glutaric acids (XXV and XXVI). Syntheses of the
acids (XXIII)–(XXV) were carried out (*64, 65, 9, 10*).

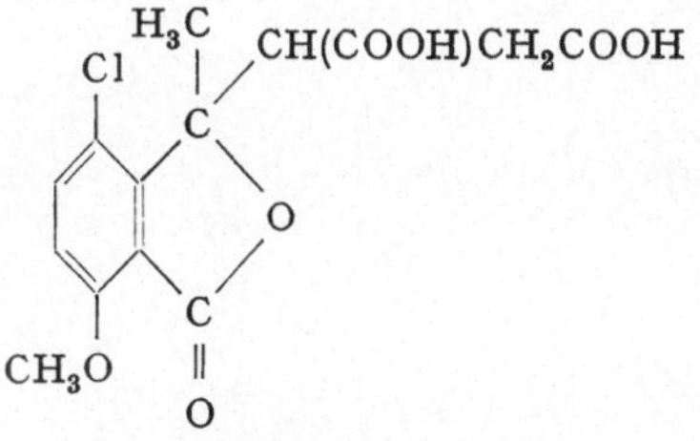

(XXV.) 4-Chloro-7-methoxy-3-methyl-
phthalide-3-succinic acid.

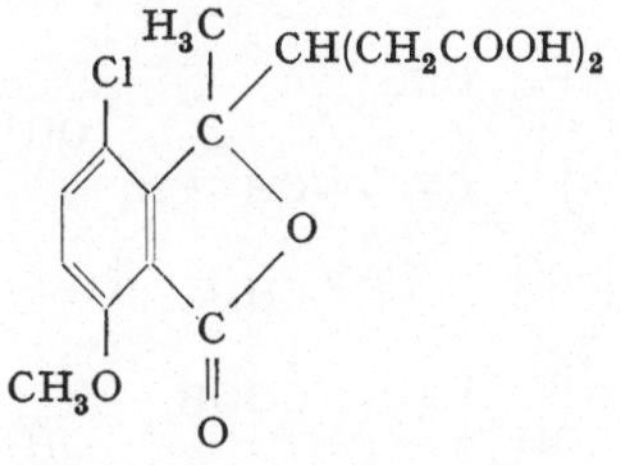

(XXVI.) 4-Chloro-7-methoxy-3-methyl-
phthalide-β-glutaric acid.

Various degrees of severity of alkaline treatment afforded a succession
of degradation products in which the bulk of the aureomycin molecule
was retained (*100, 99, 53, 101*). Dilute alkali at room temperature for
twenty-four hours converts the antibiotic to *isoaureomycin* (XXVII),

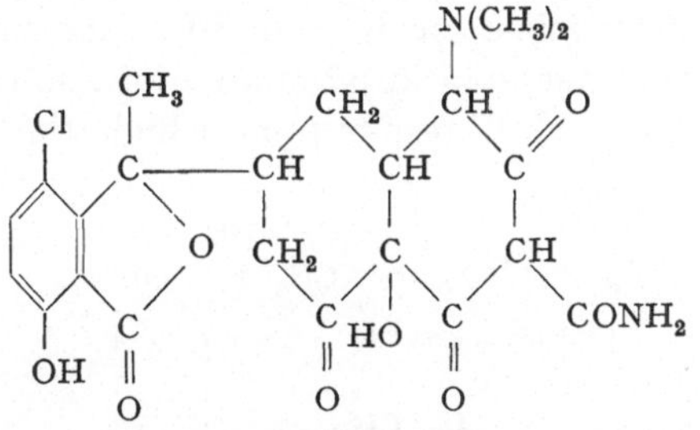

(XXVII.) Isoaureomycin.

which, on short treatment with 5 N-NaOH in the presence of sodium
hydrosulfite, is converted to the α- and β-*aureomycinic acids* (XXVIII).

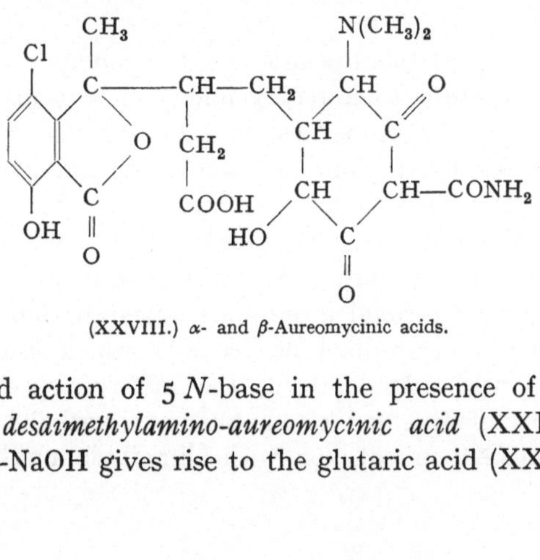

(XXVIII.) α- and β-Aureomycinic acids.

More prolonged action of 5 N-base in the presence of oxygen effects
conversion to *desdimethylamino-aureomycinic acid* (XXIX), and aerial
oxidation in N-NaOH gives rise to the glutaric acid (XXVI, with —OH

instead of —OCH_3) and a product of ring contraction, 3,4-dihydroxy-2,5-diketocyclopentane-1-carboxamide (XXX). Drastic reduction of the latter with hydrogen halides to 1,3-cyclopentandione or 1,2,4-cyclo-

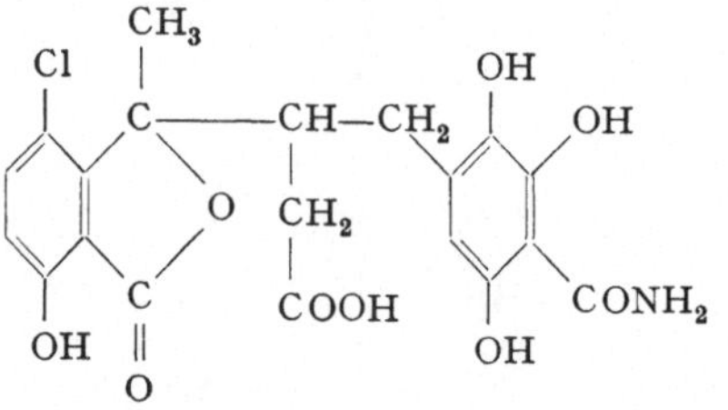

(XXIX.) Desdimethylamino-aureomycinic acid.

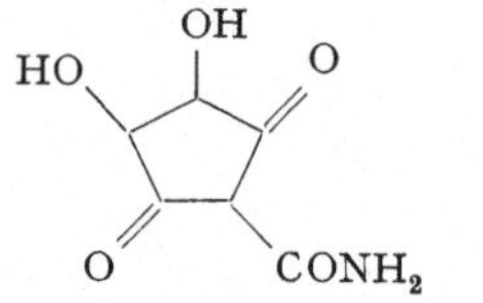

(XXX.) 3,4-Dihydroxy-2,5-diketo-cyclopentane-1-carboxamide.

pentantrione, which were synthesized (*101*), and formation of 4,5-dihydroxy-1,3-cyclopentandione on barium hydroxide hydrolysis, largely established the structure of this product. Infrared and ultraviolet spectral data disclosed the nature of the chromophoric groups present in des-dimethylamino-aureomycinic acid, and pKa measurements allowed specification of the centers of acidity; study of the substituted tetralone, *aureoneamide* (XXXI), which results from dehydration, was most useful

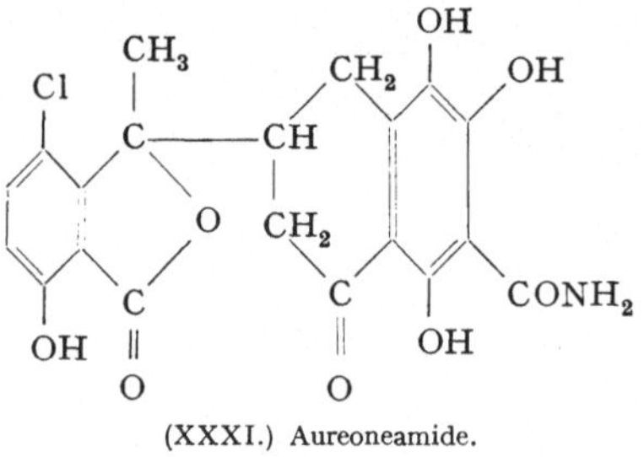

(XXXI.) Aureoneamide.

in the allocation of substituent positions in the polyhydroxylated nucleus. Assignment of structure to aureomycinic acid was made possible by its spectral and acidity properties, and by suitable interpretation of the reaction whereby dimethylamine is eliminated and desdimethylamino-aureomycinic acid formed. Again, spectral and pKa data pointed to the enolized β-diketone, non-conjugated ketone, and lactone systems of isoaureomycin, the structure (XXVII) (or one in which the dimethylamino- and angular hydroxyl groups are reversed) following from the nature of the change, viz. β-diketone cleavage, which results in formation of aureomycinic acid. Aureomycin, the infrared spectrum of which indicated it to be a conjugated ketone but not a phthalide, was therefore formulated as (XXI, p. 98) (or with dimethylamino- and hydroxyl groups reversed, as in the case of isoaureomycin). Support for this assignment was provided by the acid-catalyzed dehydration of aureomycin to *anhydro-aureomycin* (XXXII) (*98*).

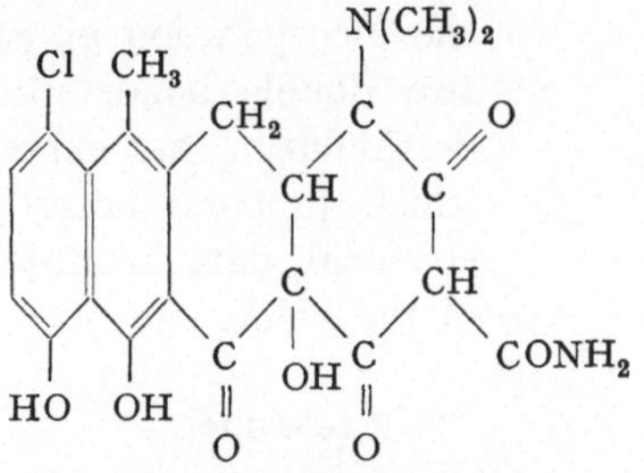

(XXXII.) Anhydro-aureomycin.

After the work on the related antibiotic terramycin had been completed, it was possible to adapt directly the developed structure to the aureomycin case (*87, 88*). The proposal (XXI) was confirmed by various physical measurements and by a series of experiments similar to many used in the course of the terramycin studies.

6. Aureothricin and Thiolutin.

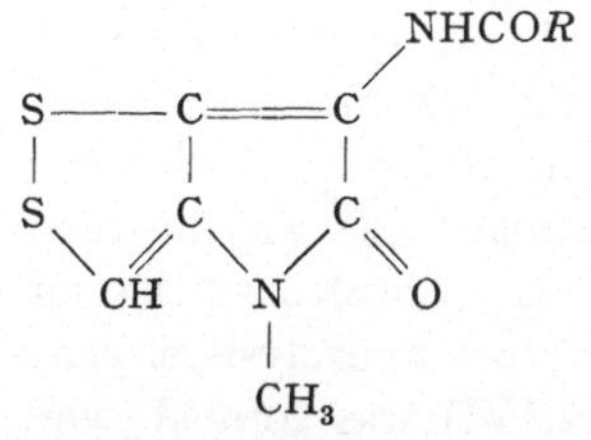

(XXXIII.) Aureothricin: $R = CH_3CH_2-$. Thiolutin: $R = CH_3-$.

Thiolutin and aureothricin are yellow, crystalline substances produced by various *Streptomyces* species, including *S. albus*. Active against a variety of fungi, ameboid parasites, Gram-positive, Gram-negative and acid-fast bacteria, the two entities are the N-acetyl and N-propionyl derivatives of a common nucleus, *pyrrothine* ($C_6H_6ON_2S_2$) (*20*). Thiolutin (*19*), $C_8H_8O_2N_2S_2$, on Raney nickel desulfurization accompanied by saturation of two double bonds, was converted to the colorless, neutral *desthiolutin*, $C_8H_{14}O_2N_2$. Removal of the N-acetyl group could be effected by treatment with dioxane-hydrochloric acid, thereby leaving the parent base of desthiolutin, $C_6H_{12}ON_2$. The action of 20% hydrochloric acid at 150° served to transform either desthiolutin or the C_6-base to a ninhydrin-positive diaminoacid, $C_6H_{14}O_2N_2$. In consideration of the single C-methyl and N-methyl groups in the system, the necessity of a five- or six-membered lactam ring in desthiolutin, as evidenced by its stability toward hydrolysis, and the required placement of two double bonds in the antibiotic itself, the structure (XXXIV) for desthiolutin was developed and then confirmed by synthesis. The molecular formula for thiolutin demands the incorporation of two sulfur atoms and two double bonds into the framework of desthiolutin, as expressed in (XXXIII), which proposal accords with spectral and chemical data acquired during the course of the work.

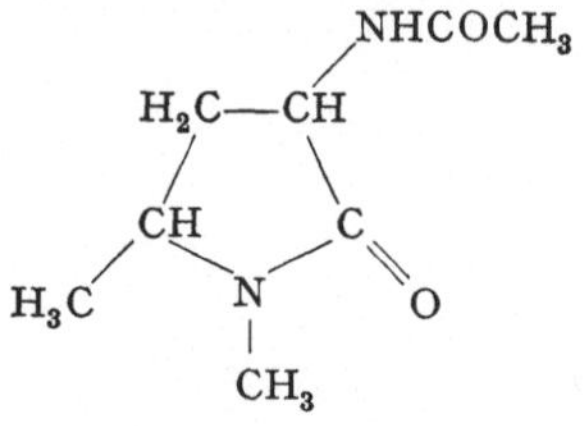

(XXXIV.) Desthiolutin.

7. Azaserine.

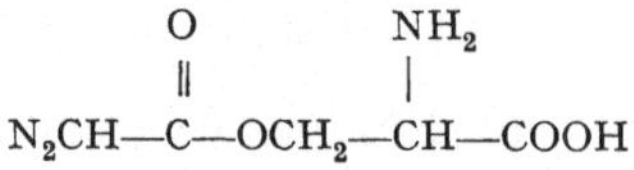

$$\text{N}_2\text{CH--}\overset{\text{O}}{\overset{\|}{\text{C}}}\text{--OCH}_2\text{--}\overset{\text{NH}_2}{\overset{|}{\text{CH}}}\text{--COOH}$$

(XXXV.) Azaserine.

Azaserine, a markedly unstable, tumor-inhibitory substance, is isolated from the culture broth filtrates of a *Streptomyces* by initial concentration, followed by adsorption of a 90% ethanol solution on alumina and elution with ethanol-water. Purification was managed by carbon column chromatography previous to crystallization (37).

The unusual chemical nature of azaserine, $C_5H_7O_4N_3$, was portended by its spectral properties: a yellow-green color and an absorption band in the infrared spectrum at 4.66 μ, indicative of a cumulative double bond system. Since nitrogen gas was liberated under acid conditions, the unsaturated group was regarded as that of the diazo type and was located by room temperature hydrolysis of azaserine at pH 2.0, whereby O-glycolyl-L-serine (XXXVI) was formed (36).

$$\text{HOCH}_2\text{--COOCH}_2\text{--CH(NH}_2\text{)COOH}$$

(XXXVI.) O-Glycolyl-L-serine.

The synthesis of azaserine has been accomplished through selective diazotization of O-glycyl-L-serine (XXXVII), which was prepared by several routes, one of which is included here (69, 70) (Chart 1).

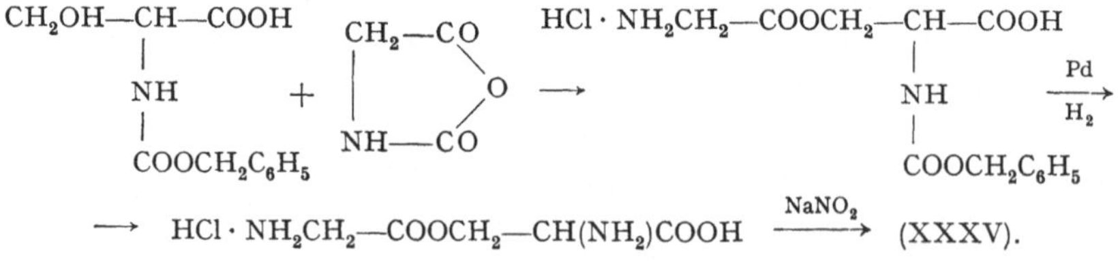

$$\text{CH}_2\text{OH--CH--COOH} \quad \text{CH}_2\text{--CO} \quad \text{HCl}\cdot\text{NH}_2\text{CH}_2\text{--COOCH}_2\text{--CH--COOH}$$

NaNO₂

$$\longrightarrow \quad \text{HCl}\cdot\text{NH}_2\text{CH}_2\text{--COOCH}_2\text{--CH(NH}_2\text{)COOH} \xrightarrow{\text{NaNO}_2} \text{(XXXV).}$$

(XXXVII.) O-Glycyl-L-serine.

Chart 1. Synthesis of Azaserine.

8. Chloromycetin (Chloramphenicol).

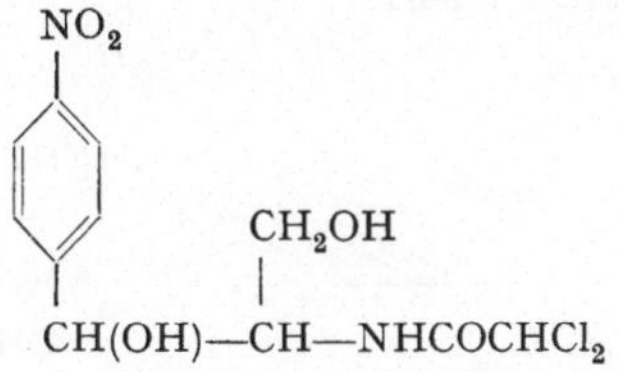

NO$_2$

CH$_2$OH

CH(OH)—CH—NHCOCHCl$_2$

(XXXVIII.) Chloromycetin.

Cultures of the actinomycete, *Streptomyces venezuelae*, which produces the useful antibiotic chloromycetin, have been located in various parts of the world: Caracas, Venezuela; Urbana, Illinois; and certain parts of Japan. After the fermentation run is complete, the desired product can be isolated by filtration of the broth and extraction with ethyl acetate. Concentration, followed by dilution with kerosene and then acid and base washes, gives a solution from which chloromycetin crystallizes at 5°, after partial removal of solvent.

Chloromycetin combats Gram-positive bacteria and is potent against bacteria producing typhoid, dysentery, as well as Rocky Mountain spotted fever. Although the drug has a rather low general toxicity, certain blood disorders (e. g., aplastic anemia) have been attributed to its prolonged administration.

Chemical studies on chloromycetin foreshadowed the succession of unusual structures which were in later years to make their appearance during the course of *Streptomyces* antibiotics investigations.

Initially, ultraviolet studies on chloromycetin, $C_{11}H_{12}Cl_2O_5N_2$, revealed the presence of a nitro group, and various chemical tests disclosed the presence of organically-bound halogen. These rare features were located more exactly by hydrolysis, which effected cleavage to dichloroacetic acid and a primary amine $C_9H_{12}O_4N_2$. Chloromycetin, the amide derived from these two components, could be reconstituted by treatment of the C_9-base with methyl dichloroacetate. With the observations that the natural product did not consume periodate, whereas the C_9-base gave rise to ammonia, formaldehyde, and *p*-nitrobenzaldehyde, chloromycetin was formulated as (XXXVIII) (75). Chemical and optical rotation data indicated that the substance belongs to the *pseudo*-ephedrine-*(threo)* series.

The chemical nature of the antibiotic makes it amenable to preparation in the laboratory, and in fact chloromycetin is produced commercially by synthesis. Several approaches have been described, the one outlined below being representative (24, 67) (*Chart 2*, p. 104).

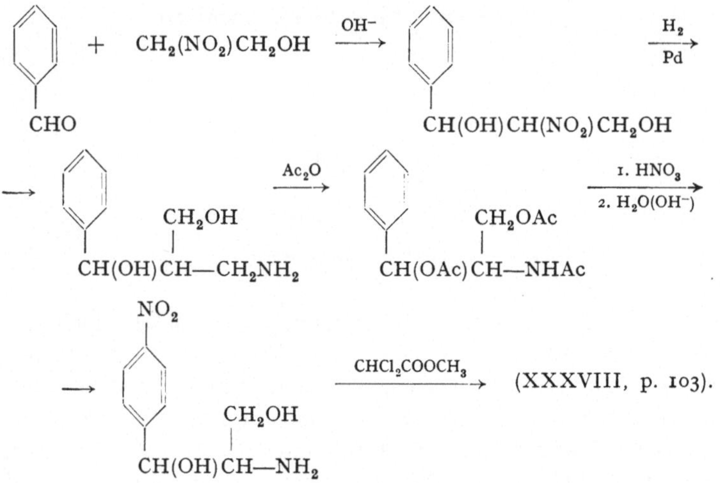

$$\text{(benzene ring)} + CH_2(NO_2)CH_2OH \xrightarrow{OH^-} \text{(benzene ring)} \xrightarrow[Pd]{H_2}$$

CHO

CH(OH)CH(NO$_2$)CH$_2$OH

$$\longrightarrow \text{(benzene ring)} \quad CH_2OH \xrightarrow{Ac_2O} \text{(benzene ring)} \quad CH_2OAc \xrightarrow[2.\ H_2O(OH^-)]{1.\ HNO_3}$$

CH(OH)CH—CH$_2$NH$_2$

CH(OAc)CH—NHAc

NO$_2$

$$\longrightarrow \text{(benzene ring)} \quad CH_2OH \xrightarrow{CHCl_2COOCH_3} \text{(XXXVIII, p. 103)}.$$

CH(OH)CH—NH$_2$

Chart 2. Synthesis of Chloromycetin.

9. Cycloserine (Oxamycin).

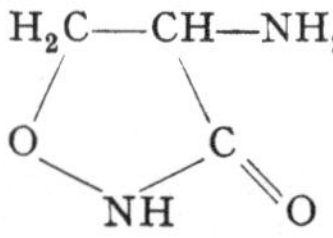

H$_2$C—CH—NH$_2$

O   C

NH   O

(XXXIX.) Cycloserine.

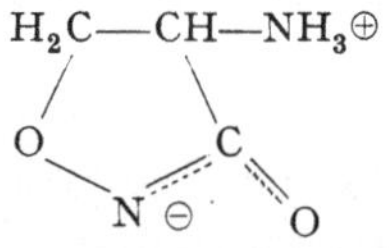

H$_2$C—CH—NH$_3$⊕

O   C

N⊖   O

(XL.) Cycloserine (zwitterion form).

In 1955, two industrial research groups disclosed the isolation and proof of structure of *D*-4-amino-3-isoxazolidone (described trivially as cycloserine or oxamycin), a *Streptomyces* metabolic product characterized by broad spectrum antibiotic activity (63, 44). The major degradation devices which allowed a structural assignment were: acid hydrolysis, which yielded hydroxylamine and serine; catalytic hydrogenolysis, which gave *D*-serine amide; and ring-opening by means of methanolic hydrogen chloride, which produced β-aminoxy-*D*-alanine methyl ester. The infrared spectral character of the antibiotic suggests that the zwitterion form (XL) contributes to the overall structure (63).

Synthesis was accomplished by starting with the methyl ester of *DL*-serine hydrochloride (84, 85) (Chart 3). Conversion to the oxazoline (XLI) was effected by means of ethyl iminobenzoate, which operation was followed by treatment with hydroxylamine-sodium ethoxide, yielding *DL*-2-phenyl-4-carbohydroxamido-2-oxazoline (XLII). Ring-opening with dry hydrogen chloride produced the hydroxamic acid (XLIII), which was subsequently ring-closed by means of alkali to the isoxazolidone (XLIV). Removal of the benzoyl group was brought about by a second treatment with hydrogen chloride, giving *DL*-βaminoxyalanine methyl ester dihydrochloride, followed by reclosure of the ring with alkali. *D*-Tartaric acid was utilized in the resolution which produced the natural *D*-isomer.

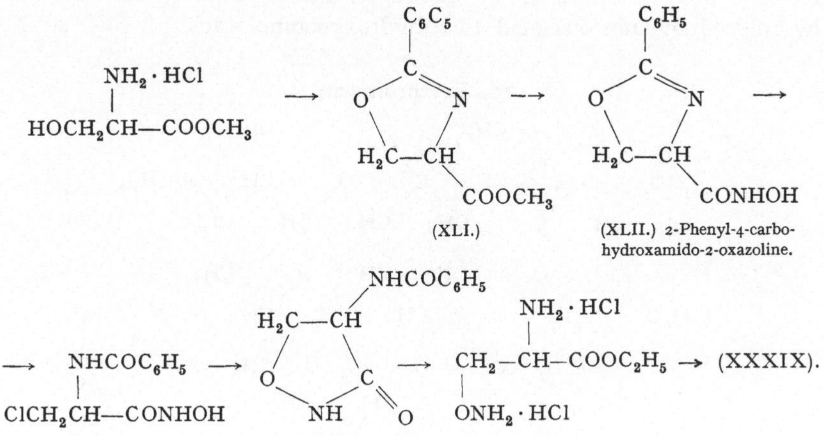

Chart 3. Synthesis of Cycloserine.

10. Elaiomycin.

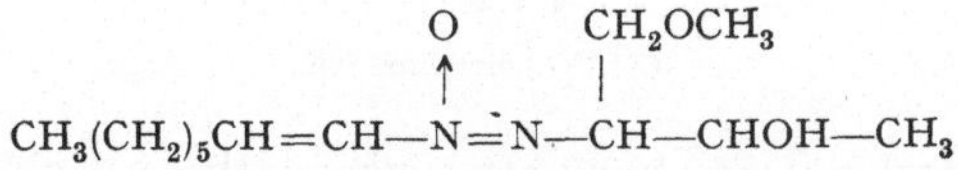

$$CH_3(CH_2)_5CH=CH-\overset{\overset{O}{\uparrow}}{N}\overset{=}{=}N-\overset{\overset{CH_2OCH_3}{|}}{CH}-CHOH-CH_3$$

(XLV.) Elaiomycin.

Elaiomycin, a *Streptomyces* metabolite exhibiting tuberculostatic properties, is one of the few natural products bearing an aliphatic azoxy linkage. The constitution of the antibiotic, an optically active, neutral oil possessing the formula $C_{13}H_{26}O_3N_2$, is implied by the following findings (*90*). Hydrogenation-hydrogenolysis, carried out in acetic acid over platinum, produced two amines: *n*-octylamine and a $C_5H_{13}O_2N$ amine. The latter fragment, isolated as the O,N-diacetyl derivative, was regarded as 4-methoxy-3-amino-2-butanol (XLVI) in that (i) both methoxyl and C-methyl groupings were known to be present; (ii) the parent antibiotic gave a positive iodoform test; and (iii) there was evidence available indicating

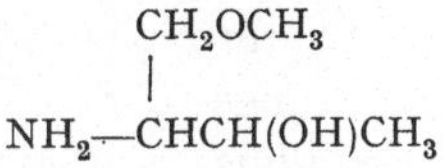

(XLVI.) 4-Methoxy-3-amino-2-butanol.

the adjacence of the hydroxyl and the nitrogen function. The O,N-diacetyl derivative of the base corresponding to structure (XLVI) was synthesized. The hydrogenolysis described above—indicative of a nitrogen-to-nitrogen linkage—along with that carried out in alcohol—which involved ready hydrogenolysis of a single oxygen and thus indicated the N-oxide function—, pointed to the presence of the azoxy linkage in elaiomycin, which assignment was supported by various spectral data. The positions of the olefinic linkage and the oxygen bound to nitrogen were established by the observations that elaiomycin was hydrolyzed by aqueous acid to α-hydroxyoctanoic acid.

11. Erythromycin.

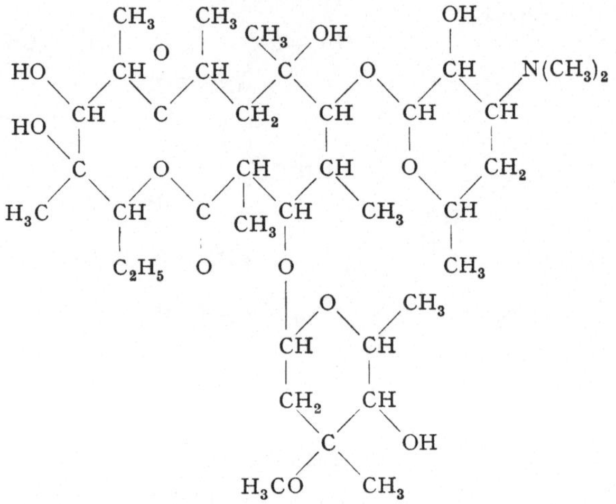

(XLVII.) Erythromycin.

From the broths of *Streptomyces erythreus* there is obtained, through a sequence of extractions and concentrations, the crystalline antibiotic erythromycin, $C_{37}H_{67}O_{13}N$, an agent active against many Gram-positive bacteria, the typhus rickettsiae as well as some large viruses. Of low toxicity, the substance is an important clinical agent, produced—as is the case with other antibiotic materials—by several pharmaceutical houses. Erythromycin, representing the "macrolide" class of natural products, apparently was the first of the antibiotics in which the large lactone ring was discerned.

Erythromycin acts as a monoacidic base, stable only in the pH range 6–8. Mild acid hydrolysis yielded two compounds, the crystalline *erythralosamine*, $C_{29}H_{49}O_8N$, and a high-boiling, neutral liquid, *cladinose*, $C_8H_{16}O_4$ (*32, 42, 109*). The nitrogen-free fragment represents one of the three rings in the erythromycin molecule, and its structure (XLVIII) was demonstrated by the following facts (*51*). Being a hemiacetal, cladinose was oxidizable to a lactone (XLIX), in which a methoxyl group is situated β to the lactone carbonyl, since methanol was eliminated readily, yielding lactone (L) on treatment with dilute alkali. Subsequent to treatment of lactone (XLIX) with aqueous base,

$$
\begin{array}{llll}
\text{HO—CH} & \text{O=C} & \text{O=C} & \\
\quad | & \quad | & \quad | & \\
\text{CH}_2 & \text{CH}_2 \quad\;\; O & \text{CH} \quad\;\; O & \text{H}_3\text{C} \\
\quad | & \quad | & \quad \| & \qquad\quad \diagdown \\
\text{H}_3\text{CCOCH}_3 \quad O & \text{H}_3\text{CCOCH}_3 & \text{H}_3\text{CC} & \qquad\qquad \text{C=CH—COOH} \\
\quad | & \quad | & \quad | & \qquad\quad \diagup \\
\text{HCOH} & \text{HC} & \text{HC} & \text{OCH} \\
\quad | & \quad | & \quad | & \\
\text{HC} & \text{HCOH} & \text{HCOH} & \text{(LI.) }\beta\text{-Formylcrotonic acid.} \\
\quad | & \quad | & \quad | & \\
\text{CH}_3 & \text{CH}_3 & \text{CH}_3 & \\
\text{(XLVIII.) Erythralosamine.} & \text{(XLIX.)} & \text{(L.)} &
\end{array}
$$

periodate oxidation generated acetaldehyde and β-formylcrotonic acid (LI), the identity of which was proved by synthesis.

When either erythromycin or erythralosamine was hydrolyzed under more drastic conditions, the aminosugar, *desosamine*, $C_8H_{17}O_3N$, was liberated. Periodate oxidation gave rise to croton-aldehyde, formic acid and dimethylamine (*21, 32*), fragments which result from oxidative cleavage and β-elimination of the elements of water from an intermediary β-hydroxybutyraldehyde or an equivalent. A positive iodoform test, reduction of Fehling's solution, and evidence that deso-samine was a hemiacetal allowed the formu-lation (LII).

$$
\begin{array}{c}
\text{N(CH}_3)_2 \\
| \\
\text{CH} \\
\diagup \quad \diagdown \\
\text{CH}_2 \qquad \text{CHOH} \\
| \qquad\qquad | \\
\text{CH} \qquad \text{CHOH} \\
\diagup \quad \diagdown \quad \diagup \\
\text{H}_3\text{C} \qquad \text{O}
\end{array}
$$
(LII.) Desosamine.

Evidence concerning the structure of the macrolactonic aglycone, *erythronolide*, $C_{21}H_{38}O_8$, was acquired by study of dihydro-erythronolide, secured by first reducing with sodium borohydride the ketonic grouping in the antibiotic (*107, 108, 38, 81*), then hydrolyzing with acid under carefully controlled conditions. The key to the structure (LIII) of

$$
\begin{array}{c}
\qquad\;\; \text{CH}_3 \quad \text{CH}_3 \\
\qquad\qquad |\;\;\;\; \text{OH} \quad | \quad \text{CH}_3 \quad \text{OH} \\
\text{HO} \diagdown \quad \text{CH} \quad | \quad \text{CH} \quad \text{C} \cdots^{b} \text{OH} \\
\quad^{a} \text{CH} \qquad \text{CH} \quad \text{H}_2\text{C} \quad \text{C} \\
\text{HO} \diagdown \\
\qquad \text{C} \qquad \text{O} \qquad \text{C} \qquad \text{CH} \\
\text{H}_3\text{C} \qquad \text{CH} \quad \text{C} \quad \text{CH} \quad \text{CH}_3 \\
\qquad\quad | \qquad \| \quad \text{H}_3 \quad | \\
\qquad \text{C}_2\text{H}_5 \quad \text{O} \qquad \text{OH}
\end{array}
$$
(LIII.) Dihydro-erythronolide.

dihydro-erythronolide lay in careful scrutiny of the products resulting from the action of periodate. The lactone itself was cleaved (a and b) to the complex ester (LIV) and the hydroxyketoaldehyde (LV). Among the various pieces of evidence required for proof of structure of these two halves are the following. Oxidation of (LIV) with peroxytrifluoracetic

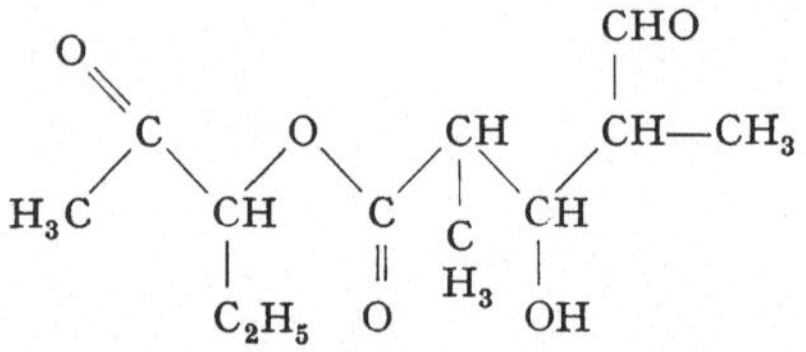

(LIV.)

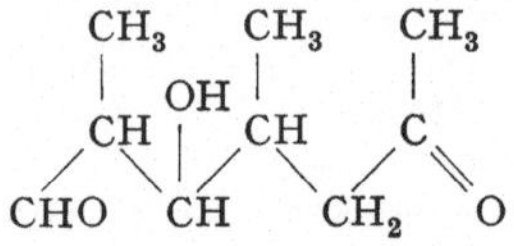

(LV.)

acid afforded the diester acid (LVI), which, on basic hydrolysis, produced *meso-α,α'*-dimethyl-β-hydroxyglutaric acid. Direct hydrolysis of (LIV) yielded 3-hydroxypentanone, the alcoholic component of the ester. Catalytic reduction of (LIV) to the triol ester (LVII), followed by hydro-

$$CH_3CO-O-CHO-COCH(CH_3)CH(OH)CH(CH_3)COOH$$
$$|$$
$$C_2H_5$$

(LVI.)

$$CH_3CH(OH)CHO-OCCH(CH_3)CH(OH)CH(CH_3)CH_2OH$$
$$|$$
$$C_2H_5$$

(LVII.)

lysis, led ˙instead to pentan-2,3-diol and the lactone (LVIII), which was identified by lithium aluminum hydride reduction to the triol. Treatment of the hydroxyketoaldehyde (LV) by means of peroxy-trifluoracetic acid resulted in oxidation of the aldehyde group as well as in transformation of the methyl ketone terminus to O-acetate; the

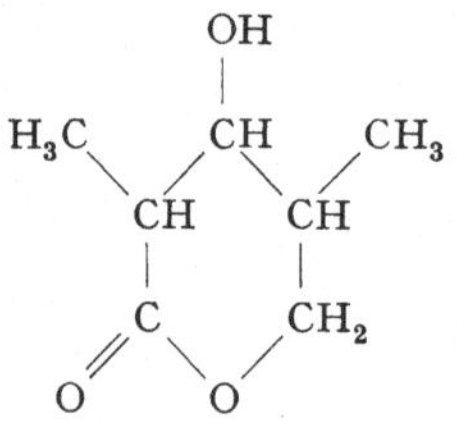

(LVIII.)

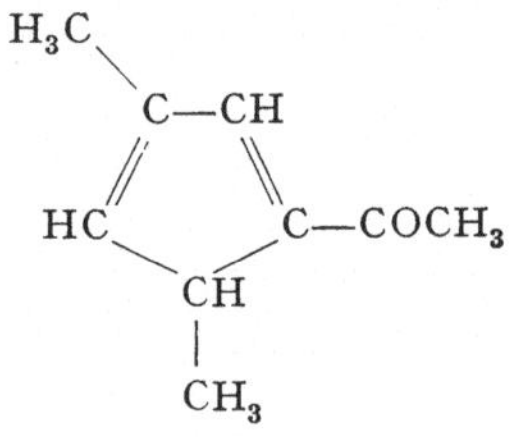

(LIX.)

intermediate ester was not isolated in that it cyclized spontaneously to the lactone (LVIII). The assigned structure (LV) was supported by base-catalyzed cyclization to a doubly-unsaturated ketone, formulated as (LIX). Techniques similar to those described above were applied to the hydroxyacid corresponding to dihydro-erythronolide, and the results confirmed the assigned structure.

More recent results (*106*) have demonstrated the position of the ketone function on the macrolide ring as well as the site of attachment of the desosamine and cladinose rings, and thus complete the elegant series of investigations which led to the establishment of the complete structure of erythromycin.

A parallel series of operations have indicated that erythromycin-B (*39*) possesses a similar type of structure; this minor antibiotic lacks the *t*-hydroxyl group present on the carbon atom adjacent to the lactone termination point in erythromycin itself.

12. Etamycin (Viridogrisein).

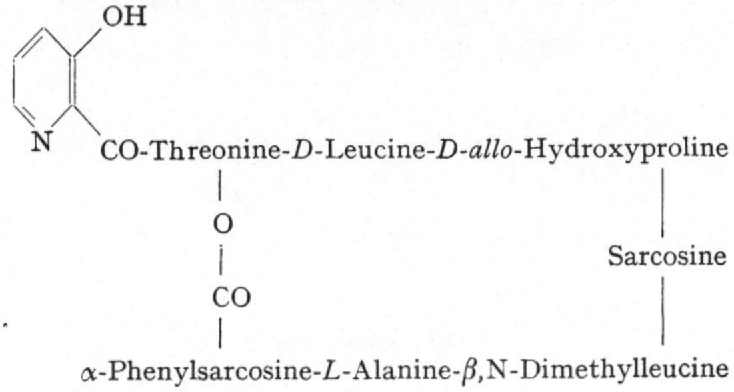

(LX.) Etamycin.

Total hydrolysis of etamycin (*6, 54, 43, 78*) gives rise to the constituent units of this polypeptide antibiotic: 3-hydroxypicolinic acid (LXI), *L*-alanine, *allo*-hydroxy-*D*-proline, *D*-leucine, threonine, sarcosine, α-phenylsarcosine (LXII) and β-N-dimethylleucine (LXIII). The two

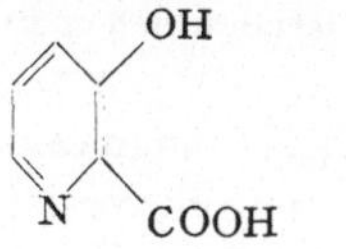

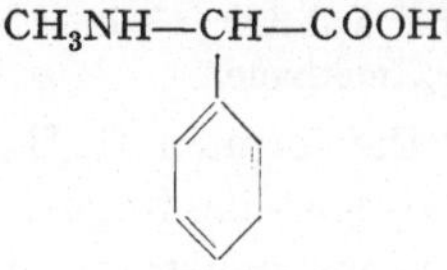

(LXI.) 3-Hydroxypicolinic   (LXII.) α-Phenylsarcosine.   (LXIII.) β-N-Dimethylleucine.
acid.

last-named amino acids had not been encountered in nature before, and the structures were proved, in the penultimate case, by synthesis; and in the last case, by hypochlorite degradation to methylamine and iso-propylmethyl-acetaldehyde. That etamycin contains a lactone ring was indicated by alkaline cleavage without loss of carbon to an anti-biotically inactive acid; the observation that, whereas the threonine unit resists oxidation in the intact antibiotic, it is virtually destroyed by chromic acid in the derived acid, indicates the placement of the lactone ether oxygen of the natural product.

Application of the EDMAN technique for stepwise degradation of polypeptides, to the product resulting from catalytic reduction of the pyridine ring, revealed the sequential relationship of the amino acids present.

13. Magnamycin (Carbomycin).

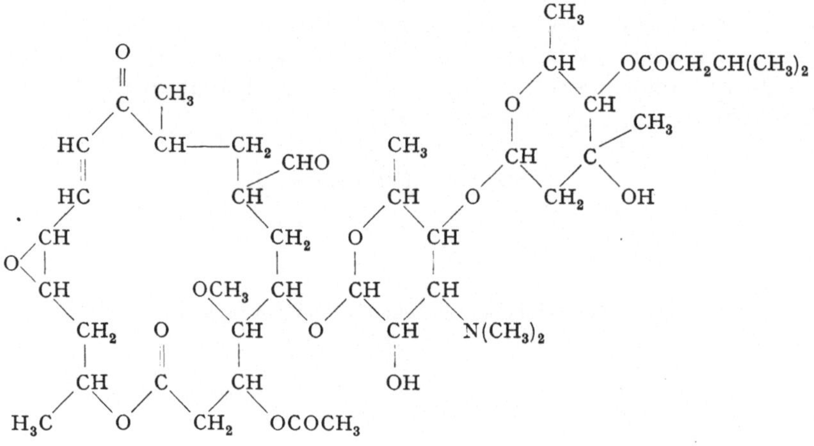

(LXIV.) Magnamycin.

Magnamycin, a metabolite of *Streptomyces halstedii* and an impressive member of the "macrolide" group of natural products, is a commercially produced antibiotic active (like erythromycin) against Gram-positive bacteria, rickettsia and large viruses.

The substance may be isolated by extraction of the fermentation broth with methyl isobutyl ketone; transference to a aqueous acidic (pH 2) medium, followed by washing with benzene; extraction into ether after adjusting the aqueous phase to pH 6.5; and evaporation of the ether solution, yielding material which may be purified by recrystallization from 50% methanol.

A monoacidic base possessing the formula $C_{42}H_{67}O_{16}N$, magnamycin on mild methanolysis is split into a sugar-like fragment, methylisovaleryl-mycaroside (LXV) and a nitrogeneous moiety, $C_{30}H_{47}O_{12}N$, referred to as *carimbose* (LXVI). Saponification of the glycoside yielded isovaleric acid and methylmycaroside, which, again treated with acid, afforded the free sugar *mycarose* (LXVII) (76). The structure of mycarose was

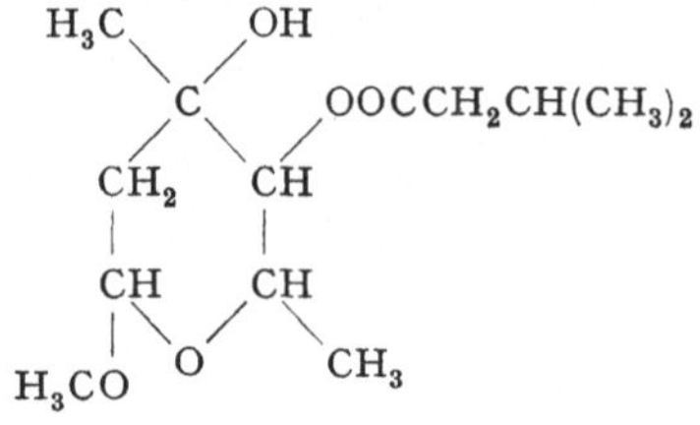

(LXV.) Methyl-isovaleryl-mycaroside.

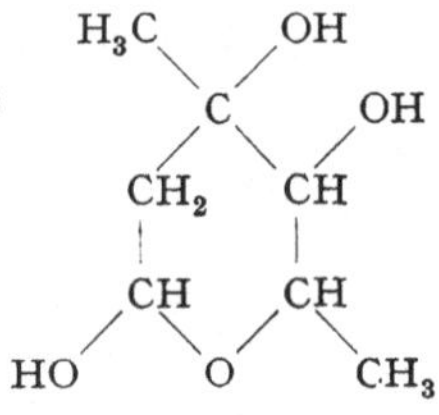

(LXVII.) Mycarose.

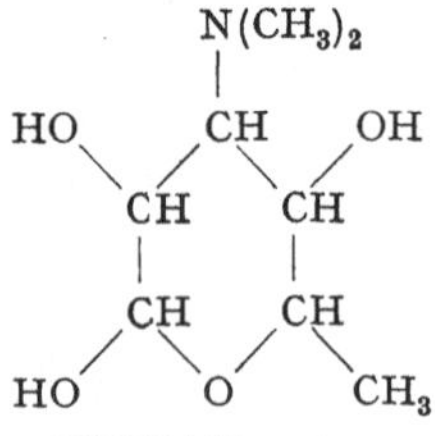

(LXVIII.) Mycaminose.

demonstrated largely by oxidation to the corresponding lactone and by periodate oxidation, which gave rise to acetaldehyde, acetoacetaldehyde and formic acid. Further acid hydrolysis of carimbose, although leading to destruction of the lactone nucleus, liberates the aminosugar *mycaminose* (47). This C_8-sugar, bearing dimethylamino and C-methyl groupings, shows the usual characteristics of a nitrogeneous monosaccharide, and could be oxidized with one mole of periodate to formic acid and a new, C_7-sugar. Further oxidation by the same reagent afforded acetaldehyde. Since mycaminose rapidly eliminated dimethylamine on treatment with alkali, the basic substituent was regarded as β to the aldehyde group of the sugar, and the structure (LXVIII) for mycaminose followed. pK measurements on magnamycin and certain related substances indicated that one of the hydroxyl groups adjacent to the nitrogen in myaminose was masked (by attachment of the mycarose unit), and this view was confirmed and defined more sharply by subjecting lithium aluminum hydride-reduced magnamycin to the action of silver oxide and methyl iodide; drastic hydrolysis of the methylated product produced the mono-O-methylated mycaminose (LXIX).

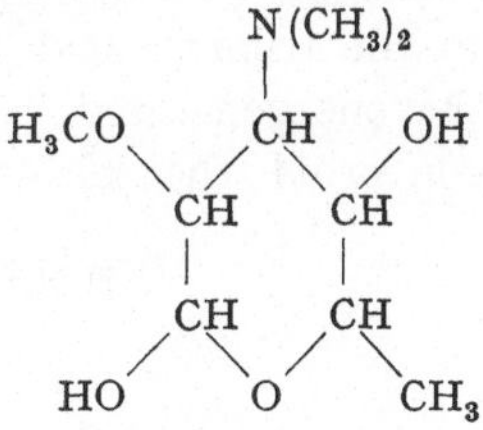

(LXIX.)

By various means it was demonstrated that carimbose possessed one each of the following functions: methoxyl, O-acetyl, aldehyde, α,β-unsaturated ketone, epoxide, and lactone. Further, the action of potassium iodide on magnamycin resulted in loss of a single oxygen atom and formation of magnamycin-B, which natural product possesses an $\alpha,\beta,\gamma,\delta$-unsaturated carbonyl system; this finding was taken to mean attachment of an epoxide ring to the β-position (LXX) of the α,β-

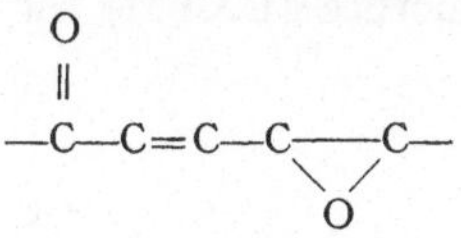

(LXX.)

unsaturated carbonyl system of magnamycin. Also, if octahydrocarimbose was treated drastically with base, dimethylamine was split out, and a $\alpha,\beta,\gamma,\delta$-unsaturated carboxylic acid system resulted. This behavior was interpreted to indicate the presence of the system (LXXI), which

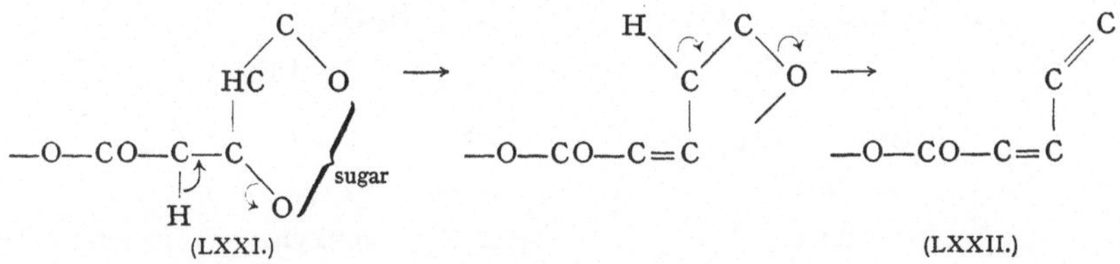

(LXXI.)   (LXXII.)

collapses as indicated by the mechanism (LXXI) → (LXXII), ultimately liberating the mycaminose carbonyl group, which development then allows further elimination of dimethylamine (*113*).

Considerable insight into the structure of the lactone nucleus was gained by study of the methoxy acid, $C_{13}H_{18}O_7$, provided by drastic treatment with aqueous potassium hydroxide of the crude periodate-permanganate oxidation product of magnamycin, magnamycin-B or carimbose. Ultraviolet spectral data again revealed the $\alpha,\beta,\gamma,\delta$-doubly unsaturated carboxylic acid system, and also showed that the product of aqueous acid treatment was an α,β-unsaturated-γ-ketoacid, HOOC—C=C—CO—, (LXXIII), resulting from loss of methanol. These interpretations were supported by oxidation experiments, including the ozonization of the C_{13}-acid, which process at the same time gave rise to the tribasic acid (LXXIV). The indicated (*) carboxyl group was the one generated by the ozonization operation, in that aqueous base converted the keto-triacid (LXXIII) to a C_{10}-keto-diacid (LXXV),

$$HOOC—CH(CH_3)CH_2CH(COOH)CH_2—\overset{*}{COOH}$$
(LXXIV.)

$$HOOC—CH(CH_3)CH_2CH(COOH)CH_2CH_2COCH_3$$
(LXXV.)

which, on Wolff-Kishner reduction, gave the expected product. The relationship of the carboxyl groups in the latter was apparent from the infrared spectrum of the derived anhydride, clearly shown to be of the glutaric acid type. These findings demonstrate the structure (LXXVI) for the C_{13}-methoxy acid; and its formation from magnamycin, interpreted on the basis of rational oxidative cleavages and elimination steps, establishes the partial structure (LXVII) for the antibiotic (*113*).

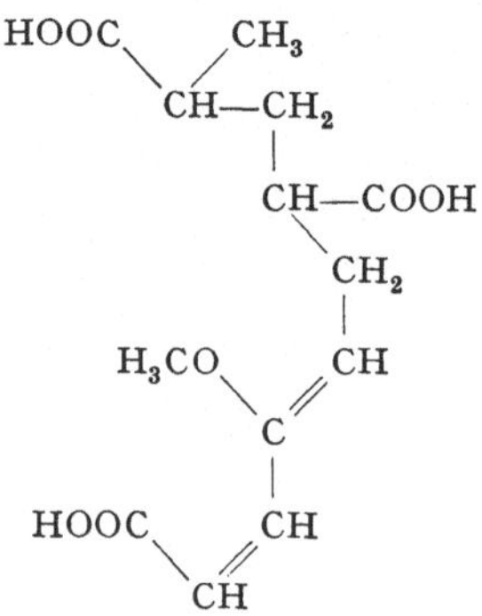

(LXXVI.) 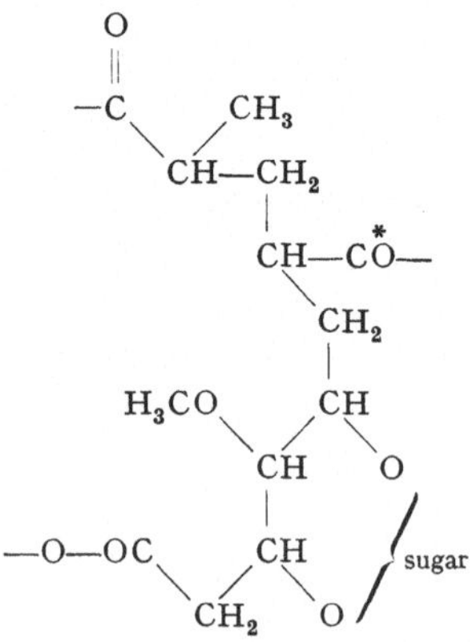   (LXXVII.)

By working with suitable transformation products of magnamycin, lactonization involving the oxidized aldehyde group and a γ-oxygen could be demonstrated, but only after removal of the sugar substituent. These results showed not only that the asterisked carbonyl group must be the aldehyde one, but also indicated that the γ-oxygen is involved in a glycosidic linkage, which in turn meant that the acetyl group present in magnamycin is attached to the oxygen β to the lactone carbonyl.

The nature of the remainder of the macrocyclic ring was uncovered by various oxidation experiments, of which the following is exemplary. Tetrahydromagnamycin-B (LXXVIII), in which the two double bonds of the parent antibiotic are saturated, yielded pimelic acid (LXXVIIIa)

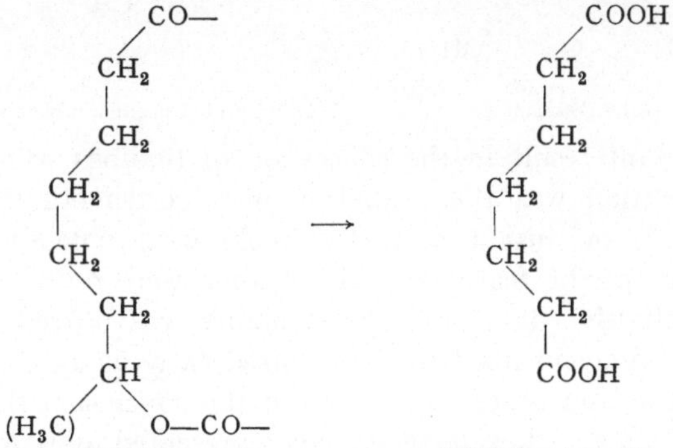

$$
\begin{array}{ll}
\text{CO—} & \text{COOH} \\
\text{CH}_2 & \text{CH}_2 \\
\text{CH}_2 & \text{CH}_2 \\
\text{CH}_2 & \text{CH}_2 \\
\text{CH}_2 & \text{CH}_2 \\
\text{CH}_2 & \text{CH}_2 \\
\text{CH} & \text{COOH} \\
\text{(H}_3\text{C)} \quad \text{O—CO—} &
\end{array}
$$

(LXXVIII.)   (LXXVIIIa.) Pimelic acid.

on oxidation with boiling nitric acid. The result indicates a straight chain sequence extending from the ketone carbonyl to the 7-carbon atom, which must be oxygenated; and the single remaining C-methyl group in magnamycin is attached, then, to this same carbon atom. By superimposing the α,β-unsaturated-γ,δ-epoxyketone system previously defined, (LXXVIII) may be expanded to (LXIV, p. 110), the complete structure of magnamycin (*113*).

14. Methymycin.

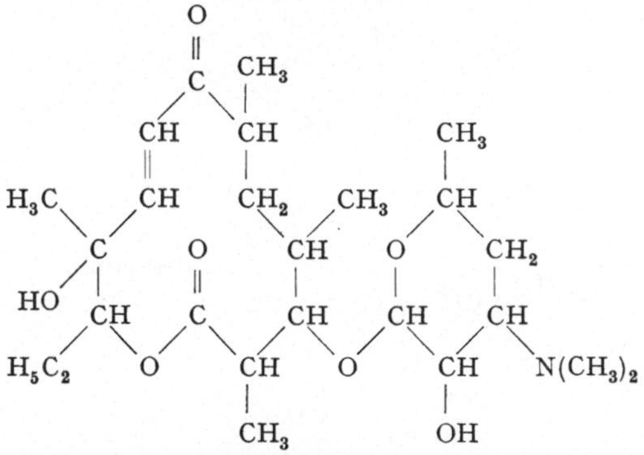

(LXXIX.) Methymycin.

The first clue as to the nature of the *Streptomyces* antibiotic, methy-mycin, $C_{25}H_{43}O_7N$, was the finding that aqueous acid hydrolysis afforded desosamine (LII), previously secured in a similar fashion from erythro-mycin (*25, 26*) (p. 107). By analogy, a macrolactone system seemed likely, and this supposition was confirmed by further study (*28*).

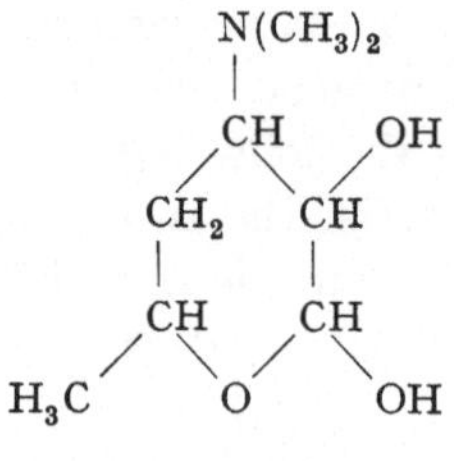

(LII.) Desosamine.

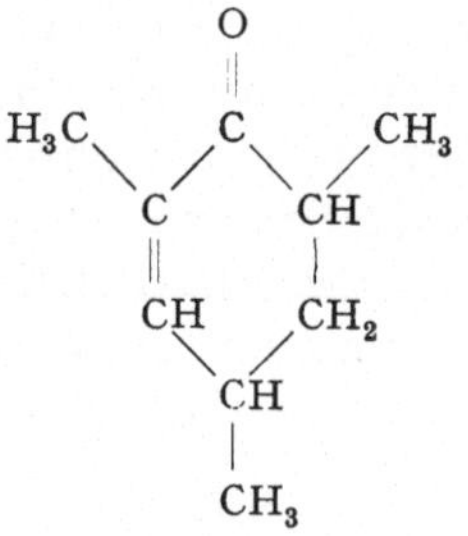

(LXXX.) 2,4,6-Trimethylcyclohex-2-en-1-one.

An important result in the succession of findings which led to the structural solution was the formation of 2,4,6-trimethylcyclohex-2-en-1-one (LXXX) on fusion of methymycin with potassium hydroxide at 360°. The specific features of this ketone were revealed by spectral and other definitive data, and the structure was proved by synthesis. Although methymycin itself was recognized early as an α,β-unsaturated ketone, spectral and other factors forced the conclusion that the cyclo-hexenone system represented by (LXXX) was not actually present in the intact antibiotic, but was generated by a cleavage and cyclization process. Careful consideration of the possible reaction routes by which the ketone might be formed, taken to-gether with the partial structure (LXXXI) deve-loped on independent grounds, led to several

(LXXXI.)

alternatives for the constitution of the lactone nucleus. Of these, (LXXXII) was shown to be correct, mainly through isolation of a series of oxidation

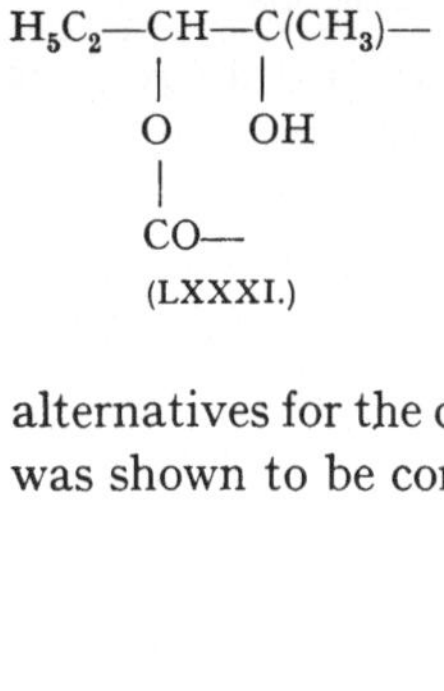

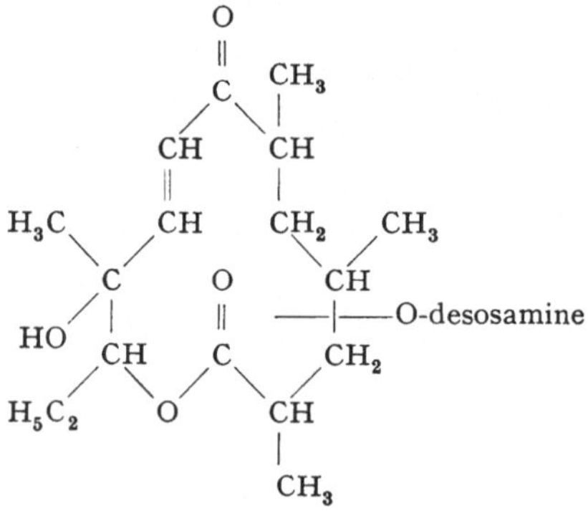

(LXXXII.)

products (LXXXIII–LXXXV) obtained by treatment of *methynolide*, the aminosugar-free product, with permanganate. The largest of the three, being an α-hydroxyacid, was oxidizable by lead tetraacetate, giving the ketoesterlactone (LXXXIV) obtained directly in the permanganate degradation. Hydrolysis of this intermediate afforded 3-hydroxy-2-pentanone and a third permanganate product, the lactonic acid (LXXXV). This acid had been previously obtained by oxidation of the antibiotics narbomycin and pikromycin, and its structure had been demonstrated by pyrolysis followed by ozonolysis, which sequence led to pyruvic acid and *meso*-α,α'-dimethylglutaric acid (*1*).

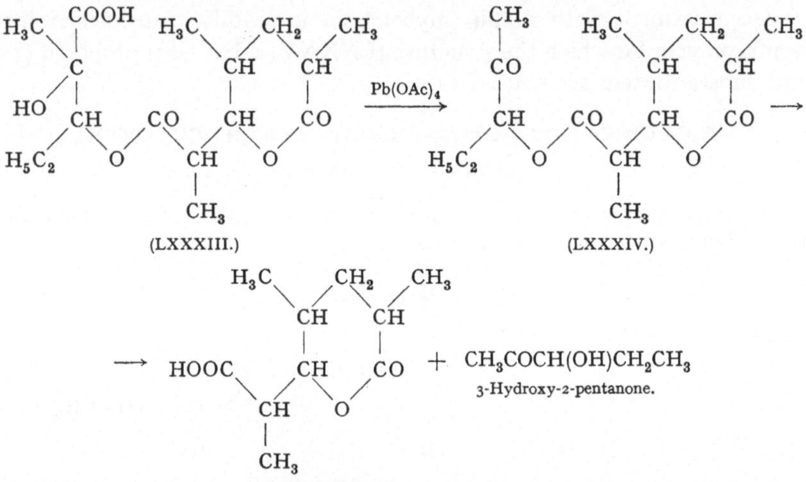

(LXXXIII.)   (LXXXIV.)

(LXXXV.)

More recent investigations have proved the structure of neomethy-mycin, an isomer and congener of methymycin, to be (LXXIX) with the hydroxyl group on the lactone ring replaced by hydrogen and the ethyl group replaced by methylcarbinol (*27*).

Although narbomycin and pikromycin as well as oleandomycin (*113*) fall into the macrolide class, results sufficient to establish unequivocally total structures for these substances have not yet been disclosed.

15. Mycomycin.

$$HC \equiv C—C \equiv C—CH = C = CH—CH = CH—CH = CH—CH_2—COOH$$

(LXXXVI.) Mycomycin.

Mycomycin is one of the few *Nocardia* antibiotics which have been subjected to chemical study (*18*). Apparently non-toxic, the substance acts upon fungi and mycobacteria as well as on both Gram-positive and Gram-negative bacteria.

Mycomycin is an optically active, highly unsaturated acid which explodes at its melting point (75°) and, even when pure, rapidly decomposes at room temperature. Catalytic reduction, requiring eight moles of hydrogen for saturation, gave rise to *n*-tridecanoic acid. Ultraviolet and infrared spectral data reflected the presence in the antibiotic of two, conjugated triple bonds, one of which, as indicated also by diagnostic tests with metallic reagents on the methyl ester, is terminal. The presence and location of the diene and allene groupings in mycomycin were revealed solely by infrared and ultraviolet measurements. Thus the optical activity of this unusual substance must arise from the allene linkage, which had not appeared before in a compound of natural origin.

On treatment with alkali, mycomycin is readily transformed into isomycomycin for which the structure (LXXXVII) has been proposed (*18*) and substantiated by synthesis (*8*).

$$H_3C-C\equiv C-C\equiv C-C\equiv C-CH=CH-CH=CH-CH_2-COOH$$

(LXXXVII.) Isomycomycin.

A general survey of acetylenic plant products has appeared recently in this Series [Bohlmann and Mannhardt (*7a*)].

16. Netropsin.

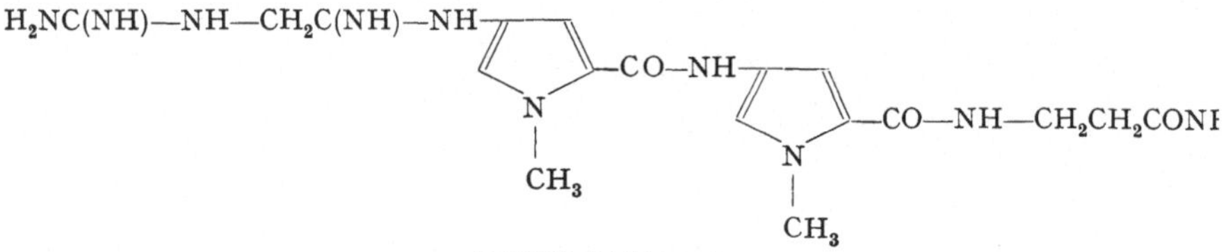

(LXXXVIII.) Netropsin.

Netropsin (Congocidine, Antibiotic T-1384), an elaboration product of *Streptomyces netropsis* (*30*), active against fungi, Gram-positive and Gram-negative bacteria, is striking on account of its high nitrogen content ($C_{18}H_{26}O_3N_{10}$). The substance, unstable as the free diacidic base, is isolated and handled in the form of several crystalline salts.

$$H_2NC(NH)NH-CH_2COOH$$

(LXXXIX.) Guanidinoacetic acid.

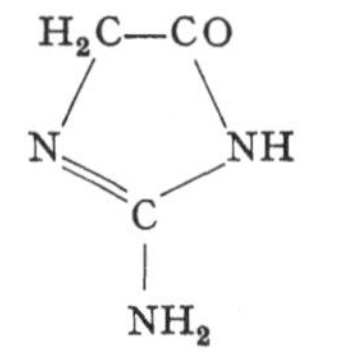

(XC.) Glycocyamidine. 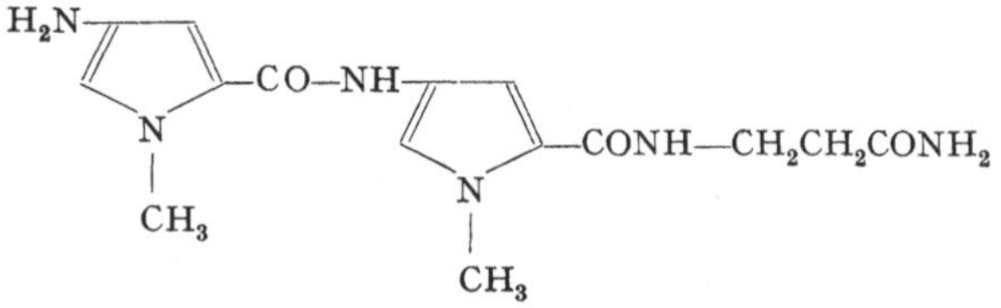   (XCI.)

By varying the nature of the reagent employed, a series of hydrolysis products can be obtained which reflects the molecular nature of the parent base. Netropsin is converted readily by mild alkaline treatment to one mole each of ammonia, guanidinoacetic acid (LXXXIX) (or glycocyamidine, XC) and the tripeptide (XCI) (*93, 94, 102, 104*). Hydrolysis of the triamide with hot dilute alkali liberates ammonia from the terminal carboxamido group, forming the carboxylic acid (XCII) corresponding to (XCI). More concentrated alkali effected further cleavage to β-alanine, 4-amino-1-methyl-2-pyrrolecarboxylic acid (XCIII) (*56*),

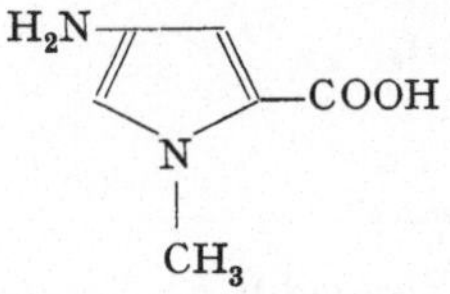

H_2N—pyrrole ring—COOH, N—CH_3

H_2N—pyrrole ring—CN, N—CH_3

(XCIII.) 4-Amino-1-methyl-2-pyrrolecarboxylic acid. 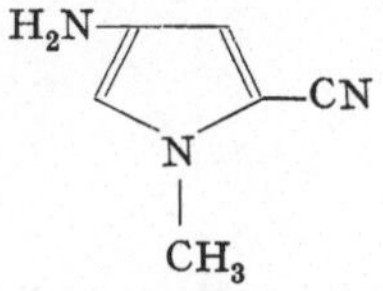   (XCIV.)

and possibly the dipeptide of the latter. Zinc dust or soda-lime distillation of the tripeptide (XCI) or the corresponding acid afforded a high yield of the nitrile (XCIV). The position of the substituents on the pyrrole ring was determined by synthesis of the aminopyrrolecarboxylic acid (*104*).

Evidence as to the mode of linkage of the constituent units was secured in several ways. Masking of the amino group of the β-alanine residue in (XCII) was indicated by the fact that no nitrogen was evolved in the van Slyke determination; on the other hand, suitable color tests clearly demonstrated the presence of the free primary aromatic amino group in the various degradation products discussed above. Also, the order of linkage in formula (XCI) was supported by the observation that, in contrast to the behavior of the simple pyrrole carboxylic acid (XCIII), the diamide acid (XCII) was not readily decarboxylated; therefore the latter substance does not possess a free aromatic carboxylic group. The point of origin of ammonia and guanidinoacetic acid was suggested by the failure of netropsin to give the EHRLICH test characteristic of the free 4-amino group on the pyrrole ring, as well as by the formation from netropsin of the guanidinoacetamido counter-

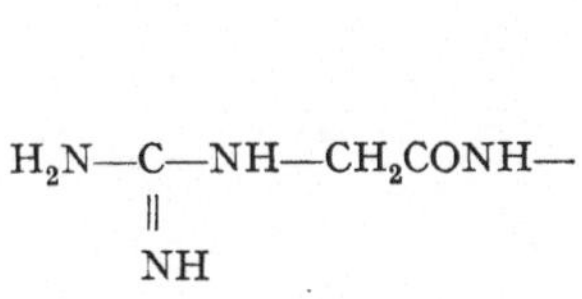

$$H_2N—C—NH—CH_2CONH—$$
$$\parallel$$
$$NH$$

(XCV.)

$$NH$$
$$\parallel$$
$$CH_2—C—NH—$$
$$NH \qquad NH_2$$
$$\diagdown C$$
$$\parallel$$
$$NH$$

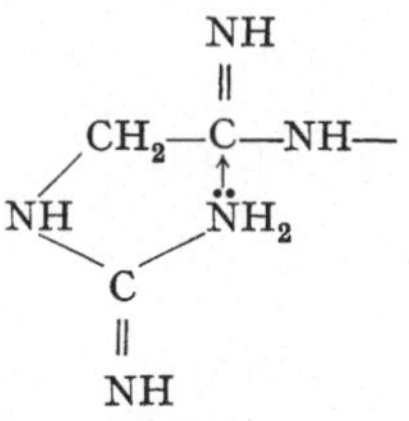

(XCVI.)

part (XCV) by spontaneous hydrolysis, suggestive of the participation mechanism (XCVI).

Confirmation of these structural assignments has been provided by the synthesis of degradation products (XCI, XCII, XCIII, and XCV).

17. Nocardamine.

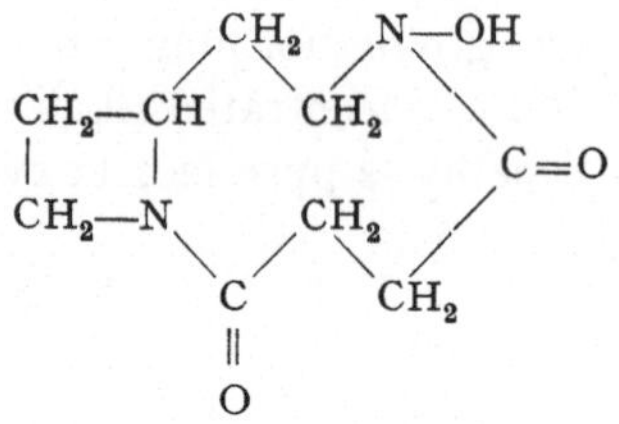

$$
\begin{array}{c}
\text{CH}_2 \quad \text{N—OH} \\
\text{CH}_2\text{—CH} \quad \text{CH}_2 \\
| \qquad | \qquad\qquad \text{C}=\text{O} \\
\text{CH}_2\text{—N} \qquad \text{CH}_2 \\
\text{C} \quad \text{CH}_2 \\
|| \\
\text{O}
\end{array}
$$

(XCVII.) Nocardamine.

Nocardamine (92), a second antibiotic derived from *Nocardia*, possesses the composition $C_9H_{14}O_3N_2$, exhibits reducing properties and features one weakly acidic, but no basic, center. Treatment with hot hydrochloric acid liberated succinic acid and at the same time led to the formation of a monoacidic base, $C_5H_{12}ON_2Cl$, which, containing organically-bound halogen, was reduced by tin and hydrochloric acid to cadaverine, $H_2N(CH_2)_5NH_2$ (XCVIII). The ready incorporation of halogen during the initial hydrolysis step was construed to indicate a small—preferably four-membered—heterocyclic ring, whereas the ready hydrogenolytic removal of oxygen from the C_5-base was regarded as evidence for its attachment to nitrogen. On this basis, the formula (XCIX) for the

$H_2NCH_2CH_2CHClCH_2CH_2NHOH$

(XCIX.)

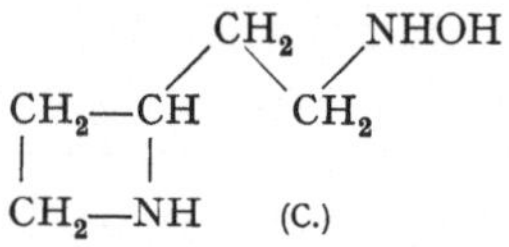

$$
\begin{array}{c}
\text{CH}_2 \quad \text{NHOH} \\
\text{CH}_2\text{—CH} \quad \text{CH}_2 \\
| \qquad | \\
\text{CH}_2\text{—NH} \quad \text{(C.)}
\end{array}
$$

acid cleavage product was proposed; and the structure (C) for the parent heterocycle follows. Nocardamine was regarded as the cyclic amide-hydroxamide of succinic acid, with the N—OH group being the site of the observed acidity.

18. Novobiocin.

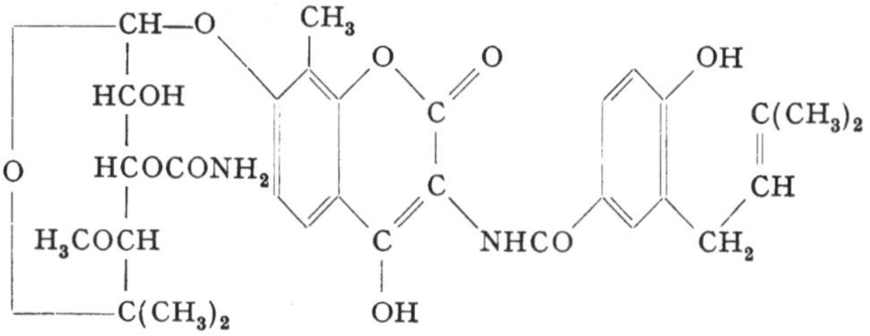

(CI.) Novobiocin.

Streptomyces niveus produces the antibiotic novobiocin (strepto-nivicin), a dibasic acid bearing the molecular formula $C_{31}H_{36}O_{11}N_2$. The molecule, consisting of three distinct moieties—sugar, substituted coumarin and substituted benzoic acid—is freed from the glycosidic unit by treatment with ethanolic hydrochloric acid, yielding the amide (CII). Refluxing acetic anhydride cleaved the latter, producing

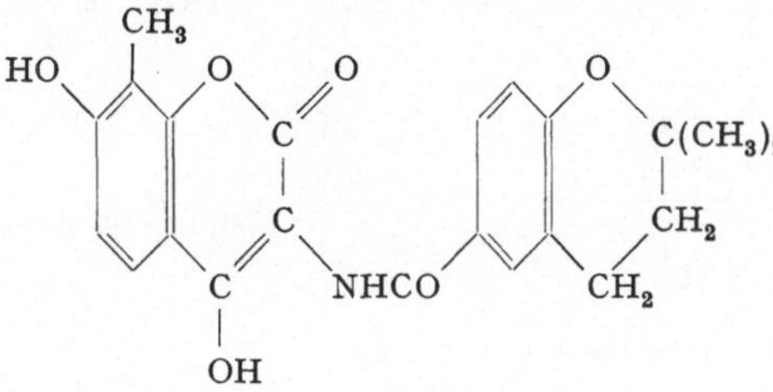

(CII.)

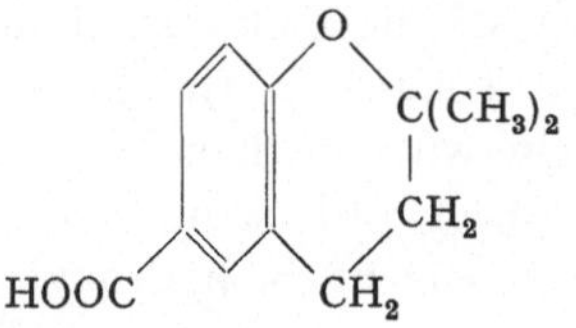

(CIII.)

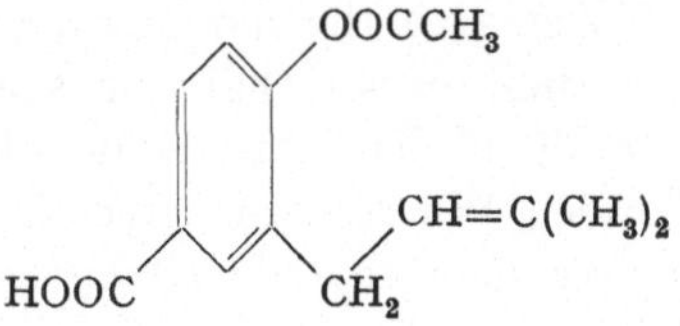

(CIV.)

the known substance, 2,2-dimethyl-6-carboxychroman (CIII) (*5r*). On the other hand, the action of acetic anhydride on the antibiotic itself gave rise to the O-acetate of an isomer of the chroman (CIII), which was shown to have the structure (CIV) by the production of acetone as a consequence of treatment with osmium tetroxide followed by sodium periodate.

Further treatment of the amide (CII) with acetic anhydride involved removal of the chromancarboxylic acid and conversion of the remainder

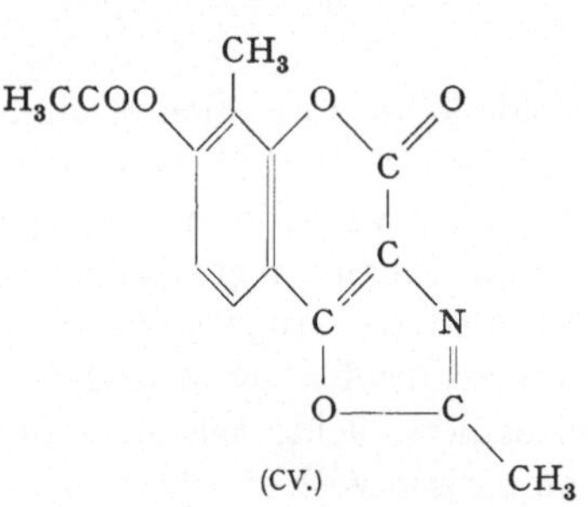

(CV.)

of the molecule to an oxazole (CV), which was transformed by methanolic hydrogen chloride to the parent coumarin (CVI). The exact structure of the latter product, which was recognized as a 3-amino-4-hydroxy-coumarin by comparison with models, was demonstrated by its conversion through alkaline cleavage to a known substance, 2,4-dihydroxy-*m*-toluic acid (CVII) (*46*).

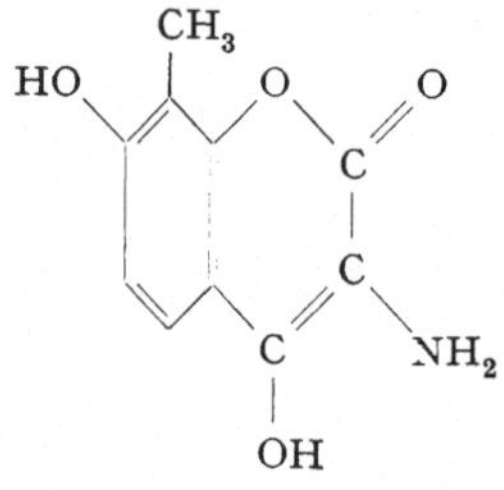

(CVI.) 3-Amino-4-hydroxycoumarin.

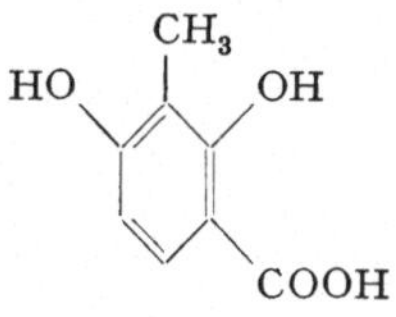

(CVII.) 2,4-Dihydroxy-*m*-toluic acid.

Turning to the monosaccharide portion of the molecule, it may be noted first of all that the sugar can be isolated in the form of its glycoside (CVIII) by fission of novobiocin with methanolic hydrogen chloride. Whereas the glycoside did not react with sodium periodate, the free heptose reduced one mole of the reagent; a hydroxyl group adjacent to the glycosidic carbon was thereby indicated. Alkaline

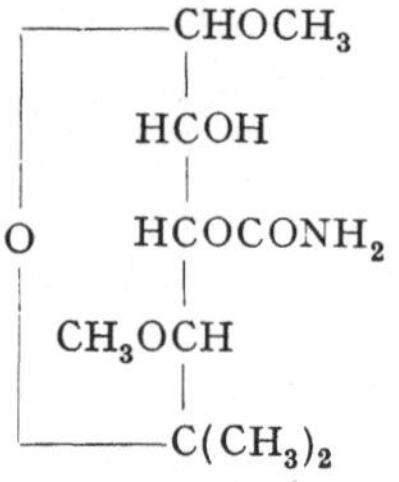

(CVIII.) 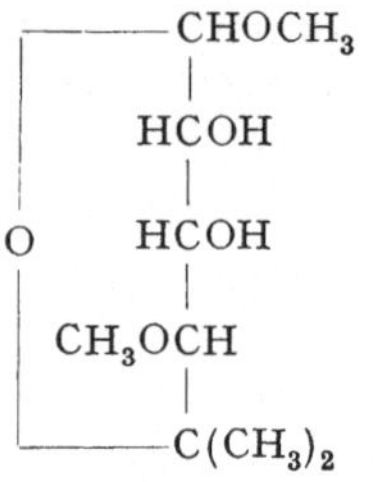   (CIX.) 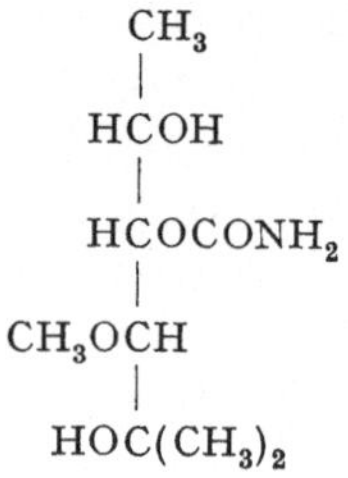   (CX.)

hydrolysis of the glycoside led to formation of ammonia, carbon dioxide, and the 4-methylglycoside (CIX), thus revealing the presence of the urethane grouping. Glyoxal was obtained on periodate oxidation of the urethane-free material. Finally, mercaptanolysis and desulfurization with Raney nickel afforded the trimethylurethane (CX), which, after alkaline hydrolysis, was oxidizable by periodate to acetaldehyde and α-methoxy-β-hydroxyisovaleric acid; the structure of the degradation acid was proved by comparison with synthetic material (*50, 57, 80, 103*).

The deductions summarized have been confirmed in part by the synthesis of amide (CII) (*83*).

19. Puromycin (Achromycin).

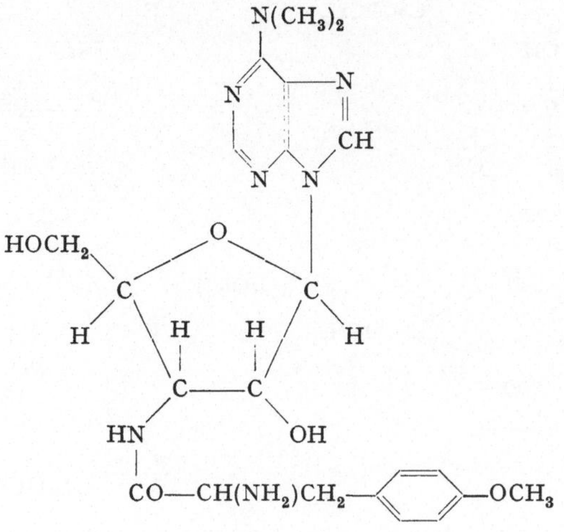

(CXI.) Puromycin.

In puromycin, an antibacterial and antiprotozoal agent isolated from *Streptomyces alboniger*, we find combined two structural units similar to the building blocks typical of living systems: ribonucleoside and α-amino acid. A diacidic base bearing the formula $C_{22}H_{29}O_5N_7$, puromycin is cleaved by ethanolic hydrogen chloride into three fragments: 6-dimethylaminopurine (CXII), O-methyl-*L*-tyrosine (CXIII) and a new aminosugar, which was shown by synthesis to be *D*-3-aminoribose (CXIV). By taking into consideration such data as the presence of one primary amino group (VAN SLYKE) and two free hydroxyl groups in puromycin, the structure (CXI) for the antibiotic could be advanced (*2, 97*).

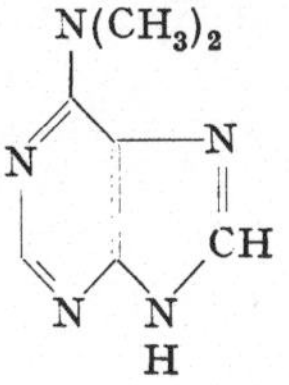

(CXII.) 6-Dimethylamino-purine. 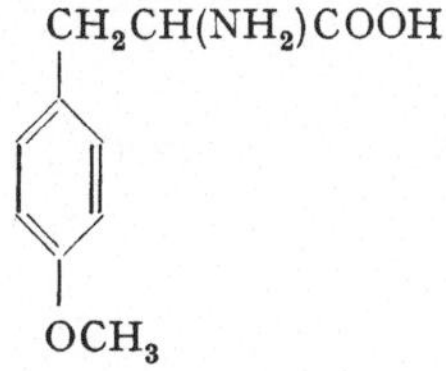   (CXIII.) O-Methyl-*L*-tyrosine. 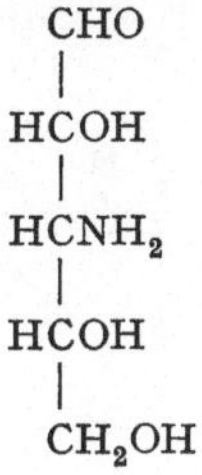   (CXIV.) *D*-3-Aminoribose.

Puromycin is one of the few *Streptomyces* antibiotics to have been attained, along with related substances, by synthesis, and the following route (*3–5*) is exemplary of the extensive efforts expended along such lines *(Chart 4)*.

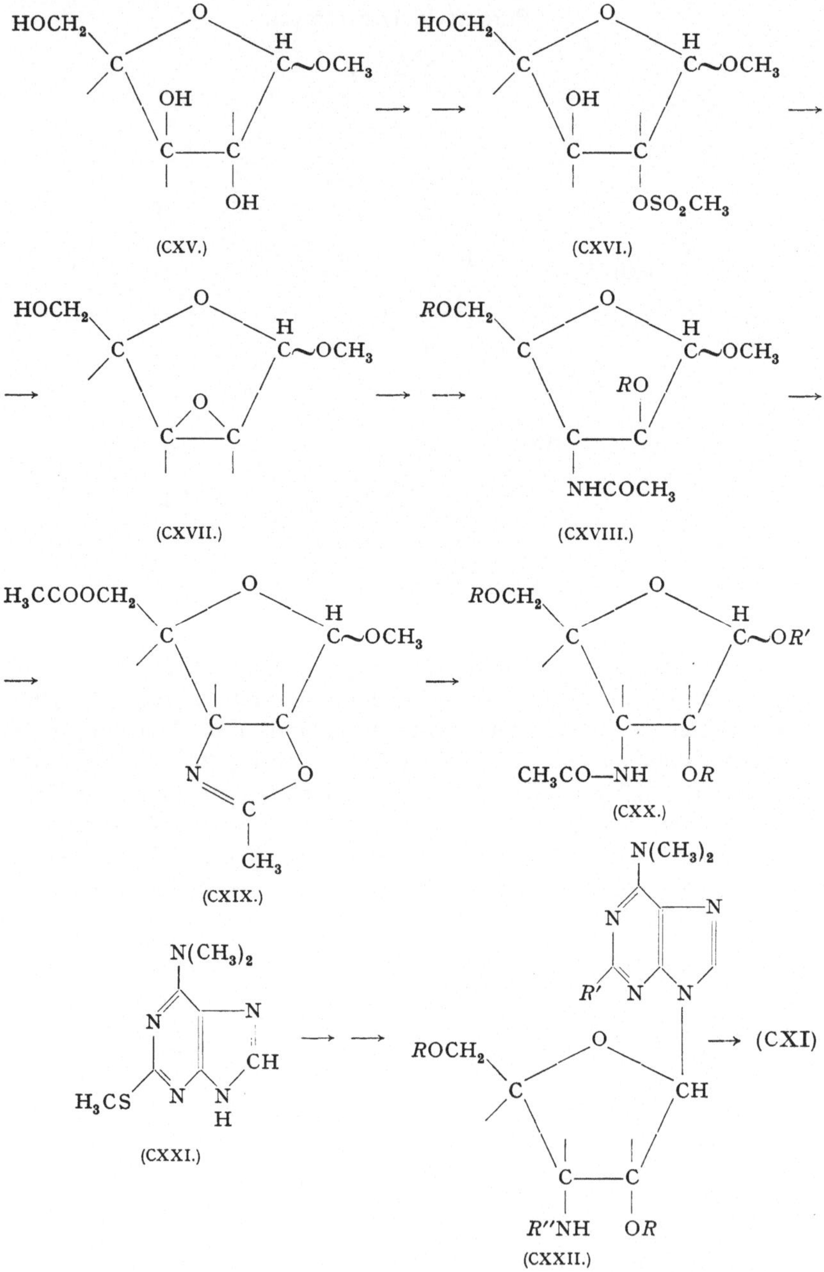

Chart 4. Synthesis of Puromycin.

(The sign ∼ indicates that the group attached is of unknown configuration.)

The acid-catalyzed reaction of methyl D-xylofuranoside (CXV) with acetone afforded the 3,5-acetonide, which was converted to the 2-O-methanesulfonate and thence deacetonated to the simple xyloside derivative (CXVI). The latter was ring-closed to methyl 2,3-anhydro-lyxofuranoside (CXVII), which was then subjected to the action of ammonia, yielding, after acetylation, 3-acetamino-3-deoxy-D-arabino-furanoside (CXVIII, $R = H$), along with the isomeric 2-acetamino-glycoside. Now that the amino group had been introduced into the proper position, conversion of the arabinose to the ribose configuration was required. This change was brought about by treatment of the 2,5-dimethanesulfonate (CXVIII, $R = CH_3SO_2$) with sodium acetate; intermediary cyclization with inversion at $C_{(2)}$ to (CXIX) was followed by hydrolytic cleavage of the oxazoline ring and removal of the 5-acetyl group, affording after acetylation, methyl-2,5-di-O-acetyl-3-acetamino-3-deoxy-D-ribofuranoside (CXX, $R = H_3CCO$, $R' = CH_3$). O-De-acetylation followed by O-benzoylation and removal of the glycosidic methoxyl group led to (CXX, $R = C_6H_5CO$, $R' = H$). After acetylation of the hemiacetal hydroxyl group, the anomeric mixture of ribosamine derivatives was allowed to react, in the presence of titanium tetrachloride, with the chloromercury derivative of 2-methylmercapto-6-dimethyl-aminopurine (CXXI), which operation resulted in the formation of the nucleoside (CXXII, $R = H_3CCO$, $R' = CH_3S$, $R'' = C_6H_5CO$). Reductive RANEY nickel desulfurization followed by base-induced debenzoylation and deacetylation provided the key intermediate (CXXII, $R = R' = = R'' = H$). Reaction with carboethoxy mixed anhydride of N-carbo-benzoxy-p-methoxy-L-phenylalanine led to N-carbobenzoxypuromycin, which, after catalytic hydrogenolysis, gave natural puromycin.

20. Sarkomycin.

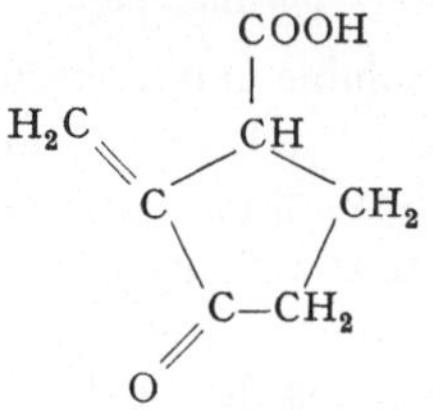

(CXXIII.) Sarkomycin.

Possessing a simple yet novel structure, the *Streptomyces* antibiotic sarkomycin is of particular interest because of its suppressive action on the EHRLICH ascites tumor in mice. Ozonolysis revealed the presence of the exocyclic methylene group, whereas catalytic reduction afforded 2-methyl-3-cyclopentanone carboxylic acid (52). Syntheses of *DL*-sarkomycin (95) and its semicarbazone (79) have been reported.

21. Streptomycin.

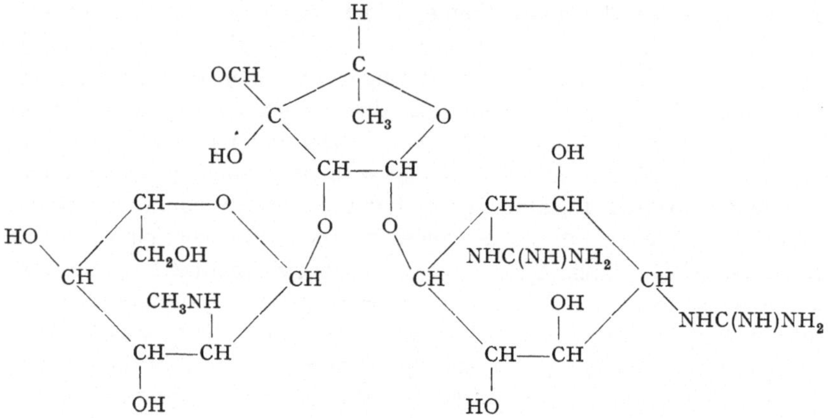

(CXXIV.) Streptomycin.

One of the earliest and best-known of the newer therapeutic agents is streptomycin, the product of a systematic search for a useful antibiotic exhibiting activity against Gram-negative and acid-fast bacteria. Originally detected in cultures of microorganisms isolated from a heavily manured field soil and from the throat of a chicken, streptomycin was in the earlier stages of its development prepared by growing *Streptomyces griseus* either according to surface culture or submerged culture methods.

Plant production is accomplished by using a submerged, aerated process, carried out in fermentors with capacities on the order of thousands of gallons. Raw materials include a nitrogen source, such as peptone or beef extract; carbohydrate, for example, glucose or starch; and various inorganic materials. In the published Merck procedure (*74*), isolation involves removal of mycelium by filtration, adsorption of streptomycin on activated carbon, elution of the active material from the carbon adsorbant, concentration of the eluates and double solvent precipitation, followed by further purification.

Optically active, water soluble and unstable outside the pH range 3–7, streptomycin is a strong base, $C_{21}H_{39}O_{12}N_7$, ordinarily handled in the form of its inorganic salts. The Sakaguchi color test indicated the presence of the guanidino group; reaction with hydroxylamine or thiosemicarbazide demonstrated the presence of the free carbonyl function. Perhaps the most important single reaction is the ready cleavage which streptomycin undergoes in the presence of acidic reagents such as methanolic hydrogen chloride, which produces *streptidine* (CXXV) (*73*) and the dimethyl-acetal of *methyl streptobiosaminide* (CXXVI) (*12, 71*). Prolonged alkaline treatment of streptidine brought about disruption of the guanidino groups, with the formation of ammonia, carbon dioxide and *streptamine* (CXXVII) (*34*). The consumption of· six moles of periodate without formation of formaldehyde denoted that streptamine was a diamino-tetrahydroxy-

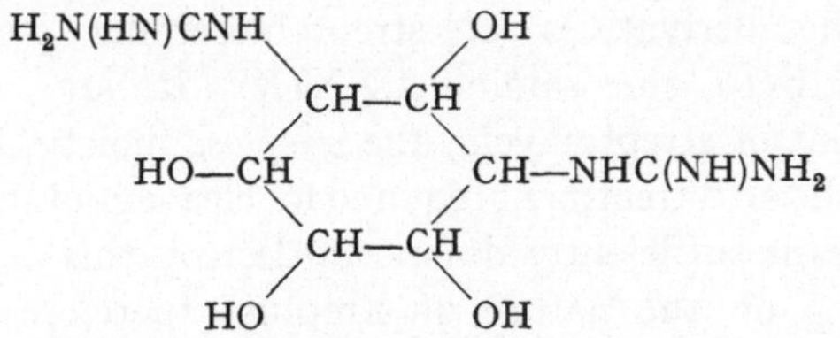

$$H_2N(HN)CNH \quad\quad OH$$
$$CH—CH$$
$$HO—CH \quad\quad CH—NHC(NH)NH_2$$
$$CH—CH$$
$$HO \quad\quad OH$$

(CXXV.) Streptidine.

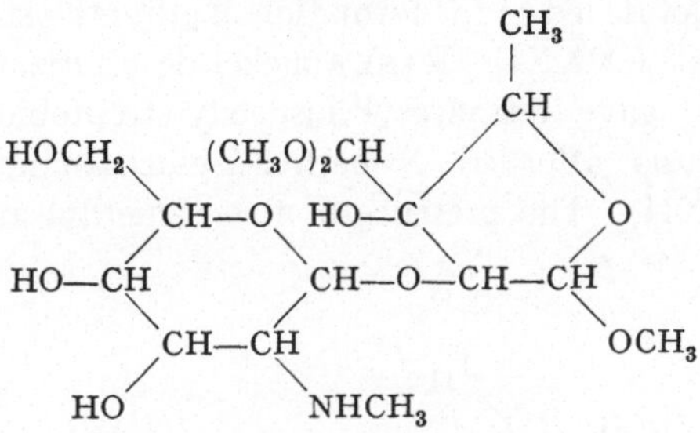

$$CH_3$$
$$|$$
$$CH$$
$$HOCH_2 \quad (CH_3O)_2CH$$
$$CH—O \quad HO—C \quad\quad O$$
$$HO—CH \quad\quad CH—O—CH—CH$$
$$CH—CH \quad\quad OCH_3$$
$$HO \quad\quad NHCH_3$$

(CXXVI.) Methyl streptobiosaminide.

cyclohexane; the consumption of two moles of periodic acid by the di-N-benzoyl derivative, along with the conversion to a di-N-benzoyl-amino-hydroxyglutaric acid (CXXVIII), distinguished the correct

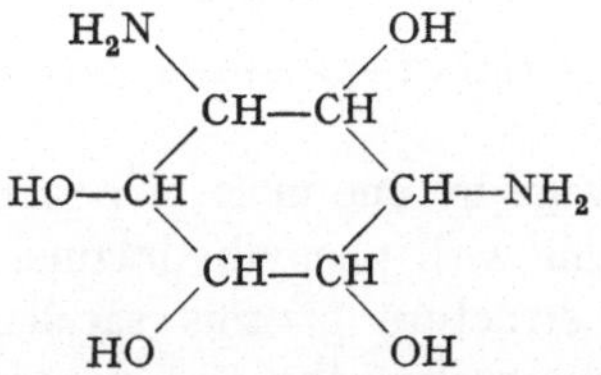

$$H_2N \quad\quad OH$$
$$CH—CH$$
$$HO—CH \quad\quad CH—NH_2$$
$$CH—CH$$
$$HO \quad\quad OH$$

(CXXVII.) Streptamine.

$$C_6H_5CONH \quad\quad OH$$
$$CH—CH$$
$$HOOC \quad\quad CH—NHOCC_6H_5$$
$$HOOC$$

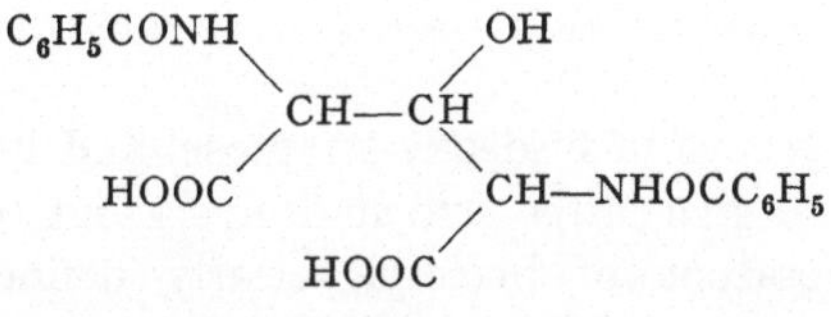

(CXXVIII.) Di-N-benzoylamino-hydroxyglutaric acid.

structure (CXXV) for streptidine from the several remaining possibilities (*16, 17*). This assignment as well as the stereochemical nature of the base was confirmed by synthesis (*111*).

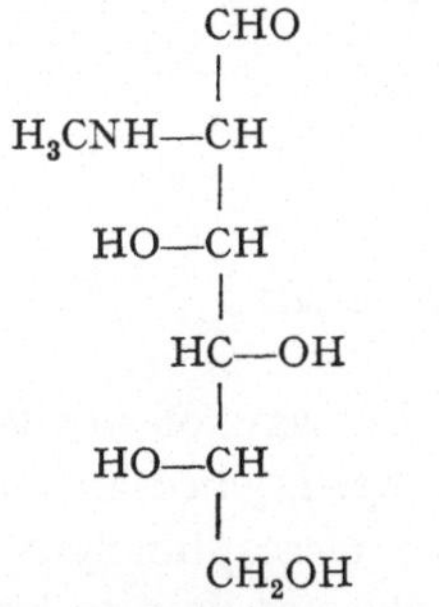

$$CHO$$
$$|$$
$$H_3CNH—CH$$
$$|$$
$$HO—CH$$
$$|$$
$$HC—OH$$
$$|$$
$$HO—CH$$
$$|$$
$$CH_2OH$$

(CXXIX.) N-Methyl-*L*-glucosamine.

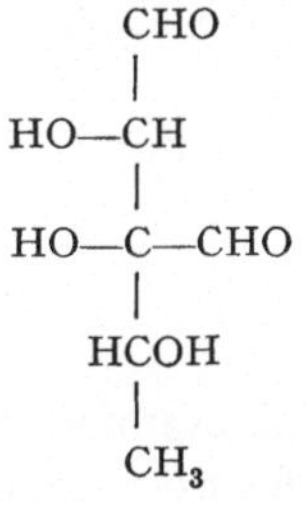

$$CHO$$
$$|$$
$$HO—CH$$
$$|$$
$$HO—C—CHO$$
$$|$$
$$HCOH$$
$$|$$
$$CH_3$$

(CXXX.)

Although the derivatives of streptobiosamine yielded on acid hydrolysis N-methyl-*L*-glucosamine (CXXIX) (*61*, *112*), the third main structural element of streptomycin, the *streptose* moiety, has never been isolated in that the acid treatment required for cleavage of streptobiosamine or its derivatives is sufficiently drastic to destroy this C_5-unit (CXXX). Evidence bearing on the nature of streptose therefore was indirectly acquired (*35*, *60*, *110*). Treatment of streptomycin with ethyl mercaptan and hydrogen chloride led to the formation of ethyl thiostreptobiosaminide diethyl mercaptal (CXXXI). RANEY nickel desulfurization of the tetra-acetyl derivative gave tetraacetyl bisdeoxy-streptobiosaminide, which on acid hydrolysis afforded N-methyl-*L*-glucosamine and bisdeoxy-streptose (CXXXII). The presence of two C-methyl and two hydroxyl

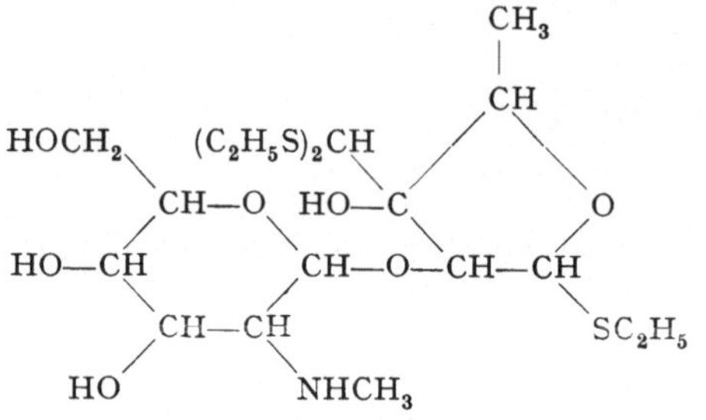

(CXXXI.) Ethyl thiostreptobiosaminide diethyl mercaptal. 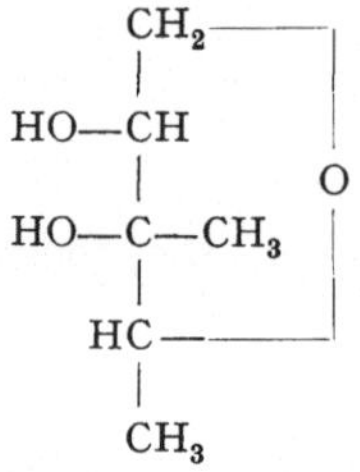   (CXXXII.) Bisdeoxy-streptose.

groups in bisdeoxy-streptose, and its cleavage by one mole of periodate to give, after acid hydrolysis and treatment with phenylhydrazine, the osazone of biacetyl, clearly define the structure of this saccharide transformation product. Apart from proving that the two sugar-like

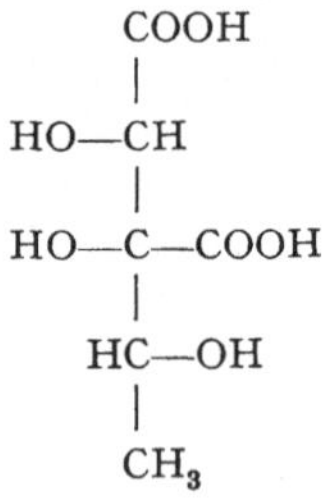

(CXXXIII.) Streptosonic acid.

units of streptobiosamine are joined by way of a glycosidic linkage involving the formyl group of N-methyl-*L*-glucosamine (as indicated by the amino sugar's survival of the mercaptanolysis-desulfurization sequence), the findings outlined limit the possibilities for the structure of streptose to those which result from incorporation of an additional hydroxyl group and the replacement of a methyl by a formyl group in (CXXXII). These assignments were made possible by the successful oxidative conversion of tetraacetyl-streptobiosamine to derivatives of the dibasic *streptosonic acid* (CXXXIII), the structure of which was evident from the simple observation that acetaldehyde was liberated on periodate oxidation. Since the pair of carboxyl groups must have arisen by oxidation of two aldehyde groups in the streptose system, the structure (CXXX) follows.

The following arguments illustrate the means by which the mode of linkage of the three moieties was elucidated. Appropriate periodate studies demonstrated the ring size of the N-methylglucosaminide system, and its attachment to the secondary hydroxyl group of the streptoside nucleus was indicated by the presence of a non-acetylatable—hence tertiary—hydroxyl group in various streptobiosamine derivatives. The facile cleavage of streptomycin under acid conditions is consistent with a glycosidic linkage joining streptidine to the streptobiosamine system. The streptidine hydroxyl group used for this linkage was identified by a stepwise conversion of streptomycin to a product, N,N'-dibenzoyl-deoxystreptamine (CXXXIV), in which the oxygen to which the strepto-biose was attached, had been replaced by hydrogen. Of the three possible structures for this deoxystreptamine derivative, the one indicated was shown to be correct by periodate oxidation to a diaminohydroxy-adipdialdehyde (CXXXV) (*62*).

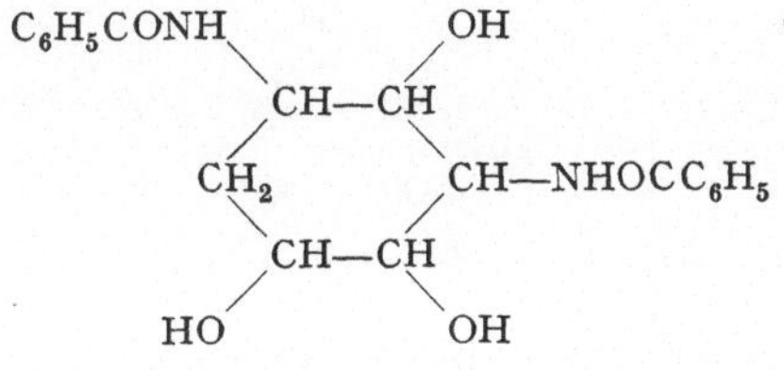

C_6H_5CONH OH

CH—CH

CH_2 CH—NHOCC_6H_5

CH—CH

HO   OH

C_6H_5CONH OH

CH—CH

CH_2 CH—NHOCC_6H_5

CHO CHO

(CXXXIV.) N,N'-Dibenzoyl-deoxystreptamine. 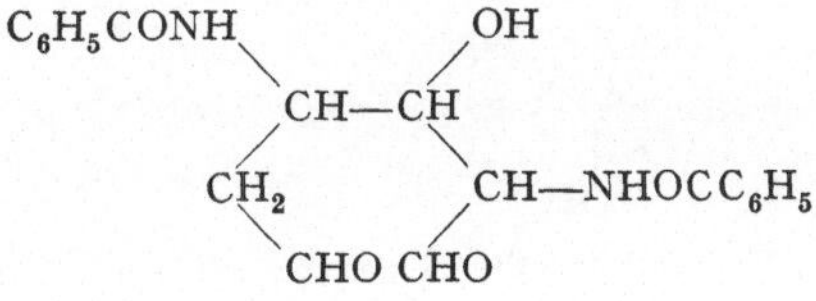   (CXXXV.) Diaminohydroxy-*L*-adipdialdehyde.

Two *Streptomyces* metabolites structurally related to streptomycin have been isolated. One of these, hydroxy-streptomycin, isolated from a strain *(S. griseocarneus)* obtained from Japanese soil, differs in that it possesses a primary carbinol, rather than the C-methyl, group in streptomycin (*7, 40, 91*). The second relative, mannosido-streptomycin (streptomycin-B), accompanies the better known antibiotic in the fermentation broth and is in fact a *D*-mannoside of streptomycin in which the additional monosaccharide is attached in the pyranose form to the 4-position of the N-methyl-*L*-glucosamine nucleus (*72, 86*).

22. Terramycin.

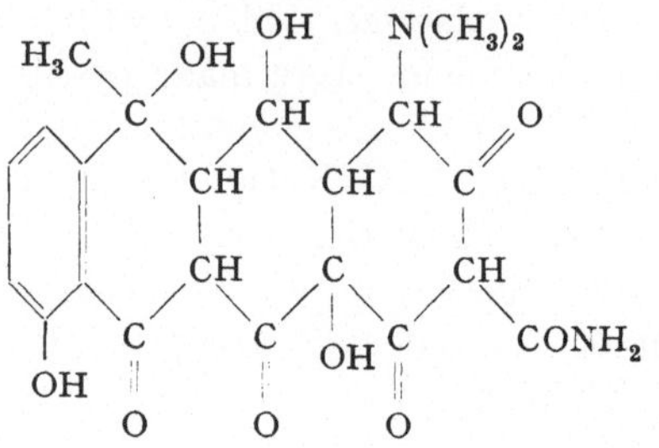

(CXXXVI.) Terramycin.

Terramycin, the "broad-spectrum" antibiotic produced by *S. rimosus*, is an amphoteric compound, $C_{22}H_{24}O_9N_2$, the first of the tetracycline-type to be assigned a complete structural formula (CXXXVI) (*48, 49*).

Terramycin is unstable in basic media, undergoing interesting degradation reactions which offered valuable clues as to the structure of the natural product. The result of one of these operations, *terracinoic acid*, possesses structure (CXXXVII). For various reasons, the acid

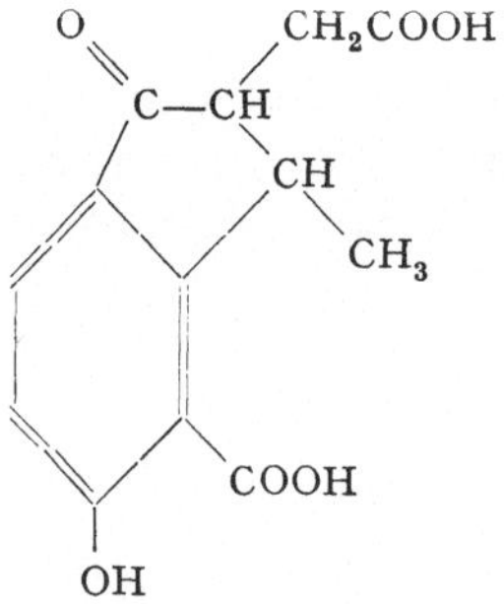

(CXXXVII.) Terracinoic acid. 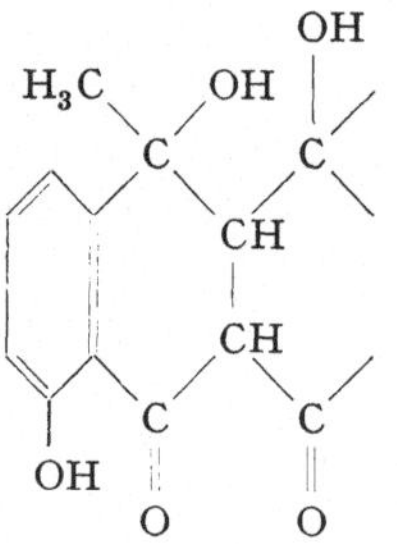   (CXXXVIII.)

could not be regarded as arising from terramycin by any direct process; and careful consideration of its probable mode of genesis along with certain other pieces of evidence led to the conclusion that the antibiotic possessed, in fact, the partial structure (CXXXVIII).

A more extensive and reliable body of evidence was associated with the series of transformation products secured through the use of successively more strongly acidic conditions. Hydrogen chloride in acetone at 5° converts terramycin to *anhydro-terramycin* (CXXXIX), whereas the action of dilute hydrochloric acid at 60° not only effects aromatization to the naphthalenoid system but also cleavage and lactonization with the result that the diastereoisomeric α- and β-*apo-terramycins* (CXL) are formed. Still more severe aqueous acid conditions

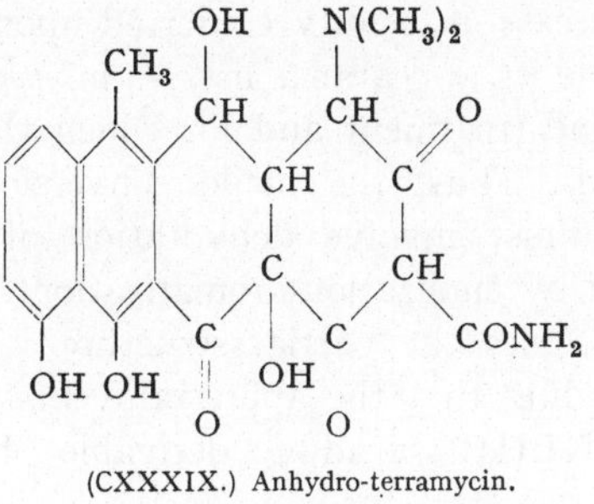

$$CH_3 \quad \overset{OH}{\underset{|}{}} \quad \overset{N(CH_3)_2}{\underset{|}{}}$$

(CXXXIX.) Anhydro-terramycin.

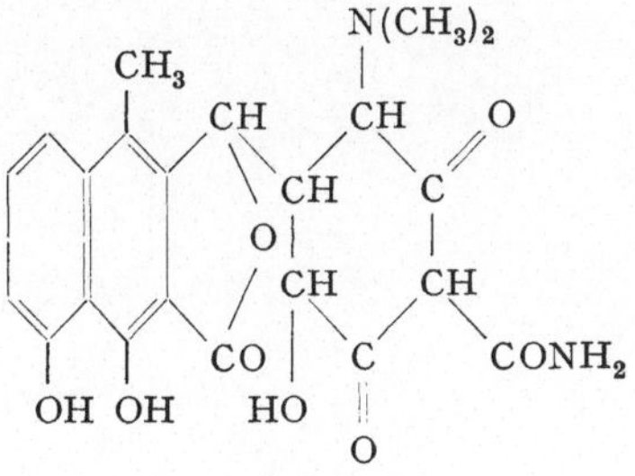

(CXL.) α- and β-Apoterramycins.

result in loss of dimethylamine and further aromatization, giving *terrinolide* (CXLI). On treatment with hot 12 *N*-sulfuric acid, terrinolide suffers loss of carboxamido group, yielding *decarboxamido-terrinolide* (CXLI, with $CONH_2$ replaced by H).

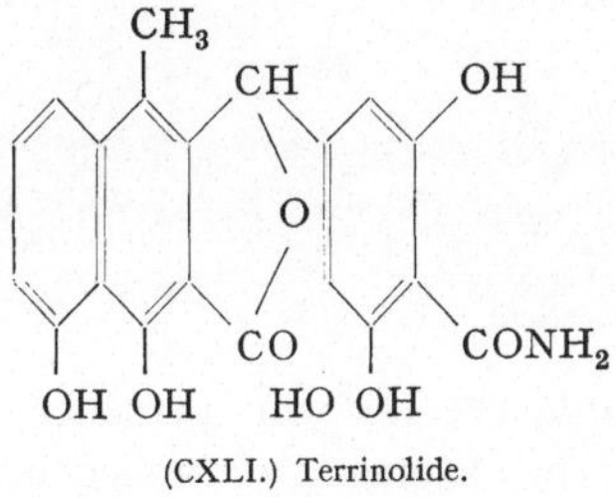

(CXLI.) Terrinolide.

The 1,8-dihydroxynaphthalene system, present in all of the members in this dehydration sequence, was characterized spectrophotometrically

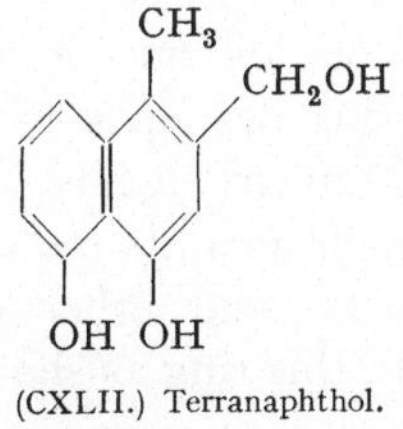

(CXLII.) Terranaphthol.

as well as by the increase in acidity conferred upon boric acid; further, the same system made its appearance in *terranaphthol*, a simpler product resulting from alkaline treatment and for which the structure (CXLII) was readily deduced. Thus, the facile conversion of terramycin to anhydro-terramycin must involve dehydration of the *t*-benzylalcohol system, accompanied by enolization-aromatization. The earlier proposal that terramycin possesses the partial structure (CXXXVIII) received support from the finding that the apoterramycins were lactones of the phthalide type (CXLIII), readily derivable from the proposed

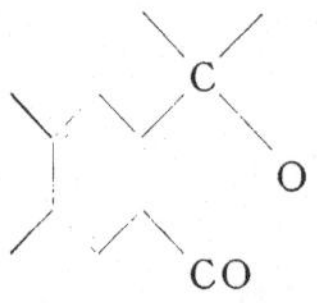

(CXLIII.) 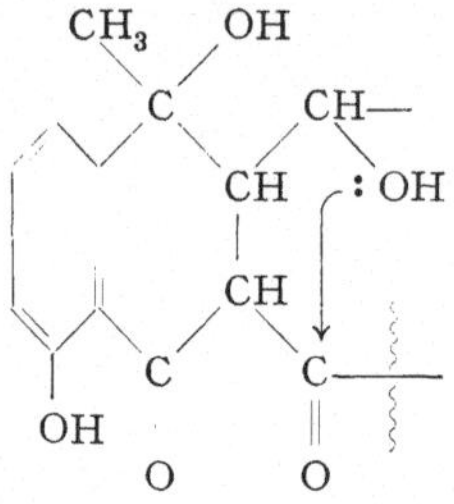   (CXLIV.)

structure (CXLIV). The presence of the 1,8-dihydroxybenzophthalide nucleus received confirmation in the alkaline fusion of apoterramycin, producing *terranaphthoic acid* (CXLV), and in the oxidation of per-O-methylated terrinolide to (CXLVI).

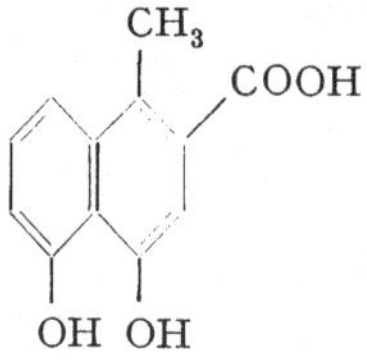

(CXLV.) Terranaphthoic acid. 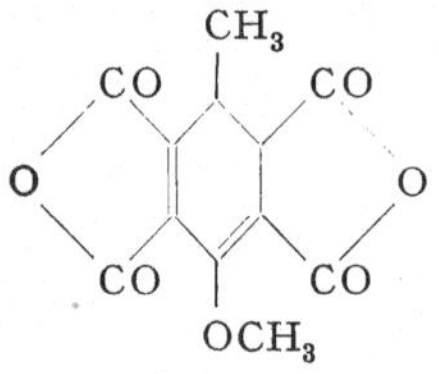   (CXLVI.)

Taking into account the known presence of carboxamido and dimethylamino units, the atoms yet unaccounted for constitute a C_6-residue, present in decarboxamido-terrinolide in the form of what appeared to be a polyhydric phenolic system. By comparison with suitable model compounds, it was shown that this system conformed to the 1,2,4-trihydroxybenzene type. Moreover, the formation of 2,5-dihydroxybenzoquinone on alkaline fusion of the apo compounds demonstrated that this C_6-ring system was also present at this stage. Thus the change, apoterramycin (CXL) → terrinolide (CXLI), must involve loss of dimethylamine from a hydroaromatic substituent, resulting in aromatization. Mechanistic and other considerations allowed the assignment of substituents in this ring as shown in (CXL), and the ready lactonization involved in formation of the apoterramycins clearly indicated that the carbonyl group of the lactone ring must have originally been situated β to a second carbonyl group. The placement of the carboxamido group flanked by hydroxyls followed from the acidity behavior of terrinolide, decarboxamido-terrinolide, and model compounds.

Although the results described above do not define the hydro-naphthacenoid system of terramycin, its presence was supported by, among others, the formation of naphthacene (CXLVI) in the zinc-dust distillation of the transformation product (CXLVII).

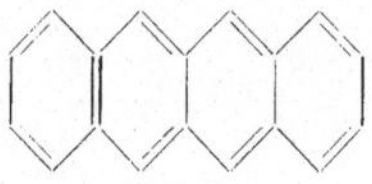

(CXLVI.) Naphthacene. 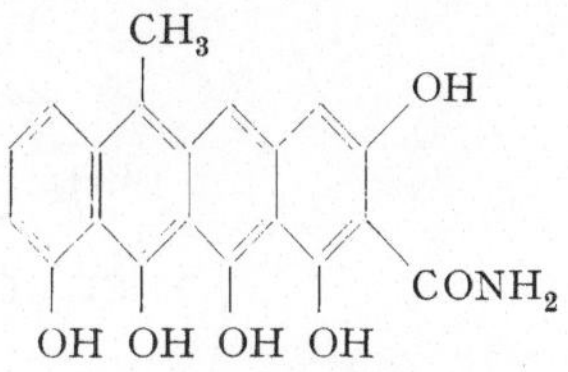   (CXLVII.)

Finally, the positions of the dimethylamino group and the angular hydroxyl group were shown to be as indicated in (CXXXVI) (rather than the structure in which the two groups are interchanged), since reductive removal of the hydroxyl group resulted in the formation of a new enolic system (CXLVIII), whereas similar replacement of the dimethylamino group did not.

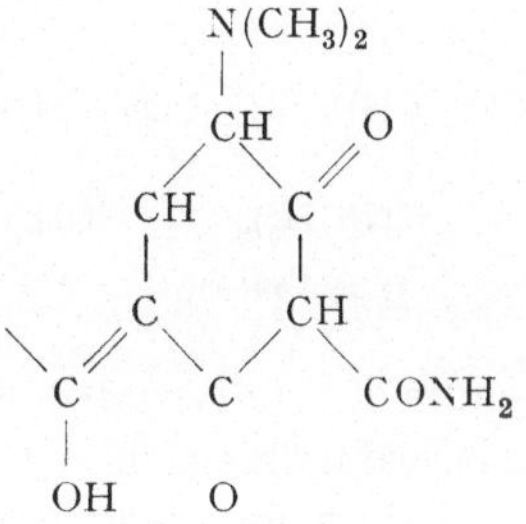

(CXLVIII.)

23. Tetracycline.

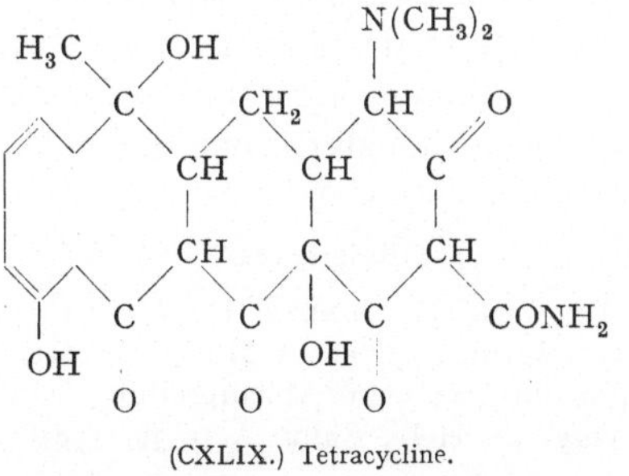

(CXLIX.) Tetracycline.

9*

Shortly after the structural studies on terramycin and aureomycin were announced, the two groups of workers simultaneously reported the preparation of tetracycline (CXLIX) by catalytic hydrogenolysis of halogen from aureomycin (chlorotetracycline, p. 98) (*11, 23*). It is noteworthy that not only does tetracycline possess antibiotic activity similar to that of the two substitution products, but also that the newer substance actually has been detected as a fermentation product, produced by a *Streptomyces* found in Texan soil, and also, in smaller amount and simultaneously with aureomycin, by *S. aureofaciens*.

24. Valinomycin.

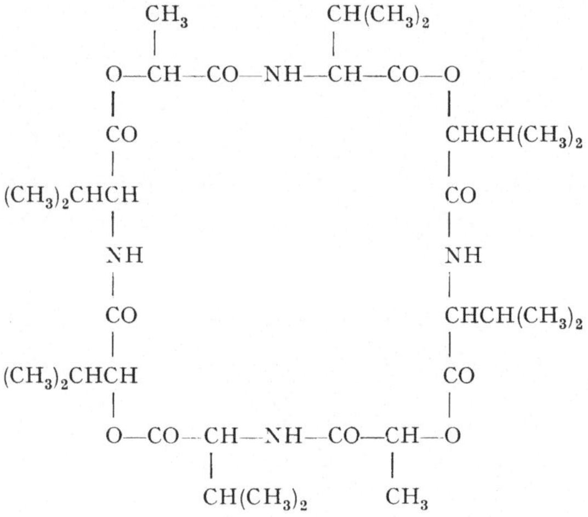

(CL.) Valinomycin.

Isolated from the mycelium of *S. fulvissimus*, valinomycin, $C_{36}H_{60}O_{12}N_4$, produces on total hydrolysis two moles each of *L*-valine, *D*-valine, *L*-lactic acid and *D*-α-hydroxyisovaleric acid. Selective cleavage of the lactone linkages, brought about by hydrolysis with barium hydroxide, gave rise to 1.7 moles each of *L*-lactyl-*L*-valine and *D*-α-hydroxyisovaleryl-*D*-valine. Since valinomycin possesses no free hydroxyl, amino or carboxyl groups, the hydrolysis results demand a cyclic structure in which the four molecules of the two amides are connected by ester linkages. Two arrangements are possible, although the one shown (CL) was favored (*14*).

References.

1. Anliker, R., D. Dvornik, K. Gubler, H. Heusser und V. Prelog: Stoff-wechselprodukte von Actinomyceten. 5. Mitt. Über das Lacton der β-Hydroxy-α,α′,γ-trimethyl-pimelinsäure, ein Abbauprodukt von Narbomycin, Pikro-mycin und Methymycin. Helv. Chim. Acta **39**, 1785 (1956).

2. BAKER, B. R. and R. E. SCHAUB: Achromycin. Synthetic Studies. III. Synthesis of 3-Amino-*D*-ribose, a Hydrolytic Product. J. Amer. Chem. Soc. 75, 3864 (1953).

3. BAKER, B. R., R. E. SCHAUB, J. P. JOSEPH and J. H. WILLIAMS: Total Synthesis of the Antibiotic Puromycin. J. Amer. Chem. Soc. 76, 4044 (1954).

4. — — — — Puromycin. Synthetic Studies. IX. Total Synthesis. J. Amer. Chem. Soc. 77, 12 (1955).

5. BAKER, B. R., R. E. SCHAUB and J. H. WILLIAMS: Puromycin. Synthetic Studies. VIII. Synthesis of 3-Amino-3-deoxy-*D*-ribofuranoside Derivatives. A Second Synthesis of 3-Amino-3-deoxy-*D*-ribose. J. Amer. Chem. Soc. 77, 7 (1955).

6. BARTZ, Q., J. STANDIFORD, J. D. MOLD, D. W. JOHANNESSEN, A. RYDER, A. MARETZKI and T. H. HASKELL: Griseoviridin and Viridogrisein: Isolation and Characterization. Antibiotics Annual 2, 777 (1954–1955).

7. BENEDICT, R. G., F. H. STODOLA, O. L. SHOTWELL, A. M. BORUD and L. A. LINDENFELSER: A New Streptomycin. Science (Washington) 112, 77 (1950).

7a. BOHLMANN, F. und H. J. MANNHARDT: Acetylenverbindungen im Pflanzenreich. Fortschr. Chem. organ. Naturstoffe 14, 1 (1957).

8. BOHLMANN, F. und H. G. VIEHE: Polyacetylenverbindungen, V. Mitt.: Synthese des Isomycomycins und ähnlicher Triacetylenverbindungen. Ber. dtsch. chem. Ges. 87, 712 (1954).

9. BOOTHE, J. H., S. KUSHNER, J. PETISI and J. H. WILLIAMS: Synthesis of Degradation Products of Aureomycin. IV. J. Amer. Chem. Soc. 75, 3261 (1953).

10. BOOTHE, J. H., S. KUSHNER and J. H. WILLIAMS: Synthesis of Degradation Products of Aureomycin. V. J. Amer. Chem. Soc. 75, 3263 (1953).

11. BOOTHE, J. H., J. MORTON II, J. P. PETISI, R. G. WILKINSON and J. H. WILLIAMS: Tetracycline. J. Amer. Chem. Soc. 75, 4621 (1953).

12. BRINK, N. G., F. A. KUEHL, Jr. and K. FOLKERS: Streptomyces Antibiotics. III. Degradation of Streptomycin to Streptobiosamine Derivatives. Science (Washington) 102, 506 (1945).

13. BROCKMANN, H., G. BOHNSACK, B. FRANCK, H. GRÖNE, H. MUXFELDT und C. SÜLING: Zur Konstitution der Actinomycine. Angew. Chem. 68, 70 (1956).

14. BROCKMANN, H. und H. GEEREN: Valinomycin. II; Antibiotica aus Actinomyceten. XXXVII. Die Konstitution des Valinomycins. Liebigs Ann. Chem. 603, 216 (1957).

15. BULLOCK, E. and A. W. JOHNSON: Actinomycin. Part. V. The Structure of Actinomycin D. J. Chem. Soc. (London) 1957, 3280.

16. CARTER, H. E., R. K. CLARK, Jr., S. R. DICKMAN, Y. H. LOO, J. S. MEEK, P. S. SKELL, W. A. STRONG, J. T. ALBERI, Q. R. BARTZ, S. B. BINKLEY, H. M. CROOKS, Jr., I. R. HOOPER and M. C. REBSTOCK: Degradation of Streptomycin and the Structure of Streptidine and Streptamine. Science (Washington) 103, 53 (1946).

17. CARTER, H. E., R. K. CLARK, Jr., S. R. DICKMAN, Y. H. LOO, P. S. SKELL and W. A. STRONG: Degradation of Streptomycin and the Structure of Streptidine and Streptamine. Science (Washington) 103, 540 (1946).

18. CELMER, W. D. and I. A. SOLOMONS: Mycomycin. III. The Structure of Mycomycin, an Antibiotic Containing Allene, Diacetylene and *cis*, *trans*-Diene Groupings. J. Amer. Chem. Soc. 75, 1372 (1953).

19. — — The Structures of Thiolutin and Aureothricin, Antibiotics Containing a Unique Pyrrolinonodithiole Nucleus. J. Amer. Chem. Soc. 77, 2861 (1955).

20. Celmer, W. D., F. W. Tanner, Jr., M. Harfenist, T. M. Lees and I. A. Solomons: Characterization of the Antibiotic Thiolutin and its Relationship to Aureothricin. J. Amer. Chem. Soc. 74, 6304 (1952).
21. Clark, R. K., Jr.: The Chemistry of Erythromycin. I. Acid Degradation Products. Antibiotics and Chemotherapy 3, 663 (1953).
22. Clark, R. K., Jr. and J. R. Schenck: Actathiazic Acid. III. The Synthesis of dl-Actathiazic Acid, Derivatives and Homologs. Arch. Biochem. Biophys. 40, 270 (1952).
23. Conover, L. H., W. T. Moreland, A. R. English, C. R. Stephens and F. J. Pilgrim: Terramycin. XI. Tetracycline. J. Amer. Chem. Soc. 75, 4622 (1953).
24. Controulis, J., M. C. Rebstock and H. M. Crooks, Jr.: Chloramphenicol (Chloromycetin). V. Synthesis. J. Amer. Chem. Soc. 71, 2463 (1949).
25. Djerassi, C., A. Bowers, R. Hodges and B. Riniker: The Partial Structure of Methymycin. J. Amer. Chem. Soc. 78, 1733 (1956).
26. Djerassi, C., A. Bowers and H. N. Khastgir: Methymycin. Reduction and Oxidation Studies. J. Amer. Chem. Soc. 78, 1729 (1956).
27. Djerassi, C. and O. Halpern: The Structure of the Antibiotic Neomethymycin. J. Amer. Chem. Soc. 79, 2022 (1957).
28. Djerassi, C. and J. A. Zderic: The Structure of the Antibiotic Methymycin. J. Amer. Chem. Soc. 78, 2907, 6390 (1956).
29. Dunitz, J. D. and J. H. Robertson: Relationship between Aureomycin and Terramycin. J. Amer. Chem. Soc. 74, 1108 (1952).
30. Finlay, A. C., F. A. Hochstein, B. A. Sobin and F. X. Murphy: Netropsin, a New Antibiotic Produced by a Streptomyces. J. Amer. Chem. Soc. 73, 341 (1951).
31. Flynn, E. H., J. W. Hinman, E. L. Caron and D. O. Woolf, Jr.: The Chemistry of Amicetin, a New Antibiotic. J. Amer. Chem. Soc. 75, 5867 (1953).
32. Flynn, E. H., M. V. Sigal, Jr., P. F. Wiley and K. Gerzon: Erythromycin. I. Properties and Degradation Studies. J. Amer. Chem. Soc. 76, 3121 (1954).
33. Ford, J. H. and B. E. Leach: Actidione, an Antibiotic from Streptomyces Griseus. J. Amer. Chem. Soc. 70, 1223 (1948).
34. Fried, J., G. A. Boyack and O. Wintersteiner: Streptomycin: The Chemical Nature of Streptidine. J. Biol. Chem. 162, 391 (1946).
35. Fried, J., D. E. Walz and O. Wintersteiner: Streptomycin. III. 4-Desoxy-L-erythrose(threose)phenylosazone from Streptobiosamine. J. Amer. Chem. Soc. 68, 2746 (1946).
36. Fusari, S. A., R. P. Frohardt, A. Ryder, T. H. Haskell, D. W. Johannessen, C. C. Elder and Q. R. Bartz: Azaserine, a New Tumor-inhibitory Substance. Isolation and Characterization. J. Amer. Chem. Soc. 76, 2878 (1954).
37. Fusari, S. A., T. H. Haskell, R. P. Frohardt and Q. R. Bartz: Azaserine, a New Tumor-inhibitory Substance. Structural Studies. J. Amer. Chem. Soc. 76, 2881 (1954).
38. Gerzon, K., E. H. Flynn, M. V. Sigal, Jr., P. F. Wiley and U. C. Quarck: Erythromycin. VIII. Structure of Dihydroerythronolide. J. Amer. Chem. Soc. 78, 6396 (1956).
39. Gerzon, K., R. Monahan, O. Weaver, M. V. Sigal, Jr. and P. F. Wiley: Erythromycin. IX. Degradation Studies of Erythromycin B. J. Amer. Chem. Soc. 78, 6412 (1956).

40. GRUNDY, W. E., J. R. SCHENCK, R. K. CLARK, Jr., M. P. HARGIE, R. K. RICHARDS and J. C. SYLVESTER: A Note on a New Antibiotic. Arch. Biochemistry **28**, 150 (1950).

41. GRUNDY, W. E., A. L. WHITMAN, E. G. RDZOK, E. J. RDZOK, M. E. HANES and J. C. SYLVESTER: Actithiazic Acid. I. Microbiological Studies. Antibiotics and Chemotherapy **2**, 399 (1952).

42. HASBROUCK, R. B. and F. C. GARVEN: The Chemistry of Erythromycin. II. Acid Degradation Products. Antibiotics and Chemotherapy **3**, 1040 (1953).

43. HEINEMANN, B., A. GOUREVITCH, J. LEIN, D. L. JOHNSON, M. A. KAPLAN, D. VANAS and I. R. HOOPER: Etamycin, a New Antibiotic. Antibiotics Annual **2**, 728 (1954–1955).

44. HIDY, P. H., E. B. HODGE, V. V. YOUNG, R. L. HARNED, G. A. BREWER, W. F. PHILLIPS, W. F. RUNGE, H. E. STAVELY, A. POHLAND, H. BOAZ and H. R. SULLIVAN: Structure and Reactions of Cycloserine. J. Amer. Chem. Soc. **77**, 2345 (1955).

45. HINMAN, J. W., E. L. CARON and C. DE BOER: The Isolation and Purification of Amicetin. J. Amer. Chem. Soc. **75**, 5864 (1953).

46. HINMAN, J. W., H. HOEKSEMA, E. L. CARON and W. G. JACKSON: The Partial Structure of Novobiocin (Streptonivicin). II. J. Amer. Chem. Soc. **78**, 1072 (1956).

47. HOCHSTEIN, F. A. and P. P. REGNA: Magnamycin. IV. Mycaminose, an Aminosugar from Magnamycin. J. Amer. Chem. Soc. **77**, 3353 (1955).

48. HOCHSTEIN, F. A., C. R. STEPHENS, L. H. CONOVER, P. P. REGNA, R. PASTERNACK, K. J. BRUNINGS and R. B. WOODWARD: Terramycin. VII. The Structure of Terramycin. J. Amer. Chem. Soc. **74**, 3708 (1952).

49. HOCHSTEIN, F. A., C. R. STEPHENS, L. H. CONOVER, P. P. REGNA, R. PASTERNACK, P. N. GORDON, F. J. PILGRIM, K. J. BRUNINGS and R. B. WOODWARD: The Structure of Terramycin. J. Amer. Chem. Soc. **75**, 5455 (1953).

50. HOEKSEMA, H., E. L. CARON and J. W. HINMAN: Novobiocin. III. The Structure of Novobiocin. J. Amer. Chem. Soc. **78**, 2019 (1956).

51. HOEKSEMA, H., J. L. JOHNSON and J. W. HINMAN: Structural Studies on Streptonivicin, a New Antibiotic. J. Amer. Chem. Soc. **77**, 6710 (1955).

52. HOOPER, I. R., L. C. CHENEY, M. J. CRON, O. B. FARDIG, D. A. JOHNSON, D. L. JOHNSON, F. M. PALERMITI, H. SCHMITZ and W. B. WHEATLEY: Studies on Sarkomycin. Antibiotics and Chemotherapy **5**, 585 (1955).

53. HUTCHINGS, B. L., C. W. WALLER, R. W. BROSCHARD, C. F. WOLF, P. W. FRYTH and J. H. WILLIAMS: Degradation of Aureomycin. V. Aureomycinic Acid. J. Amer. Chem. Soc. **74**, 4980 (1952).

54. HUTCHINGS, B. L., C. W. WALLER, S. GORDON, R. W. BROSCHARD, C. F. WOLF, A. A. GOLDMAN and J. H. WILLIAMS: Degradation of Aureomycin. J. Amer. Chem. Soc. **74**, 3710 (1952).

55. JOHNSON, A. W., A. R. TODD and L. C. VINING: Actinomycin. Part II. Studies on the Chromophoric Grouping. J. Chem. Soc. (London) **1952**, 2672.

56. JULIA, M. et N. JOSEPH: Premières études sur la structure chimique d'un nouvel antibiotique, la congocidine. C. R. hebd. Séances Acad. Sci. **243**, 961 (1956).

57. KACZKA, E. A., C. H. SHUNK, J. W. RICHTER, F. J. WOLF, M. M. GASSER and K. FOLKERS: Novobiocin. III. Cyclonovobiocic Acid, a Methyl Glycoside, and Other Reaction Products. J. Amer. Chem. Soc. **78**, 4125 (1956).

58. KORNFELD, E. C. and R. G. JONES: The Structure of Actidione, an Antibiotic From *Streptomyces griseus*. Science (Washington) **108**, 437 (1948).

59. Kornfeld, E. C., R. G. Jones and T. V. Parke: The Structure and Chemistry of Actidione, an Antibiotic from *Streptomyces griseus*. J. Amer. Chem. Soc. **71**, 150 (1949).

60. Kuehl, F. A., Jr., R. L. Clark, M. N. Bishop, E. H. Flynn and K. Folkers: Streptomyces Antibiotics. XXII. Configuration of Streptose and Streptobiosamine. J. Amer. Chem. Soc. **71**, 1445 (1949).

61. Kuehl, F. A., Jr., E. H. Flynn, F. W. Holly, R. Mozingo and K. Folkers: Streptomyces Antibiotics. V. N-Methyl-*l*-glucosamine from Streptomycin. J. Amer. Chem. Soc. **68**, 536 (1946).

62. Kuehl, F. A., Jr., R. L. Peck, C. E. Hoffhine, Jr. and K. Folkers: Streptomyces Antibiotics. XVIII. Structure of Streptomycin. J. Amer. Chem. Soc. **70**, 2325 (1948).

63. Kuehl, F. A., Jr., F. J. Wolf, N. R. Trenner, R. L. Peck, E. Howe, B. D. Hunnewell, G. Downing, E. Newstead, K. Folkers, R. P. Buhs, I. Putter, R. Ormond, J. E. Lyons and L. Chaiet: *D*-4-Amino-3-isoxazolidone, a new Antibiotic. J. Amer. Chem. Soc. **77**, 2344 (1955).

64. Kushner, S., J. H. Boothe, J. Morton II, J. Petisi and J. H. Williams: Synthesis of Degradation Products of Aureomycin. J. Amer. Chem. Soc. **74**, 3710 (1952).

65. Kushner, S., J. Morton II, J. H. Boothe and J. H. Williams: Synthesis of Degradation Products of Aureomycin. III. J. Amer. Chem. Soc. **75**, 1097 (1953).

66. Leach, B. E., J. H. Ford and A. J. Whiffen: Actidione, an Antibiotic from *Streptomyces Griseus*. J. Amer. Chem. Soc. **69**, 474 (1947).

67. Long, L. M. and H. D. Troutman: Chloroamphenicol (Chloromycetin). VI. A Synthetic Approach. J. Amer. Chem. Soc. **71**, 2469 (1949).

68. McLamore, W. M., W. D. Celmer, V. V. Bogert, F. C. Pennington and I. A. Solomons: The Structure and Synthesis of a New Thiazolidone Antibiotic. J. Amer. Chem. Soc. **74**, 2946 (1952); **75**, 105 (1953).

69. Moore, J. A., J. R. Dice, E. D. Nicolaides, R. D. Westland and E. L. Whittle: Azaserine, Synthetic Studies. I. J. Amer. Chem. Soc. **76**, 2884 (1954).

70. Nicolaides, E. D., R. D. Westland and E. L. Whittle: Azaserine, Synthetic Studies. II. J. Amer. Chem. Soc. **76**, 2887 (1954).

71. Peck, R. L., R. P. Graber, A. Walti, E. W. Peel, C. E. Hoffhine, Jr. and K. Folkers: Streptomyces Antibiotics. IV. Hydrolytic Cleavage of Streptomycin to Streptidine. J. Amer. Chem. Soc. **68**, 29 (1946).

72. Peck, R. L., C. E. Hoffhine, Jr., P. Gale and K. Folkers: Streptomyces Antibiotics. XXI. Linkage of Mannosidostreptobiosamine to Streptidine in Mannosidostreptomycin. J. Amer. Chem. Soc. **70**, 3968 (1948).

73. Peck, R. L., C. E. Hoffhine, Jr., E. W. Peel, R. P. Graber, F. W. Holly, R. Mozingo and K. Folkers: Streptomyces Antibiotics. VII. The Structure of Streptidine. J. Amer. Chem. Soc. **68**, 776 (1946).

74. Porter, R. W.: Streptomycin Engineered into Commercial Production. Chem. Eng. **53**, No. 10, 94, 142 (1946).

75. Rebstock, M. C., H. M. Crooks, Jr., J. Controulis and Q. R. Bartz: Chloroamphenicol (Chloromycetin). IV. Chemical Studies. J. Amer. Chem. Soc. **71**, 2458 (1949).

76. Regna, P. P., F. A. Hochstein, R. L. Wagner, Jr. and R. B. Woodward: Magnamycin. II. Mycarose, an Unusual Branched-chain Desoxysugar from Magnamycin. J. Amer. Chem. Soc. **75**, 4625 (1953).

77. Schenck, J. R. and A. F. De Rose: Actithiazic Acid. II. Isolation and Characterization. Arch. Biochem. Biophys. **40**, 263 (1952).

78. SHEEHAN, J. C., H. G. ZACHAU and W. B. LAWSON: The Structure of Etamycin. J. Amer. Chem. Soc. **79**, 3933 (1957).

79. SHEMYAKIN, M. M., G. A. RAVDEL, Y. S. CHAMAN, Y. B. SHVETSOV and Y. I. VINOGRADOVA: Synthesis of Racemic Sarkomycin. Chem. and Ind. **1957**, 1320.

80. SHUNK, C. H., C. H. STAMMER, E. A. KACZKA, E. WALTON, C. F. SPENCER, A. N. WILSON, J. W. RICHTER, F. W. HOLLY and K. FOLKERS: Novobiocin. II. Structure of Novobiocin. J. Amer. Chem. Soc. **78**, 1770 (1956).

81. SIGAL, M. V., Jr., P. F. WILEY, K. GERZON, E. H. FLYNN, U. C. QUARCK and O. WEAVER: Erythromycin. VI. Degradation Studies. J. Amer. Chem. Soc. **78**, 388 (1956).

82. SOBIN, B. A.: A New Streptomyces Antibiotic. J. Amer. Chem. Soc. **74**, 2947 (1952).

83. SPENCER, C. F., C. H. STAMMER, J. O. RODIN, E. WALTON, F. W. HOLLY and K. FOLKERS: Novobiocin. IV. Synthesis of Dihydronovobiocic Acid and Cyclonovobiocic Acid. J. Amer. Chem. Soc. **78**, 2655 (1956).

84. STAMMER, C. H., A. N. WILSON, F. W. HOLLY and K. FOLKERS: Synthesis of *D*-4-Amino-3-isoxazolidone. J. Amer. Chem. Soc. **77**, 2346 (1955).

85. STAMMER, C. H., A. N. WILSON, C. F. SPENCER, F. W. BACHELOR, F. W. HOLLY and K. FOLKERS: Synthesis of Cycloserine and a Methyl Analogue. J. Amer. Chem. Soc. **79**, 3236 (1957).

86. STAVELY, H. E. and J. FRIED: Streptomycin. IX. The Stepwise Degradation of Mannosidostreptomycin. J. Amer. Chem. Soc. **71**, 135 (1949).

87. STEPHENS, C. R., L. H. CONOVER, F. A. HOCHSTEIN, P. P. REGNA, F. J. PILGRIM, K. J. BRUNINGS and R. B. WOODWARD: Terramycin. VIII. Structure of Aureomycin and Terramycin. J. Amer. Chem. Soc. **74**, 4976 (1952).

88. STEPHENS, C. R., L. H. CONOVER, R. PASTERNACK, F. A. HOCHSTEIN, W. T. MORELAND, P. P. REGNA, F. J. PILGRIM, K. J. BRUNINGS and R. B. WOODWARD: The Structure of Aureomycin. J. Amer. Chem. Soc. **76**, 3568 (1954).

89. STEVENS, C. L., R. J. GASSER, T. K. MUKHERJEE and T. H. HASKELL: The Structure of Amicetin. A New Dimethylaminosugar. J. Amer. Chem. Soc. **78**, 6212 (1956).

90. STEVENS, C. L., B. T. GILLIS, J. C. FRENCH and T. H. HASKELL: The Structure of Elaiomycin, a Tuberculostatic Antibiotic. J. Amer. Chem. Soc. **78**, 3229 (1956).

91. STODOLA, F. H., O. L. SHOTWELL, A. M. BORUD, R. G. BENEDICT and A. C. RILEY, Jr.: Hydroxystreptomycin, a new Antibiotic from *Streptomyces Griseocarneus*. J. Amer. Chem. Soc. **73**, 2290 (1951).

92. STOLL, A., J. RENZ und A. BRACK: Beiträge zur Konstitutionsaufklärung des Nocardamins. Helv. Chim. Acta **34**, 862 (1951).

93. TAMELEN, E. E. VAN and A. D. G. POWELL: The Structure of Netropsin. Chem. and Ind. **1957**, 365.

94. TAMELEN, E. E. VAN, D. M. WHITE, I. C. KOGON and A. D. G. POWELL: Structural Studies on the Antibiotic Netropsin. J. Amer. Chem. Soc. **78**, 2157 (1956).

95. TOKI, K.: Synthesis of *dl*-Sarkomycin(2-Methylenecyclopentanone-3-carboxylic Acid). Bull. Chem. Soc. Japan **30**, 450 (1957).

96. WAKSMAN, S. A. and M. TISHLER: The Chemical Nature of Actinomycin, an Antimicrobial Substance Produced by *Actinomyces antibioticus*. J. Biol. Chem. **142**, 519 (1942).

97. WALLER, C. W., P. W. FRYTH, B. L. HUTCHINGS and J. H. WILLIAMS: Achromycin. The Structure of the Antibiotic Puromycin. I. J. Amer. Chem. Soc. **75**, 2025 (1953).

98. Waller, C. W., B. L. Hutchings, R. W. Broschard, A. A. Goldman, W. J. Stein, C. F. Wolf and J. H. Williams: Degradation of Aureomycin. VII. Aureomycin and Anhydroaureomycin. J. Amer. Chem. Soc. 74, 4981 (1952).

99. Waller, C. W., B. L. Hutchings, A. A. Goldman, C. F. Wolf, R. W. Broschard and J. H. Williams: Degradation of Aureomycin. IV. Des-dimethylaminoaureomycinic Acid. J. Amer. Chem. Soc. 74, 4979 (1952).

100. Waller, C. W., B. L. Hutchings, C. F. Wolf, R. W. Broschard, A. A. Goldman and J. H. Williams: Degradation of Aureomycin. III. 3,4-Di-hydroxy-2,5-dioxocyclopentane-1-carboxamide. J. Amer. Chem. Soc. 74, 4978 (1952).

101. Waller, C. W., B. L. Hutchings, C. F. Wolf, A. A. Goldman, R. W. Broschard and J. H. Williams: Degradation of Aureomycin. VI. Iso-aureomycin and Aureomycin. J. Amer. Chem. Soc. 74, 4981 (1952).

102. Waller, C. W., C. F. Wolf, W. J. Stein and B. L. Hutchings: The Structure of Antibiotic T-1384. J. Amer. Chem. Soc. 79, 1265 (1957).

103. Walton, E., J. O. Rodin, C. H. Stammer, F. W. Holly and K. Folkers: Novobiocin. V. The Configuration of the Aldose Moiety. J. Amer. Chem. Soc. 78, 5454 (1956).

104. Weiss, M. J., J. S. Webb and J. M. Smith, Jr.: The Structure of Antibiotic T-1384. Synthesis of the Degradation Fragments. J. Amer. Chem. Soc. 79, 1266 (1957).

105. Whiffen, A. J., N. Bohonos and R. L. Emerson: The Production of an Antifungal Antibiotic by *Streptomyces griseus*. J. Bacterol. 52, 610 (1946).

106. Wiley, P. F.: Personal communication to the author.

107. Wiley, P. F., K. Gerzon, E. H. Flynn, M. V. Sigal, Jr. and U. C. Quarck: Erythromycin. V. Isolation and Structure of Degradation Products. J. Amer. Chem. Soc. 77, 3677 (1955).

108. — — — — — Erythromycin. IV. Degradation Studies. J. Amer. Chem. Soc. 77, 3676 (1955).

109. Wiley, P. F. and O. Weaver: Erythromycin. The Structure of Cladinose. J. Amer. Chem. Soc. 77, 3422 (1955); 78, 808 (1956).

110. Wolfrom, M. L. and C. W. de Walt: The Configuration of Streptose. J. Amer. Chem. Soc. 70, 3148 (1948).

111. Wolfrom, M. L., S. M. Olin and W. J. Polglase: A Synthesis of Streptidine. J. Amer. Chem. Soc. 72, 1724 (1950).

112. Wolfrom, M. L., A. Thompson and I. R. Hooper: N-Methyl-L-glucosaminic Acid. J. Amer. Chem. Soc. 68, 2343 (1946).

113. Woodward, R. B.: Struktur und Biogenese der Makrolide. Angew. Chem. 69, 50 (1957).

(Received, December 30, 1957.)

Protein Synthesis in Plants.

By **JAMES BONNER**, Pasadena, California.

With 5 Figures.

Acknowledgments. The author is indebted to Dr. PAUL O. P. Ts'o for the electron micrograph of Fig. 1. He is indebted also to Professor G. C. WEBSTER for permission to use previously unpublished data which appear in Figures 3 B and 3 D. For fruitful and stimulating discussions of the subject matter of this review the author is particularly indebted to Dr. PAUL O. P. Ts'o, Dr. RICHARD S. SCHWEET and Professor HOWARD DINTZIS. The writer also wishes to acknowledge the continuing council of Dr. JEROME VINOGRAD. Work on protein synthesis at the California Institute of Technology has been supported in part by the Herman Frasch Foundation and in part by the U.S. Public Health Service.

Abbreviations. Throughout the present paper the following abbreviations will be used: ADP = adenosine diphosphate; AMP = adenosine monophosphate; ATP = adenosine triphosphate; DNA = deoxyribonucleic acid; RNA = ribonucleic acid; and TCA = trichloroacetic acid.

I. Introduction.

These are exciting times in the study of protein synthesis. We are beginning to gain some insight into the mechanism by which amino acids are assembled into the peptide chains of proteins and to achieve some understanding of the way in which information is transferred from nucleus to cytoplasm, there to be used in the construction of the many individual proteins of the cell. The flood of new information and insight concerning protein synthesis has in part come from elegant enzymology and biochemistry applied to the study of the process. In part, and in a large part, however, it has come from a better understanding of the structure of the cell and from improved methods for the separation of cytoplasm into its subcellular constituents. These studies have focused attention upon the *microsomes* as the engines of protein synthesis. This review will therefore first consider the microsomes—their general role in protein synthesis, their origin and their structure. We shall then go on to the biochemistry and enzymology of protein synthesis insofar as we understand it today. Work on animal tissues and on microbial cells, as well as on plant tissues has contributed to our knowledge of protein synthesis, and although this review is particularly directed toward an understandig of plants, it will be helpful to draw on work with other materials. The final Chapters will, however, deal exclusively with plant problems. These will concern protein synthesis in subcellular components other than microsomes as well as protein synthesis on a grosser scale in varied plant organs.

II. Isolation of Microsomes.

When plant tissue is ground at low temperature and in isotonic solution in a homogenizer or other mechanical device, the cell walls are disrupted and the contents of individual cells made available for separation. It has long been known that the cellular debris—cell walls, unground fragments of tissue, etc.—may be removed from the homogenate by filtration or by low speed centrifugation, as $500 \times g$ for a few minutes. The remaining supernatant, the "whole cytoplasm" of WILDMAN and BONNER (93) contains nuclei, chloroplasts, mitochondria, starch grains, and other particulate inclusions as well as soluble materials. Generally speaking, nuclei may be removed from the homogenate by centrifugation at higher speeds, perhaps $1000 \times g$, for a few minutes [JAGENDORF and WILDMAN (34), JAGENDORF (33)]. With tissues which

Fig. 1. Microsomes of pea stems. Freeze dried preparation. Magnification 41,000×. Thorium shadowed, shadowing angle 6.5 : 1. Preparation and photograph through the courtesy of Dr. P. O. P. Ts'o.
(White sphere at top: standard polystyrene sphere, diam., 260 mμ.)

do not contain chloroplasts, such as those of etiolated seedlings, higher centrifugal forces [Ts'o and Sato (79)] can be used to sediment nuclei since the difficult problem of separation of nuclei and chloroplasts is not encountered. Application of higher centrifugal forces, of the order of 1000 to 2000 × g for 10 minutes, results in sedimentation of chloroplasts [Granick (25), Jagendorf and Wildman (34), Clendenning (16)] which may then be purified by repeated centrifugation or by flotation in glycerol-sucrose-water mixtures [Jagendorf (33)]. Application of still higher centrifugal forces of the order of 8,000 to 12,000 × g for 10 to 15 minutes results in sedimentation of mitochondria [Millerd et al. (43), Stafford (71), Goddard and Stafford (23), Hackett (26), Millerd (41)]. The supernatant remaining after removal of these larger particulate bodies contains soluble proteins as well as a further particulate entity, the *microsome*.

Microsomes of plants were first recognized by their appearance in tissue electronmicrographs [Robinson and Brown (57)] as spherical particles of diameter of 200 to 400 Å. They may, however, be sedimented from plant homogenates, previously freed of larger particulate matter by centrifugation at approximately 100,000 × g for 60 minutes. The microsomal particles thus prepared from pea stems or roots are homogeneous in the ultracentrifuge with a sedimentation constant, S, of approximately 80 Svedberg units [Ts'o et al. (77, 78)]. In the electronmicroscope they appear as almost spherical particles of diameter 280 Å as is illustrated in *Fig. 1*. They contain 35 to 40% ribose nucleic acid (RNA), the balance of the microsomal structure being largely protein.

Microsomes have been found in tissues of animals [Slautterback (69), Palade (51), Porter (54), Sjöstrand and Hanzon (67)], and of yeast [Chao and Schachman (12)], and similar but smaller particles in bacteria [Schachman et al. (59)]. The microsomal particles of liver are associated with lipoprotein membranes of the endoplasmic reticulum from which they may be removed by treatment with desoxycholate [Littlefield et al. (37)]. The purified microsomal particles of animal tissues are rich in RNA and resemble plant microsomes in size and shape. Microsomes of yeast and of certain animal tissues, as ascites tumor [Littlefield and Keller (36)] are not clearly associated with an endoplasmic

Table 1. Distribution of RNA among the Subcellular Components of Etiolated Pea Stems. [Ts'o and Sato (79).]

Component	μgm. RNA/gm. fresh weight of tissue	% of total RNA	RNA/protein ratio in component	μgm. DNA/gm. fresh weight of tissue	% of total cell protein in fraction
Nuclei	70	14.3	0.125	30	25
Mitochondria	101	21.2	0.278	2*	16
Microsomes	294	61.8	0.62	0	19
Soluble (supernatant) .	13	2.8	0.012	0	40

* This is taken to measure only the contamination of the mitochondrial fraction by nuclear fragments.

reticulum and may be obtained directly by centrifugation of an appropriate homogenate, as with pea stem tissue.

A striking fact concerning the microsomes is that they contain within them a large proportion of all of the RNA of the cell. The distribution of cellular RNA of pea stem tissue is given in *Table 1*. Fifty to seventy per cent of all of the cellular RNA is recoverable as purified microsomal particles. Of the remainder, the bulk is present in the nucleus and in the mitochondria. A smaller amount is present as soluble, non-microsomal, RNA.

III. Structure of the Microsomes.

The availability of relatively homogeneous microsomal preparations [CHAO and SCHACHMAN (*12*), Ts'o et al. (*77*)] has made it possible to approach the problem of the internal structure of these particles. It has been already noted that microsomes from a variety of sources are remarkably similar and are all characterized by sedimentation constants of 78–80 S, RNA contents of ca. 40% and, as seen in the electron microscope, shapes of oblate spheroids of major diameter ca. 280 Å. The physical characteristics of the intensively studied microsomes of pea seedlings and of yeast are assembled in *Table 2*. The microsomal molecular weight calculated on the basis of a spherical model with 40% hydration is 4 to 4.5 × 10^6 and the total molecular weight of all RNA contained within the particle, correspondingly, of the order of 1.6 to 1.8 × 10^6.

The microsome consists of *subunits*, and these subunits are knitted together by magnesium ions [CHAO (*11*), Ts'o et al. (*78*)]. Ts'o et al.

Table 2. Physical and Chemical Characteristics of Plant Microsomes. Pea stem microsomes after Ts'o et al. (*77*), and yeast microsomes after CHAO and SCHACHMAN (*12*).

Quantity	Yeast	Pea stem
Sedimentation constant, S_{20}	80 S	80 S
Partial specific volume	.67	.67–.68
Intrinsic viscosity....................	.05	.11
RNA/Protein	.4–.44	.39–.42
Diameter (electron microscope)........	240 Å	280 Å (diameter) 180 Å (height)
Shape (electron microscope)	spheroids	oblate spheroids
Molar particle weight (gm.)...........	4.1 × 10^6	4–4.5 × 10^6
Ions required for binding subunits in particle	Mg, Ca	Mg, Ca
Dissociation products after removal of divalent ions: sedimentation constants	60 S, 40 S, 27 S	60 S, 40 S, 27 S

have shown by direct analysis that in fact microsomes contain bound magnesium and calcium in the proportion of 6 to 1. The two ions together are present in sufficient quantity to combine with about half of the phosphate groups of the microsomal RNA. That they are in fact combined with the RNA of the particle rather than with the protein is indicated qualitatively by the fact that magnesium goes with the nucleotide fraction when the particles are hydrolized by the enzyme RNAase.

The magnesium of the particle may be removed by exchange for other cations in appropriately concentrated salt solutions or by chelation with versene. If about one-half of the microsomal magnesium is removed, the original 78–80 S particles cleave reversibly to two new particles of sedimentation constants 60 S and 40 S, respectively. These microsomal fragments, both of which contain RNA and protein in the same proportion as the original particles, are produced in the number ratio 1 : 1. The 60 S particle corresponds in mass to approximately two-thirds, the 40 S particle to one-third, of the 78–80 S microsome. It would appear then that partial removal of microsomal magnesium results in cleavage of one original particle to yield a two-thirds fragment and a one-third fragment.

Further removal of magnesium results in further degradation of the microsome, to yield particles of 40 S and 27 S. The 27 S particles, mass one-sixth that of the original microsome, like the 40 S particles contain RNA and protein in the proportion of the original microsome. For each original particle which is degraded there appear two 40 S and two 27 S fragments.

It is apparent then that the microsome is made up of *ribonucleoprotein units*, the smallest of which is not more than one-sixth of the original particle. The 27 S or one-sixth fragment contains one-sixth of the RNA of the original particle and this places an upper limit of ca. 280,000 ($1/6 \times 1.6 \times 10^6$) on the molecular weight of the microsomal RNA. Direct determination of the molecular weight of microsomal RNA (liver, freed of protein by detergent) has yielded a value of 250,000 [HALL and DOTY (27)]. The RNA chain in each 27 S microsomal subunit is then less than 10^3 nucleotides in length. This in turn may well set the upper limit on the amount of information which the microsome can convey.

IV. Role of the Microsomes in Protein Synthesis.

Attention was first focused on the microsome in connection with protein synthesis because of the fact that labeled amino acid supplied to living tissue or organisms appears most rapidly in the microsomal

Table 3. Distribution of Labeled Amino Acid Among Cellular Components in Short Time Incorporation Experiments. Excised bean hypocotyl data after WEBSTER (87). Excised tobacco leaf experiment after STEPHENSON et al. (72). Tissue incubated in labeled metabolite and then fractionated.

Component	Specific activity relative to specific activity of soluble proteins	
	Bean hypocotyl, glutamate C^{14}, 1 hr.	Tobacco leaf, leucine C^{14}, 12 min.
Soluble proteins ...	1.0	1.0
Nuclei............	0.48	—
Chloroplasts.......	—	0.80
Mitochondria......	0.14	1.88
Microsomes........	3.82	3.40

fraction. This fact was first noted by BORSOOK et al. (6) with liver microsomes and has subsequently been found repeatedly with a variety of tissues. *Table 3* gives the distribution of activity among the sub-

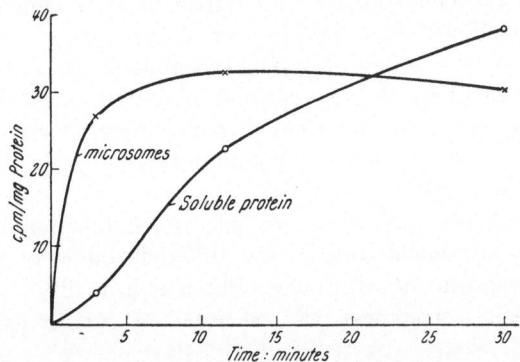

Fig. 2. Incorporation of C^{14}-valine into proteins of excised whole cells of mouse ascites tumor. After incubation the tissues were separated into subcellular components. According to LITTLEFIELD and KELLER (36). [From: J. Biol. Chem. 224, 13 (1957).]

cellular components after short term incubation of two different plant tissues with a C^{14}-labeled amino acid. It is clear that the microsomes form the most rapidly labeled fraction of the cell. With time, however, the pattern of labeling changes. This has been done more elegantly with animal tissues than with plants, and *Fig. 2* presents data taken from the work of LITTLEFIELD and KELLER (36) with intact ascites tumor tissue. Although the microsomes become rapidly labeled, they approach a plateau after a relatively short period. The soluble proteins of the cell, on the other hand, become labeled more slowly but rise ultimately to a specific activity much higher than that of the microsomes.

The level of labeling of microsomes in the steady state (Fig. 2) is very low: of the order of 0.1–1% of the protein of the particle becomes labeled [LITTLEFIELD and KELLER (*36*), LITTLEFIELD et al. (*37*), WEBSTER (*87*)]. Furthermore, the incorporation of labeled amino acid into the microsomal fraction is transitory. Labeled amino acid may be again washed out by supplying the tissue, after a short incorporation period, with an excess of unlabeled amino acid [LITTLEFIELD et al. (*37*)]. This is not true of the proteins of the soluble cytoplasmic fraction or of the reticulum. The kinetic evidence may then be interpreted by, and is in agreement with, the hypothesis that the microsomes assemble amino acids into protein molecules which are then shed from the microsomes to appear as soluble proteins. The growing protein would appear to constitute not more than 1% of the microsomal protein; that is, the microsomal machine is large compared to its product.

A second and more qualitative observation relating microsomes to protein synthesis is the fact that ability to synthesize protein may be greatly reduced by brief treatment of living tissue with ribonuclease [BRACHET (*7*)]. Such treatment removes much of the cellular RNA and would be expected to destroy the integrity of the microsomes since these are ribonuclease sensitive [Ts'o et al. (*77*)].

Finally, that microsomes do in fact assemble amino acids to protein, is directly indicated by the experiments of CAMPBELL et al. (*10*). They found that appropriate microsomal preparations produce on incubation with amino acids protein (serum albumin in this case) which is detectable by serological methods.

The picture developed above of microsome function is a working hypothesis, in agreement with many different kinds of experimental findings, in disagreement with none. But it is in reality only a picture, useful because its correctness may be tested by further experiments.

V. Biology of the Microsomes.

1. Nuclear Microsomes.

Let us now leave for the moment the function of microsomes and ask ourselves about the origin of these ribonucleoprotein particles. Where in the cell do microsomes arise? The first pertinent fact is that electron microscopy reveals objects within the nucleus which appear to resemble microsomes in size and shape [DE ROBERTIS (*20*)]. They are in fact concentrated in the nucleolus, a spherical RNA-rich body found in the nuclei of many types of cells. In addition, microsome-like particles have been isolated directly from the nucleus by OSAWA et al. (*50*) and by Ts'o and SATO (*79*). The work of OSAWA et al. was done with the thymus gland. It was found, in agreement with ALLFREY and MIRSKY (*1*), that the thymus nucleus contains two kinds of RNA, viz. that associated with protein and readily extractable into neutral phosphate buffer, and that found in the residue after further extraction with $1\,M$ sodium chloride. This latter fraction may correspond to the nucleolar RNA [ALLFREY and MIRSKY (1)]. The readily soluble ribonucleoprotein has been found by OSAWA et al. (50) to be identical with cytoplasmic microsomes in respect to precipitation properties, electrophoretic behavior, RNA content, and base ratio of the RNA. The work of Ts'o and SATO (79) was done with nuclei isolated from pea stems. It was found that such nuclei contain 10–15% of their RNA in the form of ribonucleoprotein, identical with cytoplasmic microsomes in respect to RNA content and sedimentation constant. These particles are not removed from the nuclei by washing but are released by homogenization and by freezing and thawing.

2. Origin of Microsomes.

It is apparent then that microsomes are contained within the nucleus. The next question is: are they in fact made within the nucleus? That they are synthesized within the nucleus is suggested by the vast mass of work on the kinetics of labeling of RNA. Thus, for example, cells which contain no nucleus incorporate appropriate precursors into RNA sluggishly. This is true of reticulocytes [SHIMURA and BORSOOK (63)], of enucleated amoebae (ŠKREB-GUILCHER (68)], and to a lesser extent of *Acetabularia* [BRACHET et al. (9)]. Cells or tissues which contain nuclei, on the contrary, rapidly incorporate into RNA precursors such as P^{32}-labeled orthophosphate or C^{14}-labeled nucleosides. Over short time periods the RNA of the nucleus may possess 50 times the specific activity of the cytoplasmic RNA [JEENER and SZAFARZ (35)]. This is true for example of pea stem tissue [SATO et al. (58a)]. That the RNA of the microsomes of the nucleus is actually of higher specific activity than the RNA of the cytoplasmic microsomes after a short time labeling period, has been shown directly only for the case of the thymus [OSAWA et al. (50)]. In this experiment P^{32}-labeled phosphate was supplied to an intact animal. After a period of three hours the animal was sacrificed, the thymus gland removed, and nuclear and cytoplasmic microsomes, respectively, isolated. The RNA of the nuclear microsomes was found to be approximately twice as high in specific activity as the microsomes of the cytoplasm. The relation is, then, that expected for a precursor-product relationship.

It is of interest to consider whether the protein which is associated with microsomes is, like the ribonucleic acid, synthesized in the nucleus or whether the RNA alone is there synthesized and the RNA molecules clothed with protein only after escape through the nuclear membrane. That the protein of the microsomal particles, like the RNA, is synthesized in the nucleus is indicated indirectly by the work of ALLFREY, MIRSKY

and OSAWA (2). These important experiments, which will be returned
to below, have shown not only that isolated nuclei synthesize protein
rapidly but, in addition, that a portion of the protein so synthesized
is associated with RNA as ribonucleoprotein. Such ribonucleoprotein,
"Fraction I" of ALLFREY et al. (2) has been identified by OSAWA et
al. (50) and by Ts'o and SATO (79) as microsomal as noted above. These
two observations then support the view that both RNA and protein
of the microsomal structure are synthesized within the nucleus itself.

3. Transfer of Microsomal RNA.

Let us now consider the evidence which bears upon the question
of whether or not microsomes are able to escape from the nucleus to
the cytoplasm. One pertinent observation is that of WATSON (82) who
has found that the nuclear membrane has holes in it. These holes, of
diameter approximately 500 Å, are sufficiently large to make it possible
in principle for microsomes to escape through them. Direct demonstration
of the escape of microsomes from nucleus to cytoplasm has not yet been
made. It may, however, be inferred from the elegant work of GOLDSTEIN
and PLAUT (24) on the escape of nuclear RNA to the cytoplasm of
amoebae. In these experiments amoebae were supplied with P^{32}-labeled
orthophosphate. Their nuclei became highly labeled. The labeled nuclei
thus obtained were excised and transplated to the cytoplasm of unlabeled
amoebae. Over a period of hours the P^{32}-containing material of the
labeled donor nucleus escaped to the cytoplasm of the previously unlabeled
receptor amoeba. The material contained in the labeled donor nucleus
as well as that which escaped to the receptor cytoplasm were shown
to be RNA itself rather than precursors of RNA or other low-molecular
weight compounds. It was further shown that the RNA which escapes
from the labeled donor nucleus does not enter into a second unlabeled
nucleus present in the same receptor cytoplasm, that is, transfer of
RNA from nucleus to cytoplasm is apparently a one-way process.

A second elegant experiment with amoebae is that of BRACHET (8)
in which amoebae are treated with ribonuclease. This results in depletion
of both cytoplasmic and nuclear RNA. The RNAase destroys not only
soluble RNA but also the RNA combined in the microsomal structure
[Ts'o et al. (77)]. The RNAase-treated amoebae of BRACHET were
then removed to fresh medium. It was found that RNA reappears first
in the nucleus and only subsequently in the cytoplasm. The kinetics
of the reappearance of RNA is thus also in agreement with the hypothesis
that synthesis of RNA occurs in the nucleus and is subsequently
transferred to the cytoplasm. Similar kinetic experiments with other
systems such as that of McMASTER and TAYLOR (40) with *Drosophila*
salivary gland chromosomes lead to the same conclusion.

4. Synthesis of the Microsomal Components.

The facts adduced above suggest, then, that microsomes may very well be manufactured within the nucleus. This in turn leads us to the question of how the components of the microsome are formed. This problem has been attacked in a straightforward and beautiful way by ALLFREY, MIRSKY and OSAWA (2) using isolated thymus nuclei. The facts adduced by ALLFREY et al. are numerous and complex but can be summarized as follows:

(a) Isolated nuclei synthesize both RNA and protein and these in a ratio of roughly 1 : 1. About two-thirds of the RNA synthesized in the isolated thymus nucleus over short time periods is in the form of the "Fraction I" of ALLFREY et al. (2) which, as discussed above, we may now tentatively identify with microsomes [OSAWA et al. (50)].

(b) Synthesis of protein in the nucleus is abolished by pretreatment with DNAase. Such treatment also abolishes the ability of the isolated nucleus to synthesize RNA. And in addition, DNAase treatment prevents the formation of nuclear microsomes ("Fraction I"). We may, therefore, conclude that microsomal synthesis is governed in some way by DNA of the nucleus. Interestingly enough, the synthesis of RNA or of microsomal RNA is not affected by pretreatment of the nucleus with ribonuclease. The synthesis of microsomes does not therefore appear to depend upon microsomes themselves.

(c) The major protein synthesis within the nucleus is that which leads to the formation of ribonucleoprotein, "Fraction I", the nuclear microsomes. The protein formed is evidently that which clothes the RNA of the microsomes. Little is known concerning the biochemistry of formation of the microsomal protein. It is evident, however, that RNA synthesis and protein synthesis go hand in hand in the nucleus. Thus, the synthesis of microsomal protein in the nucleus is inhibited by inhibitors of RNA synthesis such as the purine nucleoside antagonist 5,6-dichloro-β-D-ribofuranosyl-benzimidazole (2). In this fact one may for the first time obtain a feeling for the reason why RNA synthesis and protein synthesis have been so long regarded as being somehow related and geared to each other.

The synthesis of microsomal protein in the nucleus insofar as we understand it from the work of ALLFREY, MIRSKY and OSAWA (2) appears to differ in principle from the synthesis of protein by microsomes in two respects. Thus, the protein synthesis by microsomes is not dependent upon DNA and is not inhibited by DNAase (desoxyribonuclease). It is inhibited by RNAase, that is, it depends upon the integrity of the microsome. And, in addition, the synthesis of protein by microsomes, as we shall see below, does not depend upon, and is not in any as yet demonstrated way dependent on, the simultaneous synthesis of RNA. We can therefore begin to draw a distinction between the synthesis of microsomal protein in the nucleus on the one hand and the synthesis of protein by microsomes in the cytoplasm on the other. This distinction will become sharper and clearer below.

Among the facts which might be brought out as indicating the distinction between microsomal structural protein and synthesis of protein by microsomes are differences in amino acid composition. Microsomal structural protein possesses its own amino acid composition which is characteristic of the microsome and different from that of the protein being formed. This has been shown by Shimura et al. (64) to be the case for the microsomes of the silk gland as compared with that of the silk which they produce. It is also true for the protein of microsomes of the reticulocytes as contrasted with that of globin. And finally, interestingly enough, the amino acid composition of microsomal protein in the three cases in which it has been determined are remarkably similar as between different species and as between plants and animals (Table 4). All are characterized by a high content of the basic amino acids lysine and arginine.

Table 4. Amino Acid Composition of Microsomal Protein from Varied Sources. Rabbit reticulocytes and pea seedlings after Ts'o et al. (76); guinea pig liver after "Fraction B" of Simkin and Work (65, 66).

Amino acid	gms. amino acid per 100 gms. of dry protein		
	Pea stem	Rabbit reticulocytes	Guinea pig liver
Alanine	5.4	5.4	5.3
Aspartate	9.6	8.8	9.5
Arginine..................	9.2	11.8	8.3
Cystine..................	0.3	1.1	—
Glutamate................	10.7	11.5	12.0
Glycine	8.3	7.1	4.7
Histidine	2.9	2.8	2.5
Isoleucine	6.4	5.7	4.2
Leucine	8.2	8.7	10.2
Lysine	12.2	12.7	9.3
Methionine	2.0	2.0	2.0
Phenylalanine............	4.8	4.4	5.6
Proline..................	5.2	4.7	7.5
Serine...................	2.5	1.8	3.9
Threonine	4.9	4.5	5.2
Tryptophan..............	1.5	1,2	—
Tyrosine	7.0	6.5	4.0
Valine	7.6	7.2	5.8

5. Kinetics of Microsomal Labeling.

Finally, on the basis of the picture drawn above, we can begin to interpret the various types of kinetics, obtained in labeling experiments in which C14-labeled amino acids are supplied to various kinds of cells or tissues, and the specific activity of the protein of various components

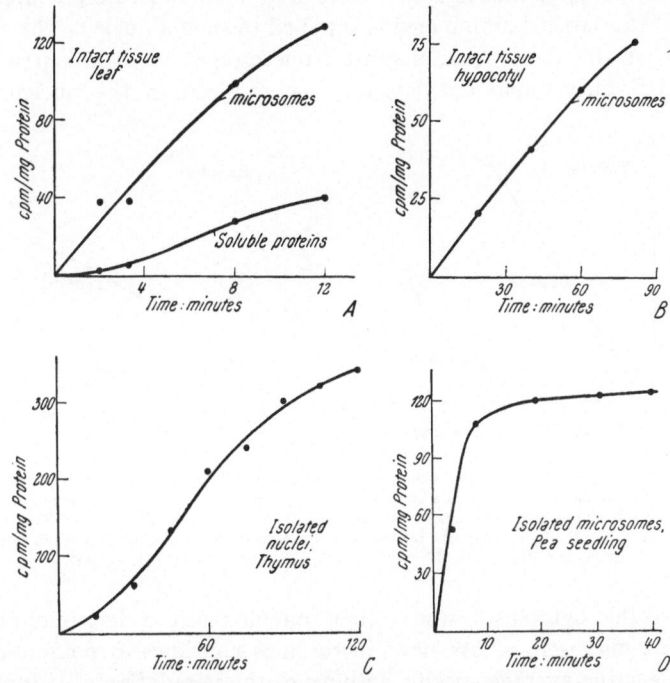

Fig. 3. Kinetics of incorporation of C14-labeled amino acids into various cellular components under various circumstances.

3 A. Incorporation of C14-leucine into microsomal and soluble proteins of excised tobacco leaf discs. After STEPHENSON et al. (72). [From: Arch. Biochem. Biophys. 65, 194 (1956).]

3 B. Incorporation of C14-glutamate into microsomal protein of excised bean hypocotyl tissue. Unpublished data of G. C. WEBSTER.

3 C. Incorporation of C14-alanine into protein of isolated thymus nuclei. After ALLFREY, MIRSKY and OSAWA (2). [From: J. Gen. Physiol. 40, 451 (1957).]

3 D. Incorporation of C14-leucine into protein of isolated microsomes in vitro. System of WEBSTER (88). Unpublished data of G. C. WEBSTER.

of the cell or tissue measured at successive time intervals. In many in vivo experiments, such as those *Fig. 3 A* and *3 B*, microsomal protein becomes more and more highly labeled with time over considerable time periods. In the case of in vitro experiments in which microsomes are incubated in an appropriate mixture and allowed to incorporate labeled amino acid, the kinetics are very different. The microsomes rapidly approach a constant level of activity at which they remain for extended periods of time *(Fig. 3 D)*. In the synthesis of nuclear micro-

somal protein in vitro in the thymus nuclei experiments of ALLFREY
et al. (2) the specific activity of nuclear protein continues to increase
over long time periods *(Fig. 3 C)*.

The interpretation which the present writer gives to these several
types of kinetics is the following. In the nucleus, microsomal structural
protein is being manufactured. More and more microsomes are being
formed. If a labeled amino acid is supplied to such a nucleus, the specific
activity of the newly manufactured microsomes is high. In a living
tissue, the microsomal entities, once synthesized in the nucleus, leak

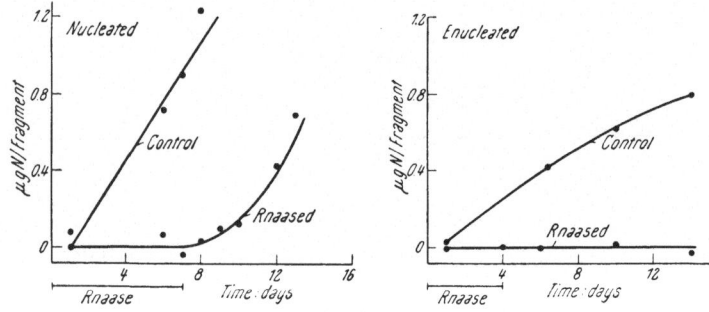

Fig. 4. Synthesis of protein in nucleated and enucleated fragments of *Acetabularia*. In these larger elongated
cells the nucleus remains at one end and enucleated (nucleus-free) fragments may be produced by merely
snipping off the nucleus-containing portion. In each case the upper curve is that for non-RNAase treated
fragments, the lower that for fragments treated with RNAase during the time period indicated. After
STICH and PLAUT (73). [From: J. Biophys. Biochem. Cytol. 4, 119 (1958).]

out into the cytoplasm where they mingle with a large number of
unlabeled microsomes. As new microsomes continue to pour out from
the nucleus the average specific activity of the population of cytoplasmic
microsomes gradually increases and continues to increase over long
periods of time. This is evidently the state of affairs in the tobacco
leaf amino acid incorporation experiment of STEVENSON et al. (Fig. 3 A).
In experiments on incorporation of amino acid into microsomes in vitro
(Fig. 3 D) microsomal structural protein is not being synthesized, since
nuclei are absent. Only the protein which is being synthesized upon
the microsomal structure becomes radioactive. The specific activity
of the microsomal particle quickly reaches a maximum level—a level
which will be determined by the number of protein molecules which are
simultaneously under assembly by each microsomal particle. This is
evidently small since as noted earlier only a few tenths to one per cent
of microsomal amino acid becomes labeled in the steady state condition.
To understand the kinetics of a labeling experiment we must know
whether new microsomes are being synthesized in the system or whether
the labeling is due merely to the synthesis of protein by microsomes.

One last elegant experiment brings together the concepts developed above of the interaction of nucleus and cytoplasm: synthesis of RNA by the nucleus, and synthesis of protein in the cytoplasm by a RNA-containing system. This experiment is that of STICH and PLAUT (73) with nucleated and enucleated fragments of the green alga, *Acetabularia*. Nucleated fragments of this creature continue to synthesize protein at a steady rate over a period of 14 days. Enucleated fragments also synthesize protein, although the rate gradually decreases over a similar time period *(Fig. 4)*. Nucleated fragments were treated with RNAase for a period of 7 days. During this period synthesis of protein was reduced to zero. Within two days after the removal of the fragments from RNAase, however, protein synthesis returned to a steady rate equal to that of non-RNAase treated fragments. RNAase-treated enucleated fragments, on the contrary, did not recover ability to produce protein after removal from RNAase solution. Clearly, the ability of *Acetabularia* to synthesize protein depends upon an RNAase-sensitive material. In the presence of the nucleus the RNA, essential to protein synthesis, may be regained by the plant. In the enucleated fragments, on the contrary, the corresponding RNA of the cytoplasm, once removed, cannot be again restored, and protein synthesis is forever abolished.

VI. Protein Synthesis as a Catenary Sequence.

The events of protein synthesis as now revealed by work on animal tissues, plant tissues, and microorganisms, can be visualized as a catenary sequence of reactions. In this sequence, amino acids are first converted to acyl-AMP-amino acid complexes, a step known as "activation", and first observed for amino acids by HOAGLAND (28). In a second step the acyl-activated amino acid is transferred to an acceptor, viz. soluble cytoplasmic RNA. This RNA is different from that of the microsomal RNA which is inactive as a direct acceptor of acyl-activated amino acids. The initial observation that an RNA acts as an acceptor of acyl-activated amino acids was made by HOLLEY (32) and actual separation of the amino acid-RNA complex was carried out by HOAGLAND et al. (30). In a further step, amino acid is transferred from soluble RNA to micro-some and there linked in peptide form as has been shown directly by HOAGLAND et al. (30). Events on the microsomal surface are still obscure. We might visualize, however, the possibility that the acyl-activated amino acids upon transfer to the microsomal structure, participate in the formation of a growing peptide chain and that when this chain is finished, the protein thus formed is in some unknown manner expelled from the microsome which is then ready to carry out its task anew.

We will now take up, one by one, these successive steps in protein biosynthesis.

1. Amino Acid Activation.

That amino acid activation is a prerequisite to protein synthesis is evident from thermodynamic considerations. The ΔF of formation of the peptide bond between alanine and glycine, for example, is 4,130 calories per mol at 37° (5). Although various enzymatic mechanisms for the activation of amino acids are known (see below), it appears probable that those concerned with protein synthesis are of the type illustrated in reaction (A):

$$\text{Enzyme} + \text{ATP} + \text{amino acid} \rightleftarrows \text{enzyme-AMP-amino}$$
$$\text{acid} + \text{pyrophosphate} \qquad\qquad \text{(A)}$$

This reaction, in which enzyme, ATP and amino acid participate to form a tripartite complex with the liberation of inorganic pyrophosphate, is similar in principle to that involved in acetate and other fatty acid activation [Berg (3)]. In this complex the carboxyl group of the amino acid is linked to the 5'-phosphate of AMP as a mixed anhydride [De Moss et al. (19)] and oxygen is transferred from the amino acid carboxyl group to AMP [Hoagland et al. (31)]. The reaction may be studied and followed quantitatively by virtue of the fortunate circumstance that the enzyme-AMP-amino acid complex reacts spontaneously and non-enzymatically with hydroxylamine to form the corresponding amino acid hydroxamate which may then be estimated colorimetrically. Alternatively, the reaction may be followed by making use of an exchange reaction. Enzyme, ATP, and amino acid are incubated in the presence of P^{32}-labeled pyrophosphate. Although no net reaction occurs, reversal of reaction (A) results in incorporation of the labeled pyrophosphate into ATP. The reaction is allowed to run for a specified period, ATP is recovered and its radioactivity determined. This is the method of choice for the study of amino acid activation.

Hoagland's initial observation (28) of amino acid activation by the mechanism described above was made with preparations from rat liver. Similar enzymes have been found in a variety of animal tissues, in yeast and a variety of microorganisms, and in higher plants [Webster (88), Clark (15)]. That there exist individual amino acid activating enzymes, each specific to one amino acid, is indicated by the work of Davie et al. (18) who have isolated an enzyme specific for the activation of tryptophan, and by the work of Schweet et al. (60, 62) who have isolated in pure form an enzyme specific to the activation of tyrosine. The most extensive study on amino acid activation by plant systems is that of Clark (15) who has demonstrated the presence of an activating enzyme in pea epicotyls, *Avena* coleoptiles, asparagus tips, and leaves of rye, grass, tobacco and spinach. In the case of the spinach enzyme, studied more extensively, activity was found toward seventeen of the twenty amino acids which occur in proteins. These effects are evidently due to individual amino acid activating enzymes, since they were partially separable from one another. In addition, activation of one amino acid by a crude preparation is in general additive to activation of a second or more amino acids by the same preparation; that is, individual amino acids do not compete with each other for an individual activation site.

The amino acid activating enzymes, both in plant [CLARK (15)] and animal [HOAGLAND et al. (29)] tissues, are present in the soluble cytoplasmic fraction, which remains after removal of all particulate matter including the microsomes. The enzymes may be precipitated from the soluble supernatant fraction by adjusting the pH of the supernatant to 5, in which case the precipitate carries not only the amino acid activating activity but also soluble RNA [HOAGLAND et al. (29)]. The mixture, which is soluble in buffer at pH 7.5 to 8, is known as "pH 5 enzyme", according to the original nomenclature of HOAGLAND (28). Activating enzyme preparations obtained from acetone powders of plant material are on the other hand free of soluble RNA activity [CLARK (15)].

WEBSTER (87) has demonstrated the presence of amino acid activating activity in crude (once sedimented) preparations of microsomes. That this property is only loosely associated with microsomes is shown by the fact that it is removed by repeated washing of the particles [CLARK (15)].

2. Transfer to Soluble RNA.

Amino acid activation as considered above involves the formation of enzyme-AMP-acylamino acid complexes. Pyrophosphate is released, AMP is not. It has been shown above that the acyl-activated amino acid can be removed from the enzyme-AMP complex non-enzymatically by hydroxylamine to form the amino acid hydroxamate. In nature, however, an RNA acts as the acceptor of the acyl-activated amino acid. This reaction was first studied by HOLLEY (32) who showed that in the presence of an RNAase-sensitive acceptor, amino acid, amino acid activating enzyme, and ATP, labeled AMP can be exchanged into ATP, according to the formulation of the sum of reactions (A) plus (B):

$$\text{Enzyme-AMP-amino acid} + \text{acceptor RNA} \rightleftharpoons$$
$$\rightleftharpoons \text{acceptor RNA-amino acid} + \text{enzyme} + \text{AMP} \qquad \text{(B)}$$

$$\text{Enzyme} + \text{amino acid} + \text{ATP} + \text{acceptor RNA} \rightleftharpoons$$
$$\rightleftharpoons \text{Enzyme} + \text{amino acid-acceptor RNA} + \text{AMP} + \text{pyrophosphate} \quad \text{(A+B)}$$

The study of the AMP exchange reaction is made difficult because of the circumstance that crude enzyme preparations also contain adenyl-kinase that catalyzes reaction (C), which reversibly converts ADP

$$2\,\text{ADP} \rightleftharpoons \text{ATP} + \text{AMP} \qquad \text{(C)}$$

to ATP + AMP. Label initially supplied as AMP may therefore be recovered in the form of ATP in the presence of adenyl-kinase. Hence, it is necessary to use for the study of reaction (B) enzyme freed of adenyl-kinase. It is necessary, too, to demonstrate that the exchange of AMP into ATP is amino acid dependent. This was done by HOLLEY (32).

A further way to study exchange of activated amino acid from amino acid activating enzyme to an RNAase sensitive material is that originally suggested by HOAGLAND et al. (29) and by OGATA and NOHARA (47). An amino acid activating enzyme system is permitted to react with free amino acid and ATP in the presence of soluble RNA. At the expiration of the reaction period the reaction is stopped with 6% TCA and the precipitate, consisting of protein and nucleic acid, washed with perchloric acid in alcohol. A portion of the labeled amino acid is found to be precipitated by the TCA, bound in some way to the insoluble matter. Such incorporation of amino acid into insoluble material by the amino acid activating enzyme-RNA system is dependent upon the presence of the RNA and is destroyed if the RNA is destroyed by RNAase. The results of an experiment of this type taken from CLARK (15) are given in *Table 5*. In this experiment the amino acid activating enzyme-soluble RNA system was prepared from the soluble supernatant of spinach homogenate after removal of all particles including microsomes.

Table 5. Transfer of C^{14}-Leucine into an RNAase Sensitive Acceptor in the Presence of Amino Acid Activating Enzyme. Transfer measured by appearance of TCA-precipitable C^{14}-leucine. The complete system contains amino acid activating enzyme from spinach (containing soluble RNA), ATP, creatine phosphate, creatine-ATP transphorylase, $MgCl_2$ buffer and KF (37°). After CLARK (15).

System and time	Counts per min. (cpm) per mg. transformed to TCA-precipitable form
Complete, 0 min.	1.5
Complete, 15 min.	15.0
Complete, 15 min. but preincubated with RNAase ..	0.3

The characterization of the soluble RNA which acts as acceptor of activated amino acid has been carried out more completely by HOAGLAND et al. (30), by SCHWEET et al. (61), and by OGATA and NOHARA (47) with systems of animal origin. In all cases it has been shown that the activated amino acid is transferred to RNA and that the resulting complex may be isolated as an acyl amino acid-RNA complex, free or essentially free of protein. The work of SCHWEET et al. (61) has further indicated that the RNA entities responsible for accepting different amino acids are different from one another. It may be presumed therefore that at least twenty different RNAs are involved in acceptance of the twenty different individual amino acids concerned in protein synthesis.

3. The Microsome as Acceptor of Amino Acids.

It has been shown by HOAGLAND et al. (*30*) in the case of a system prepared from liver that the amino acid-RNA complex described above is capable of transferring labeled amino acid to microsome and incorporating the amino acid into peptide linkage. This transfer may be done with one amino acid and in the absence of others, that is, the simultaneous presence of all twenty amino acids is not necessary for incorporation into microsomes. The direct demonstration of transfer of amino acid from amino acid soluble RNA complex to microsome has not yet been made with a plant system. We do know, however, that plant microsomal preparations possess the ability to incorporate amino acids, provided that the system includes amino acid activators as well as AMP-ATP exchange activity (activity for transfer of activated amino acid to RNA). This has been done extensively by WEBSTER (*88*) with microsomal particles prepared from etiolated pea seedlings. The once washed particles contain not only microsomal material but also amino acid activating activity which as noted above is present as an impurity and can be removed by further washing of the preparation. The crude microsomal preparation also possesses activity for exchange of AMP into ATP in the presence of added amino acid. This exchange is clearly dependent on RNA since it is decreased or abolished by ribonuclease.

The particulate system of WEBSTER (*88*) is capable of incorporating glutamate into protein, the time course resembling that of Fig. 3 D (p. 151). Ability to incorporate glutamate into protein is dependent upon the presence of ATP and magnesium ions, is increased by the addition to the system of further amino acid activating enzyme preparation (including

Table 6. Incorporation of C^{14}-Labeled Glutamate into Protein by Microsomal Particles of Yeast or Pea Stems. After WEBSTER (*88*). Incorporation for 30 min. at 38°. Complete system contains buffer, 0.1 μmol. ATP, 0.01 μmol. $MgSO_4$, 5 μmol. creatine phosphate, 0.1 mg. crystalline creatine-ATP transphorylase, 0.1 μmol. guanosine triphosphate, 10 mg. pH 5 enzyme, 1 μmol. C^{14}-glutamate (300,000 counts per min.) and 3 mg. microsomal particles (unwashed).

Components of system	Glutamate incorporation cpm/mg. protein	
	Pea stem	Yeast
Complete	239	155
Minus microsomes	5	15
Minus pH 5 enzyme	203	53
Minus ATP	12	9
Minus guanosine triphosphate..............	101	128

soluble RNA) and requires the presence of guanosine triphosphate. These relations are shown in *Table 6*. Disruption of the particles by acetone treatment, by versene, or by detergents such as sodium dodecylsulfate abolishes or essentially abolishes the ability to incorporate amino acid into protein although it leaves unaffected activity towards amino acid activation and toward AMP-ATP exchange. Interestingly enough, sonic vibration of the preparation is reported not to inhibit ability to incorporate amino acid into protein (*88*).

One important question bearing upon the transfer of amino acid from soluble RNA to microsome concerns whether or not the RNA moiety is similarly incorporated. The experiments of SHIMURA and BORSOOK (*63*) have already indicated that synthesis of protein in the reticulocyte (a nucleus-free system) does not involve synthesis or turnover of RNA. On the other hand, WEBSTER's crude particulate preparation does possess some ability to incorporate labeled AMP into RNA (*88*). Whether the incorporation is into soluble or microsomal RNA has not been determined. In any case incorporation of AMP into RNA in the pea seedling preparation is not dependent upon the simultaneous presence of amino acids and it is not evident that it has to do directly with the act of protein synthesis.

4. Alternative Pathways of Peptide Synthesis.

It has in the past been often considered that the synthesis of amide or simple peptide linkages might offer model systems for the study of peptide linkages in protein. It is now reasonably clear that this is not the case. The synthesis of the amide bonds of glutamine and asparagine proceed by an independent pathway different in principle from that used in protein synthesis. The same is true of the synthesis of the tripeptide, glutathione. In all of these cases, as in protein synthesis, ATP serves as the energy donor for the initial activation of amino acid. This was demonstrated first by ELLIOTT (*21*) for the synthesis of glutamine in a homogenate of lupine seedlings. WEBSTER (*84*) has shown that acetone powders of bean seedlings carry out the synthesis of glutamine according to the overall stoichiometry of reaction (D).

Glutamate $+ NH_3 + ATP \rightleftarrows$ glutamine $+ ADP +$ orthophosphate. (D)

It is to be noted that the cleavage of ATP in this case is one which yields ADP and orthophosphate rather than AMP and pyrophosphate as in the case of the amino acid activating enzymes of protein synthesis. In addition, the enzyme has been shown by WEBSTER to be a mitochondrial rather than a soluble one (*84, 85*). The enzyme has been purified from peas over 4,000-fold by VARNER and WEBSTER (*80, 90*). The purified enzyme is capable not only of the synthesis of glutamine from glutamate, ammonia and ATP, according to reaction (D), but also of reactions which we may term glutamyl transfers. The enzyme, glutamine synthetase, is in fact identical with the glutamyl transferase of STUMPF and LOOMIS (74). Thus, it is capable of exchanging the amide group of glutamine for a labeled amide group or of replacing this group by hydroxylamine to form glutamylhydroxamate. For these transfer reactions the presence of ADP and orthophosphate are required. The intermediate common to glutamyl-glutamine synthesis and glutamyl transfer appears to be one consisting of enzyme, glutamyl, ADP, and orthophosphate.

Asparagine synthesis, a classical problem of plant physiology, has been elucidated by the work of WEBSTER and VARNER (92). An enzyme found in lupine seedlings as well as in wheat germ brings about the synthesis of asparagine from aspartic acid, ATP, and ammonia in a manner completely analogous to that of glutamine synthesis, as indicated in reaction (E).

$$\text{Aspartic acid} + \text{ammonia} + \text{ATP} \rightleftarrows$$
$$\rightleftarrows \text{asparagine} + \text{ADP} + \text{orthophosphate} \qquad\qquad \text{(E)}$$

This enzyme, asparagine synthetase, is separate and distinct from glutamine synthetase. It can, like the latter, carry out transfer reactions involving the transfer of the aspartyl residue to acceptors other than ammonia.

WEBSTER (83) as well as WEBSTER and VARNER (89, 91) have thoroughly elucidated the pathway of glutathione synthesis in preparations of bean hypocotyls and of wheat germ. The reaction proceeds in two steps, reactions (F) and (G):

$$\text{Glutamate} + \text{cysteine} + \text{ATP} \rightleftarrows$$
$$\rightleftarrows \text{glutamyl-cysteine} + \text{ADP} + \text{orthophosphate} \qquad\qquad \text{(F)}$$

$$\text{Glutamyl-cysteine} + \text{glycine} + \text{ATP} \rightleftarrows$$
$$\rightleftarrows \text{glutathione} + \text{ADP} + \text{orthophosphate} \qquad\qquad \text{(G)}$$

Each step involves the formation of an active intermediate, viz. enzyme-glutamyl-ADP-orthophosphate in the first case, and enzyme-glutamylcysteinyl-ADP-orthophosphate in the second case. In reaction (F) cysteine acts as the acceptor for glutamyl from the complex. In reaction (G) glycine acts as the acceptor for glutamyl-cysteinyl from the complex. The determination of the amino acid sequence in glutathione involves specificity both as to substrate and as to acceptor on the part of the two individual enzymes. In principle it would be possible to build up long peptide chains of specific amino acid sequence by this sort of mechanism. Since many different amino acid sequences are involved in proteins, however, it is clear that a vast number (a maximum of $n-1$, where n equals the number of amino acid residues in the peptide chain) of different individual enzymes would be required if the glutathione model were to be applied to synthesis of a whole protein. From whence would all these specific enzymes come?

VII. Protein Synthesis in Mitochondria and Chloroplasts.

It is generally agreed that the mitochondria are entities which multiply by division [MILLERD and BONNER (42), HACKETT (26)]. Although the enzymatic constitution of the mitochondrion is undoubtedly genetically controlled by the nucleus, there are in addition mitochondrial characteristics which are inherited cytoplasmically. This is true, for example, of the poky character of *Neurospora* discovered by MITCHELL, MITCHELL et al. (44, 45). The mitochondrion behaves then to some extent as though it were an autonomous entity in the cytoplasm of its host cell. The constituent enzymes of the mitochondrion are characteristic, highly organized, and probably unique to the mitochondrion. How are these enzymes synthesized? Are they produced by microsomes and then subsequently assembled into the mitochondrial structure? Or does

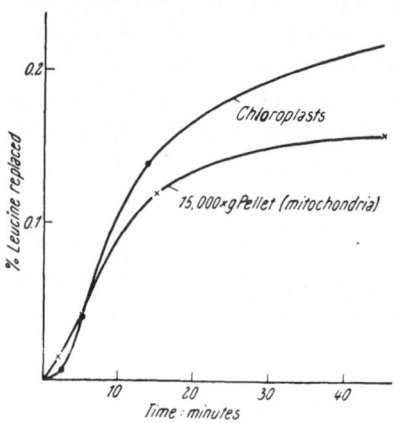

Fig. 5. Incorporation of C^{14}-leucine into isolated chloroplasts and mitochondria of tobacco leaves. After STEPHENSON et al. (72). [From: Arch. Biochem. Biophys. 65, 194 (1956).]

the mitochondrion possess a protein synthesizing mechanism of its own? This question cannot be answered, although some pertinent facts are available. In the first place, if mitochondria synthesize protein according to the microsomal model, it should be expected that mitochondria would contain RNA. They do. On the basis of a critical survey of the literature, HACKETT (26) has concluded that plant mitochondria contain of the order of 5% of their weight in the form of RNA. Ts'o and SATO (79) come to a similar conclusion. Mitochondria of young pea stem tissue possess a RNA/protein ratio of about 1 to 4; those of older tissue a ratio of about 1 to 8. On the basis that the mitochondria contain about 30% protein, these values would bracket HACKETT's (26).

Although mitochondria rapidly incorporate into their structure precursors of nucleic acids such as labeled orthophosphate, it still remains that the site of synthesis of the mitochondrial RNA is unknown. It is clear, however, that isolated mitochondria can incorporate amino acids. This has been shown for example by WEBSTER (86) who has incubated isolated mitochondria of bean hypocotyls with labeled glutamate. In the presence of ATP the glutamate is incorporated into the mitochondrial structure. Similar observations have been made on an isolated mitochondrial fraction of tobacco leaves by STEPHENSON et al. (*72*) *(Fig. 5)*.

Chloroplasts too are self-propagating entities of the plant cell [VON WETTSTEIN (*81*)]. In higher plants the chloroplasts develop from proplastids, and multiplication of the chloroplast system takes place at the proplastid level. Mature chloroplasts contain only a small amount of nucleic acid, perhaps of the order of 0.1–1% [JAGENDORF and WILDMAN (*34*), JAGENDORF (*33*)]. This nucleic acid is probably primarily RNA. Although chloroplast preparations commonly contain a trace of DNA, it is quite possible that the latter is in reality associated with nuclear fragments since it is difficult to separate nuclei quantitatively from chloroplasts. Since the mature chloroplast is several orders of magnitude larger than the proplastid, it is conceivable that the nucleic acid content of the latter is high and comparable to that of nucleus or microsome. This is, however, still unknown.

Protein synthesis by chloroplasts takes place particularly rapidly during the growth of the chloroplast from the proplastid [LYTTLETON (*38*)]. Even mature chloroplasts, however, possess the ability to incorporate labeled amino acids into their own structure. This has been shown, for example, by STEPHENSON et al. (*72*) who incubated isolated tobacco leaf chloroplasts with labeled leucine (Fig. 5). It was found that no external energy source is required for this incorporation. The rate of the reaction is, however, increased in light.

Resolution of the problem of the origin of mitochondrial and chloroplastic protein will obviously be an important further problem of protein synthesis in plants.

VIII. Protein Synthesis in Varied Plant Organs.

1. Developing Fruits and Seeds.

A center of massive protein formation in the plant is the developing seed. This is true particularly of seeds which contain protein as their reserve material. The reserve proteins of seeds are unique and characteristic to them, as for example the seed globulins. Thus, the pea seed contains as its reserve material two principal globulins, vicilin, $S = 8$, and leguminin, $S = 12.5$ [DANIELSSON (*17*)]. These two globulins are found in the seed of the pea plant and nowhere else. The seeds contain also a lesser amount of the so-called seed albumins a and b of sedimentation constants approximately 1 and 4 [DANIELSSON (*17*)]. The proteins are formed in the developing seed from soluble materials transported into the fruit, although the exact nature of these transport materials is unknown.

McKee et al. (*39*), who have followed the time course of the total protein content in the developing pea seed, have shown that the protein increases some thirty-fold in a period of 26 days, the rate of increase slowing down as the seed approaches maturity. Soluble nitrogenous materials increase to a maximum when the seed has attained about half its final dry weight and slowly decrease thereafter. RAACKE (*55*) has followed the individual component proteins during the course of development of the pea seed. During the early stages, all of the protein of the seed consists of low-molecular weight albumins. The globulins begin to appear when the seed has attained about one-fourth of its final dry weight. Once this stage is reached, globulins are formed even if the seed is detached from the plant. An interesting feature of the developing pea seed stressed by SNELLMAN and DANIELSSON (*70*) and by RAACKE (*55*) is the fact that this organ contains a considerable amount of peptide nitrogen. These peptides, like the soluble amino nitrogen of the seed, increase in quantity up to the time that the fruit is about half grown. Thereafter the peptide content decreases abruptly. It has been suggested by DANIELSSON (*17*) and by RAACKE (*55*) that the peptides constitute an intermediate between amino acids and the finished seed globulins. No compelling evidence on this point is available, however, and it is just as possible that the soluble peptides constitute products of protease action or arise in some other way not directly related to protein synthesis.

Although the cotyledons of leguminous seeds contain abundant RNA [OOTA and OSAWA (*48*)], this RNA has not been unequivocally identified with a specific subcellular component [OOTA and OSAWA (*49*)]. Neither has the behavior of RNA during the development of the seed been followed. It would appear that the developing legume seed, containing as it does only two principal protein components, would provide an excellent system for the study of the synthesis of individual proteins and of the relationship of RNA to protein synthesis.

2. Leaves.

An interesting facet of plant protein metabolism has to do with the behavior of excised leaves. The growing leaf synthesizes protein rapidly. As the leaf attains maturity its protein level attains a constant value. In senescence the leaf again loses its protein.

If a leaf is detached from a plant its behavior is drastically changed. Protein synthesis slows down or stops completely, to be replaced by protein loss. An extensive literature on the protein relationships of excised leaves has developed [BONNER (*4*)]. Thus, it has been established that protein loss upon excision is most marked in mature or aging leaves and is less marked in young leaves [ENGELBRECHT (*22*), PEARSALL and BILLIMORIA (*52, 53*)]. Protein loss in excised leaves takes place both in light and in darkness, although it is more marked in the dark. Protein loss in excised leaves cannot be inhibited by nutrient solutions which supply needed minerals, including nitrogen, to the leaf. Protein loss can, however, be replaced by protein synthesis if the leaf forms adventitious roots. Rooted leaves may be kept alive for long periods of time and show steady increases in total protein [CHIBNALL (13), ENGELBRECHT (22), MOTHES and ENGELBRECHT (46)].

The behavior of leaves upon excision is unlike that of other plant organs. Thus, isolated roots can be grown indefinitely in culture and continue to build up proteins from carbohydrate and inorganic nitrogen. It has therefore been suspected that protein loss in excised leaves might be due to some amino acid deficiency in these organs. It has been suggested, for example, that leaves may be unable to make all amino acids and may depend upon roots for a continuing supply of some of these essential metabolites [WOOD and CRUICKSHANK (94), BONNER (4)]. This hypothesis has, however, been shown to be untrue by ROGERS (58). ROGERS floated excised leaves, stems, or roots of bean plants on nutrient solutions containing C^{14}-labeled sucrose or C^{14}-labeled acetate. He showed that the soluble amino acids of leaves all become labeled from one or the other of these two substrates and that the specific activity of the amino acids thus synthesized by the isolated leaf is as high as that of the amino acids synthesized by roots.

ROGERS did, however, find very great differences in rate of incorporation of label into protein as between leaves and roots. Thus, over a 12-hour period after excision, leaves incorporated their labeled amino acids into protein only one-seventh as rapidly as did the roots. In similar experiments, RACUSEN and ARONOFF (56) have compared rates of incorporation of C^{14}-labeled CO_2 into amino acids and leaf protein of soybean leaves which were either attached to the plant or excised from it. They noted that rate of incorporation of label into leaf protein decreases with time after excision, although ability of the excised leaf to incorporate label into soluble amino acids remains at a high level.

It would appear therefore that the lesion in the isolated leaf which results in loss of protein is not related to any inability of the leaf to synthesize its own amino acids. It appears to have to do rather with some breakdown in the protein synthesizing mechanism and the mechanism for incorporation of amino acids into protein. However, this breakdown is not complete, for CHIBNALL and WILTSHIRE (14) have shown that excised bean leaves which are undergoing a net loss of protein, nonetheless retain the ability to incorporate some N^{15} (supplied as NH_3) for periods up to several days. Interestingly enough, this inability to synthesize normal leaf protein does not apply to virus protein. Thus, it has been shown by TAKAHASHI (75) that excised leaves of tobacco, even though they are losing protein rapidly, can nonetheless be infected and synthesize tobacco mosaic virus.

The nucleic acid metabolism of excised leaves has been investigated but little. It would be of interest to know, for example, whether the RNA of the microsomes disappears after excision of the leaf; whether the leaf retains or does not retain ability to produce new RNA, and so on. Studies of this kind might well give us important insight into the mechanism of regulation of the leaf protein level as well as into the mechanism of virus synthesis.

IX. Conclusion.

At long last, the problem of how proteins are synthesized in living creatures appears to have become a soluble one. Much is still left to learn but a real start has been made. Microsomes appear to be the engines of protein synthesis, at least so far as the soluble cytoplasmic proteins are concerned. Microsomes appear to be made in the nucleus and they may well be the agency by which information is carried from nuclear DNA to cytoplasm. According to this view, differentiation may consist in enrichment of the cell or tissue in particular kinds of microsomes at the expense of others. In any case, our rapidly increasing knowledge of the mechanism of protein synthesis promises to correspondingly increase our understanding of the biology of the living organism.

References.

1. ALLFREY, V. G. and A. E. MIRSKY: Some Aspects of Ribonucleic Acid Synthesis in Isolated Cell Nuclei. Proc. Nat. Acad. Sci. (USA) **43**, 821 (1957).
2. ALLFREY, V. G., A. E. MIRSKY and S. OSAWA: Protein Synthesis in Isolated Cell Nuclei. J. Gen. Physiol. **40**, 451 (1957).
3. BERG, P.: Participation of Adenyl-acetate in the Acetate-activating System. J. Amer. Chem. Soc. **77**, 3163 (1955).
4. BONNER, J.: Plant Biochemistry. New York: Academic Press. 1950.
5. BOORSOOK, H.: Enzymatic Syntheses of Peptide Bonds. In: D. GREENBERG, Chemical Pathways of Metabolism, Vol. II, p. 173. New York: Academic Press. 1954; cf. Fortschr. Chem. organ. Naturstoffe **9**, 212 (1952).
6. BOORSOOK, H., C. L. DEASY, A. J. HAAGEN-SMIT, G. KEIGHLEY and P. H. LOWY: The Uptake in vitro of C^{14}-Labeled Glycine, L-Leucine, and L-Lysine by Different Components of Guinea Pig Liver Homogenate. J. Biol. Chem. **184**, 529 (1950).
7. BRACHET, J.: Effects of Ribonuclease on the Metabolism of Living Root-tip Cells. Nature (London) **174**, 876 (1954).
8. — Action of Ribonuclease and Ribonucleic Acid on Living Amoebae. Nature (London) **175**, 851 (1955).
9. BRACHET, J., H. CHANTRENNE et F. VANDERHAEGHE: Recherches sur les interactions biochimiques entre le noyau et le cytoplasme chez les organismes unicellulaires. II. *Acetabularia Mediterranea*. Biochim. Biophys. Acta **18** 544 (1955).

10. CAMPBELL, P. N., O. GREENGARD and B. A. KERNOT: Amino Acid Incorporation into Serum Albumin in Microsome Preparations from Regenerating Rat Liver. Biochemic. J. 68, 18 P (1958).

11. CHAO, FU-CHUAN: Dissociation of Macromolecular Ribonucleoprotein of Yeast. Arch. Biochem. Biophys. 70, 426 (1957).

12. CHAO, FU-CHUAN and H. K. SCHACHMAN: The Isolation and Characterization of a Macromolecular Ribonucleoprotein from Yeast. Arch. Biochem. Biophys. 61, 220 (1956).

13. CHIBNALL, A. C.: Protein Metabolism in Rooted Runner-Bean Leaves. New Phytologist 53, 31 (1954).

14. CHIBNALL, A. C. and G. H. WILTSHIRE: A Study with Isotopic Nitrogen of Protein Metabolism in Detached Runner-Bean Leaves. New Phytologist 53, 38 (1954).

15. CLARK, J. M., Jr.: Studies on Amino Acid Activation and Protein Synthesis. Thesis, Calif. Instit. Technology, 1958.

16. CLENDENNING, K. A.: Biochemistry of Chloroplasts in Relation to the Hill Reaction. Annu. Rev. Plant Physiol. 8, 137 (1957).

17. DANIELSSON, C. E.: Seed Globulins of the Gramineae and Leguminosae. Biochemic. J. 44, 387 (1949).

18. DAVIE, E. W., V. V. KONINGSBERGER and F. LIPMANN: The Isolation of a Tryptophan-Activating Enzyme from Pancreas. Arch. Biochem. Biophys. 65, 21 (1956).

19. DE MOSS, J. A., S. M. GENUTH and G. D. NOVELLI: The Enzymatic Activation of Amino Acids via their Acyl-Adenylate Derivatives. Proc. Nat. Acad. Sci. (USA) 42, 325 (1956).

20. DE ROBERTIS, E.: Electron Microscope Observations on the Submicroscopic Morphology of the Meiotic Nucleus and Chromosomes. J. Biophys. Biochem. Cytol. 2, 785 (1956).

21. ELLIOTT, W. H.: Studies on the Enzymic Synthesis of Glutamine. Biochemic. J. 49, 106 (1951).

22. ENGELBRECHT, L.: Über den Stickstoff-Stoffwechsel isolierter Organe. Die Kulturpflanze, Beiheft 1, 86 (1956).

23. GODDARD, D. R. and H. A. STAFFORD: Localization of Enzymes in the Cells of Higher Plants. Ann. Rev. Plant Physiol. 5, 115 (1954).

24. GOLDSTEIN, L. and W. PLAUT: Direct Evidence for Nuclear Synthesis of Cytoplasmic Ribose Nucleic Acid. Proc. Nat. Acad. Sci. (USA) 41, 874 (1955).

25. GRANICK, S.: Quantitative Isolation of Chloroplasts from Higher Plants. Amer. J. Bot. 25, 558 (1938).

26. HACKETT, D. P.: Recent Studies on Plant Mitochondria. Int. Rev. Cytology 4, 143 (1955).

27. HALL, B. and P. DOTY: Physical-chemical Studies of the Ribonucleic Acid of Microsomal Particles. Program and Abstracts, Biophysical Society, 1958 Meeting, Cambridge, Mass., p. 15.

28. HOAGLAND, M. B.: An Enzymic Mechanism for Amino Acid Activation in Animal Tissues. Biochim. Biophys. Acta 16, 288 (1955).

29. HOAGLAND, M. B., E. B. KELLER and P. C. ZAMECNIK: Enzymatic Carboxyl Activation of Amino Acids. J. Biol. Chem. 218, 345 (1956).

30. HOAGLAND, M. B., P. C. ZAMECNIK and M. L. STEPHENSON: Intermediate Reactions in Protein Biosynthesis. Biochim. Biophys. Acta 24, 215 (1957).

31. HOAGLAND, M. B., P. C. ZAMECNIK, N. SHARON, F. LIPMANN, M. P. STULBERG and P. D. BOYER: Oxygen Transfer to AMP in the Enzymic Synthesis of the Hydroxamate of Tryptophan. Biochim. Biophys. Acta 26, 215 (1957).

32. HOLLEY, R. W.: An Alanine-dependent, Ribonuclease-inhibited Conversion of AMP to ATP, and its Possible Relationship to Protein Synthesis. J. Amer. Chem. Soc. 79, 658 (1957).

33. JAGENDORF, A. T.: Purification of Chloroplasts by a Density Technique. Plant Physiol. 30, 138 (1955).

34. JAGENDORF, A. T. and S. G. WILDMAN: The Proteins of Green Leaves. VI. Centrifugal Fractionation of Tobacco Leaf Homogenates and some Properties of Isolated Chloroplasts. Plant Physiol. 29, 270 (1954).

35. JEENER, R. and D. SZAFARZ: Relations Between the Rate of Renewal and the Intracellular Localization of Ribonucleic Acid. Arch. Biochemistry 26, 54 (1950).

36. LITTLEFIELD, J. W. and E. B. KELLER: Incorporation of C14-Amino Acids into Ribonucleoprotein Particles from the Ehrlich Mouse Ascites Tumor. J. Biol. Chem. 224, 13 (1957).

37. LITTLEFIELD, J. W., E. B. KELLER, J. GROSS and P. C. ZAMECNIK: Studies on Cytoplasmic Ribonucleoprotein Particles from the Liver of the Rat. J. Biol. Chem. 217, 111 (1955).

38. LYTTLETON, J. W.: Protein of Pasture Plants. Cytoplasmic Protein of White Clover and Italian Ryegrass. Biochemic. J. 64, 70 (1956).

39. McKEE, H. S., R. N. ROBERTSON and J. B. LEE: Physiology of Pea Fruits. I. The Developing Fruit. Austral. J. Biol. Sci. 8, 137 (1955).

40. McMASTER-KAYE, R. and J. H. TAYLOR: Evidence for Two Metabolically Distinct Types of Ribonucleic Acid in Chromatin and Nucleoli. J. Biophys. Biochem. Cytol. 4, 5 (1958).

41. MILLERD, A.: Mitochondria and Microsomes. In: Handbuch der Pflanzenphysiologie, Vol. II, p. 573. Berlin-Göttingen-Heidelberg: Springer-Verlag. 1956.

42. MILLERD, A. and J. BONNER: Biology of Plant Mitochondria. J. Histochem. Cytochem. 1, 254 (1953).

43. MILLERD, A., J. BONNER, B. AXELROD and R. S. BANDURSKI: Oxidative and Phosphorylative Activity of Plant Mitochondria. Proc. Nat. Acad. Sci. (USA) 37, 855 (1951).

44. MITCHELL, M. B. and H. K. MITCHELL: A Case of "Maternal" Inheritance in Neurospora crassa. Proc. Nat. Acad. Sci. (USA) 38, 442 (1952).

45. MITCHELL, M. B., H. K. MITCHELL and A. TISSIERES: Mendelian and Non-Mendelian Factors Affecting the Cytochrome System in Neurospora crassa. Proc. Nat. Acad. Sci. (USA) 39, 606 (1953).

46. MOTHES, K. und L. ENGELBRECHT: Über den Stickstoffumsatz in Blattstecklingen. Flora 143, 428 (1956).

47. OGATA, K. and H. NOHARA: The Possible Role of the Ribonucleic Acid (RNA) of the pH 5 Enzyme in Amino Acid Activation. Biochim. Biophys. Acta 25, 659 (1957).

48. OOTA, Y. and S. OSAWA: Migration of "Storage PNA" from Cotyledon into Growing Organs of Bean Seed Embryo. Experientia 10, 254 (1954).

49. — — Relation between Microsomal Pentose Nucleic Acid (PNA) and Protein Synthesis in the Hypocotyl of Germinating Bean Embryo. Biochim. Biophys. Acta 15, 162 (1954).

50. OSAWA, S., K. TAKATA and Y. HOTTA: Some Aspects of the Relation between Nuclear and Cytoplasmic Ribonucleic Acids. Biochim. Biophys. Acta 25, 656 (1957).

51. PALADE, G. E.: A Small Particulate Component of the Cytoplasm. J. Biophys. Biochem. Cytol. 1, 59 (1955).

52. PEARSALL, W. H. and M. C. BILLIMORIA: Effects of Age and of Season upon Protein Synthesis in Detached Leaves. Ann. Botany (N. S.) **2**, 317 (1938).

53. — — The Influence of Light upon Nitrogen Metabolism in Detached Leaves. Ann. Botany (N. S.) **3**, 601 (1939).

54. PORTER, K. R.: Electron Microscopy of Basophilic Components of Cytoplasm. J. Histochem. Cytochem. **2**, 346 (1954).

55. RAACKE, I. D.: Protein Synthesis in Ripening Pea Seeds. I. Analysis of Whole Seeds. Biochemic. J. **66**, 101 (1957).

56. RACUSEN, D. W. and S. ARONOFF: Metabolism of Soybean Leaves. VI. Exploratory Studies in Protein Metabolism. Arch. Biochem. Biophys. **51**, 68 (1954).

57. ROBINSON, E. and R. BROWN: Cytoplasmic Particles in Bean Root Cells. Nature (London) **171**, 313 (1953).

58. ROGERS, B.: Studies on the Amino Acid Metabolism of Higher Plants. Thesis, Calif. Instit. Technology, 1955.

58 a. SATO, C., J. R. PILCHER and P. O. P. Ts'o: Phosphate Metabolism in Pea Shoots. Plant Physiol. **32**, XII (1957).

59. SCHACHMAN, H. K., A. B. PARDEE and R. Y. STANIER: Studies on the Macromolecular Organization of Microbial Cells. Arch. Biochem. Biophys. **38**, 245 (1952).

60. SCHWEET, R. S.: Purification and Properties of Tyrosine Activating Enzyme. Federat. Proc. (Amer. Soc. exp. Biol.) **16**, 244 (1957).

61. SCHWEET, R. S., F. C. BOVARD, E. ALLEN and E. GLASSMAN: The Incorporation of Amino Acids into Ribonucleic Acids. Proc. Nat. Acad. Sci. (USA) **44**, 173 (1958).

62. SCHWEET, R. S., R. W. HOLLEY and E. H. ALLEN: Amino Acid Activation in Hog Pancreas. Arch. Biochem. Biophys. **71**, 311 (1957).

63. SHIMURA, K. and H. BORSOOK: Absence of RNA Synthesis in Reticulocytes (unpublished); cf. Abstr. Int. Biochem. Kongr. Vienna (1958).

64. SHIMURA, K., J. SATO, S. SUTO and A. KIKUCHI: Amino Acid Composition of Ribonucleoprotein of Silk Gland. J. Biochemistry **43**, 217 (1956).

65. SIMKIN, J. L. and T. S. WORK: Protein Synthesis in Guinea-pig Liver. Biochemic. J. **65**, 307 (1957).

66. — — Incorporation of Radioactive Amino Acids into Proteins of the Microsome Fraction of Guinea-pig Liver in a Cell-free System. Biochemic. J. **67**, 617 (1957).

67. SJÖSTRAND, F. S. and V. HANZON: Membrane Structure of Cytoplasm and Mitochondria in Exocrine Cells of Mouse Pancreas as Revealed by High Resolution Electron Microscopy. Exptl. Cell Res. **7**, 393 (1954).

68. ŠKREB-GUILCHER, Y.: Influence de la ribonucléase sur la teneur en adénosine-triphosphate (ATP) et la consommation d'oxygène des amibes vivantes. Biochim. Biophys. Acta **17**, 599 (1955).

69. SLAUTTERBACK, D. B.: Electron Microscopic Studies of Small Cytoplasmic Particles (Microsomes). Exptl. Cell Res. **5**, 173 (1953).

70. SNELLMAN, O. and C. E. DANIELSSON: An Experimental Study of the Biosynthesis of the Reserve Globulins in Pea Seeds. Exptl. Cell Res. **5**, 436 (1953).

71. STAFFORD, H. A.: Intracellular Localization of Enzymes in Pea Seedlings. Physiol. Plantarum **4**, 696 (1951).

72. STEPHENSON, M. L., K. V. THIMANN and P. C. ZAMECNIK: Incorporation of C^{14}-Amino Acids into Proteins of Leaf Disks and Cell-Free Fractions of Tobacco Leaves. Arch. Biochem. Biophys. **65**, 194 (1956).

73. STICH, H. and W. PLAUT: The Effect of Ribonuclease on Protein Synthesis in Nucleated and Enucleated Fragments of *Acetabularia*. J. Biophys. Biochem. Cytol. **4**, 119 (1958).

74. STUMPF, P. K. and W. D. LOOMIS: Observations on a Plant Amide Enzyme System Requiring Manganese and Phosphate. Arch. Biochemistry 25, 451 (1950).

75. TAKAHASHI, W. N.: Changes in Nitrogen and Virus Content of Detached Tobacco Leaves in Darkness. Phytopathol. 31, 1117 (1941).

76. Ts'o, P. O. P., J. BONNER and H. DINTZIS: On the Similarity of Amino Acid Composition of Microsomal Nucleoprotein Particles. Arch. Biochem. Biophys. 76, 225 (1958).

77. Ts'o, P. O. P., J. BONNER and J. VINOGRAD: Microsomal Nucleoprotein Particles from Pea Seedlings. J. Biophys. Biochem. Cytol. 2, 451 (1956).

78. — — — Physical and Chemical Properties of Microsomal Nucleoprotein Particles from Pea Seedlings. Plant Physiol. 32, XIII (1957).

79. Ts'o, P. O. P. and C. SATO: Distribution of Ribonucleic Acid and Protein among Subcellular Components of Pea Epicotyls. Plant Physiol. (in press).

80. VARNER, J. E. and G. C. WEBSTER: Mechanism of Enzymatic Synthesis of Glutamine. Abstr., Amer. Soc. Plant Physiol., Gainesville, 1954, p. 33.

81. VON WETTSTEIN, D.: Genetics and the Submicroscopic Cytology of Plastids. Hereditas 43, 303 (1957).

82. WATSON, M. L.: The Nuclear Envelope. Its Structure and Relation to Cytoplasmic Membranes. J. Biophys. Biochem. Cytol. 1, 257 (1955).

83. WEBSTER, G. C.: Peptide-Bond Synthesis in Higher Plants. I. The Synthesis of Glutathione. Arch. Biochem. Biophys. 47, 241 (1953).

84. — Enzymatic Synthesis of Glutamine in Higher Plants. Plant Physiol. 28, 724 (1953).

85. — Enzymatic Synthesis of Gamma-Glutamyl-Cysteine in Higher Plants. Plant Physiol. 28, 728 (1953).

86. — An Energy Dependent Incorporation of Amino Acids into the Protein of Plant Mitochondria. Plant Physiol. 29, 202 (1954).

87. — Incorporation of Radioactive Amino Acids into the Proteins of Plant Tissue Homogenates. Plant Physiol. 30, 351 (1955).

88. — Amino Acid Incorporation by Intact and Disrupted Ribonucleoprotein Particles. J. Biol. Chem. 229, 535 (1957).

89. WEBSTER, G. C. and J. E. VARNER: Peptide-Bond Synthesis in Higher Plants. II. Studies on the Mechanism of Synthesis of γ-Glutamylcysteine. Arch. Biochem. Biophys. 52, 22 (1954).

90. — — On the Mechanism of the Enzymatic Synthesis of Glutamine. J. Amer. Chem. Soc. 76, 633 (1954).

91. — — Peptide-Bond Synthesis in Higher Plants. III. The Formation of Glutathione from γ-Glutamylcysteine. Arch. Biochem. Biophys. 55, 95 (1955).

92. — — Aspartate Metabolism and Asparagine Synthesis in Plant System. J. Biol. Chem. 215, 91 (1955).

93. WILDMAN, S. G. and J. BONNER: The Proteins of Green Leaves. I. Isolation, Enzymatic Properties and Auxin Content of Spinach Cytoplasmic Proteins. Arch. Biochemistry 14, 381 (1947).

94. WOOD, J. G. and D. H. CRUICKSHANK: The Metabolism of Starving Leaves. 5. Changes in Amounts of Some Amino Acids During Starvation of Grass Leaves; and their Bearing on the Nature of the Relationship between Proteins and Amino Acids. Austral. J. Exptl. Biol. Med. Sci. 22, 111 (1944).

(Received, April 1, 1958.)

The Electron Gas Theory of the Color of Natural and Artificial Dyes: Problems and Principles.

By **HANS KUHN**, Marburg a. d. Lahn.

With 22 Figures.

Contents.

As indicated in the Contents, this survey includes a general discussion of Problems and Principles. It is intended to present a treatment, along similar lines, of some important Applications in one of the forthcoming Volumes.

Acknowledgement. The writer is grateful to Dr. WALTER HUBER and Miss C. HARKORT for help in the preparation of the manuscript, and to Dr. E. W. HUGHES for critical reading of the text.

I. Some Experimental Facts Concerning Color and Structure.

As is well known, if the solution of a dye is irradiated with white light, a part of the visible spectrum is absorbed and the solution appears in the color of the transmitted light. Since most dyes show a single, strong and sufficiently narrow absorption band in the visible region, a relationship between the visible color and the wave length of the main absorption peak can be given as expressed in *Table 1* (*28, 89*).

Table 1. Visible Color and Wave Length Position of the Main Band.

λ_{max} (mμ)	Visible color	Excitation energy $\Delta E \cdot 10^{12}$ (erg) (cf. p. 179)
400	greenish yellow	5.0
425	yellow	4.7
450	orange......................	4.4
490	red.........................	4.1
510	purple	3.9
530	violet......................	3.7
550	reddish blue (indigo)	3.6
590	blue	3.4
640	bluish green................	3.1
730	green	2.7

The color of a dye solution can thus be predicted if its absorption spectrum is known (*1, 12, 76*), and it can be approximately predicted if the position of the main peak is given.

The observed relation between the chemical structure and the position of the main absorption band of dyes has been investigated for more than 70 years. The results have been reviewed extensively (*27, 66, 70*) and hence, only some particular cases will be discussed below.

1. Polyacetylenes, Polyenes, Symmetrical Cyanine Dyes, and Some Similar Compounds.

Figs. 1a and b show the wave length positions of the main absorption peak in the homologous series of the polyacetylenes (I) (*9, 10, 2, 3, 20*) and of the all-*trans* polyenes (II) (*11*). This peak is shifted towards longer wave lengths with increasing number of conjugated double or triple

$$R—(C\equiv C)_j—R \qquad\qquad R—(CH=CH)_j—R$$

$$\text{(I.)} \qquad\qquad\qquad\qquad \text{(II.)}$$

bonds, and the shift caused by one additional double or triple bond decreases with increasing number of j. The shift is smaller in the poly-acetylene series than in the polyene series, and thus a polyacetylene absorbs at shorter waves than a polyene containing the same number (j) of conjugated bonds.

By attaching an electron donating group (an "auxochrome" according to WITT's early color theory) (*85, 86*) to one end of a polyene and an electron accepting group (an "anti-auxochrome" in the color theory of DILTHEY and WIZINGER) (*24, 25, 87–89*) to the other end, the absorption

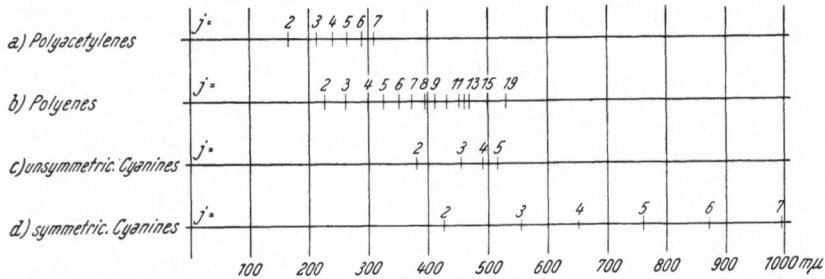

Fig. 1. Wave length location of the main absorption maxima of polyacetylenes (I) (Fig. 1a), polyenes (II) (Fig. 1b), unsymmetrical cyanines (V) (Fig. 1c), symmetrical cyanines (IV) (Fig. 1d): j is the number of triple or double bonds in the main resonating system [number of conjugated triple or double bonds in polyacetylenes (I) and polyenes (II); number of double bonds in the resonating chain between the two N atoms in unsymmetrical and symmetrical cyanines (V, IV)]. Solvent: methanol.

band is shifted towards the long wave region. A colorless polyene can thus be converted into a colored substance. An especially simple relationship exists between color and chemical structure when, e. g. the donor $(CH_3)_2\overline{N}$—, and the acceptor —$CH{=}\overset{\oplus}{N}(CH_3)_2$, or the

donor [structure] and the acceptor —C [structure] are attached. In general, if such donor and acceptor are chosen that the resulting molecule can be written in two mirror image limiting structures, dyes of symmetrical cyanine type are obtained. For example, (IIIa) or (IVa) can be converted into (IIIb) or (IVb) by interchanging single and double bonds *(Chart 1)*.

This relationship is demonstrated in *Fig. 1d* for the homologous series of cyanine dyes (IVa)–(IVb) (*13–15*); j is the number of double bonds in the chain between the two N atoms.

Lengthening the chain by the inclusion of an additional —CH=CH— group shifts the wave length of the main absorption band by about 100 mμ towards the longer wave length region, independently of the initial chain length. This simple experimental relationship was found first by KÖNIG (*40–43*) and was later verified in a great number of instances by BROOKER (*13–15*, *17–18*), SCHWARZENBACH (*77*), and others (*31, 8, 29, 80*).

$$(CH_3)_2\overset{-}{N}—CH=CH—CH=CH—CH=CH—CH=CH—CH=\overset{\oplus}{N}(CH_3)_2$$
(IIIa.)

$$(CH_3)_2\overset{\oplus}{N}=CH—CH=CH—CH=CH—CH=CH—CH=CH—\overset{-}{N}(CH_3)_2$$
(IIIb.)

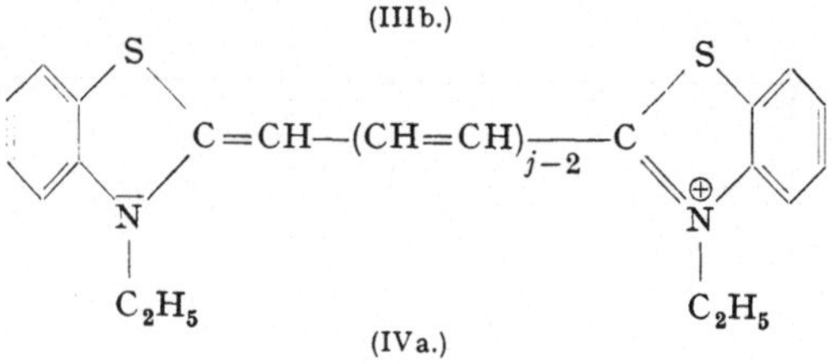

(IVa.)

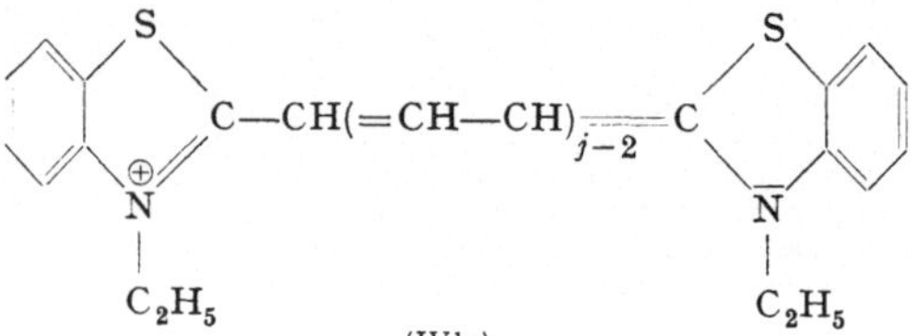

(IVb.)

Chart 1. Symmetrical Cyanines.

A type intermediate between the extreme cases of a polyene and a symmetrical cyanine (viz. an unsymmetrical cyanine) is obtained if the donor mentioned, 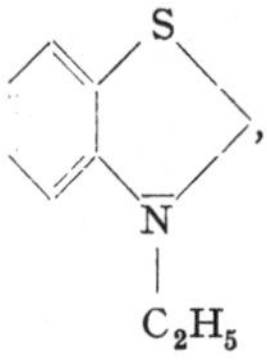 , is attached to one end of a polyene molecule, and instead of the acceptor, 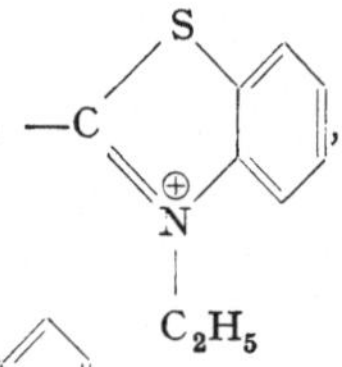 , the less strongly electron-attracting acceptor, 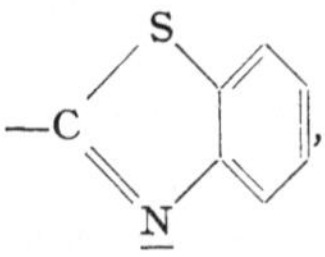 , is attached to the other end (*13–15*). The resulting compound (V) is a resonance hybrid between the two limiting formulas (Va) and (Vb) which can be derived from each other by interchanging the single and double bonds. However, these two structures are not equivalent, since (Va) contributes more than (Vb) to the ground state.

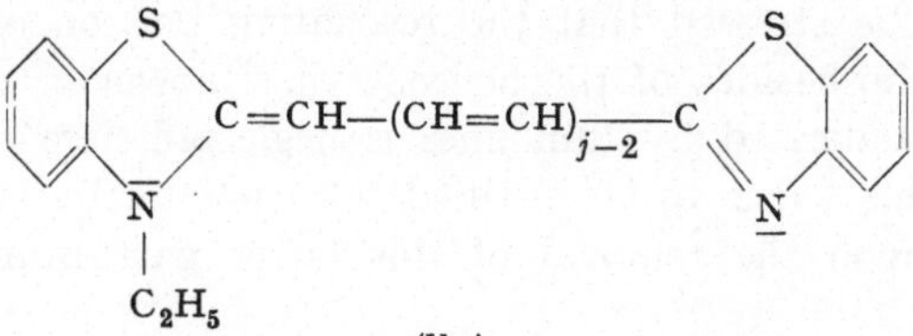

(Va.)

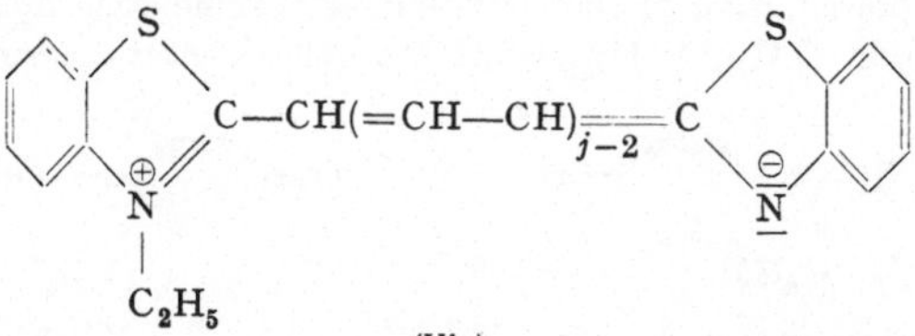

(Vb.)

Unsymmetrical Cyanines.

Fig. 1c gives the wave length of the main spectral maximum of (V) for $j = 2$ to 5 (*13–15*) and shows that the behavior of this type is intermediate between that of polyenes and of the symmetrical cyanine dyes; for any given value of j, (V) absorbs at longer waves than the polyene (II) and at shorter ones than the symmetrical cyanine (IV).

WIZINGER (*89*) has given many examples of such intermediate cases, and BROOKER (*13–15*, *17*, *18*) has systematically varied the electron donating power of the auxochrom and the electron attracting power of the antiauxochrom, thus covering the whole range between compounds similar to polyenes and those similar to the symmetrical cyanines. When the two limiting structures appeared to contribute to the ground state equally or nearly so, the same spectral characteristics as in symmetrical cyanines were observed; and, with increasing difference between the respective contributions of the two limiting structures, an increasing shift of the band towards shorter wave lengths was found, i. e. a more and more pronounced polyene type of spectral behavior.

A number of dyes can be classified either as symmetrical cyanine-like or polyene-like substances or belonging to an intermediate type. For example, MICHLER's hydrol blue can roughly be considered as a symmetrical cyanine, since the two equivalent limiting structures (VIa) and (VIb) are interconvertible by interchanging single and double bonds in the chain between auxochrom and antiauxochrom as indicated by heavy lines.

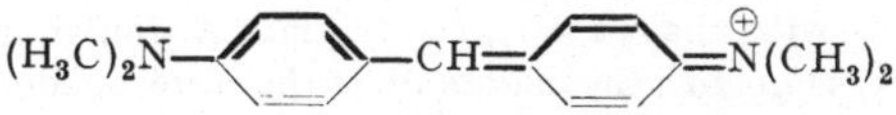

(VIa.)

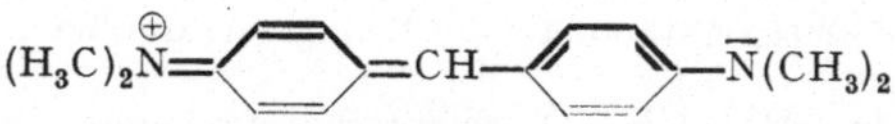

(VIb.)

It should be stressed that the resonating portion of the molecule extends over both sides of the benzene ring; however, the part of the chromophore indicated by thin lines is neglected here for the sake of simplicity. This seems to be justified since practically no spectral shift takes place upon the removal of this latter part from the molecule (VI → III)*.

A similar representative of the symmetrical cyanine type dye is the red basic form of benzaurine (VIIa, VIIb); while the yellow neutral form (VIIIa, VIIIb)

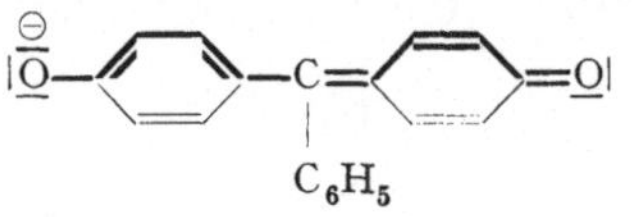

(VIIa.) 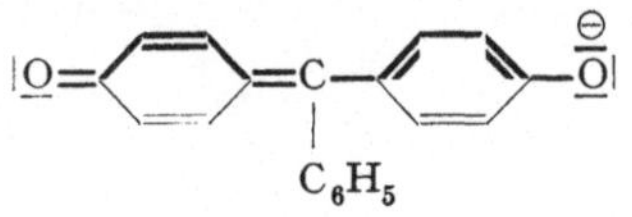   (VIIb.)

belongs to the unsymmetrical cyanine type, because (VIIIb) contributes much less than (VIIIa); thus, it absorbs at much shorter waves than (VII) [for the

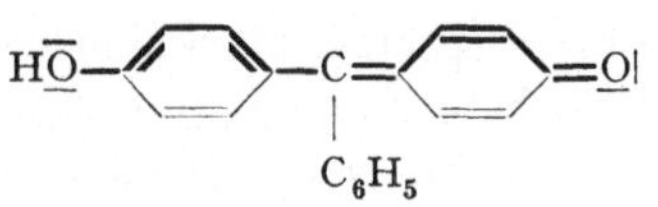

(VIIIa.) 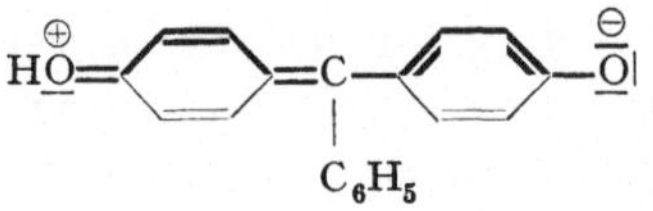   (VIIIb.)

bluish red substance (VII), $\lambda_{max} = 555$ mμ (19), while the corresponding figure for the yellow form (VIII) is 435 mμ (67)].

The blue basic form (IX) (84) of pelargonidine can roughly be considered as a symmetrical cyanine type substance, since (IXa) and (IXb) are contributing nearly equally to the ground state of the molecule. One finds here very much the same situation as in the case of (VII). However, the main resonating portion,

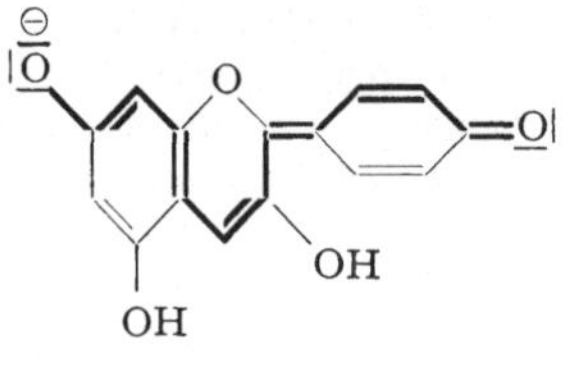

(IXa.) 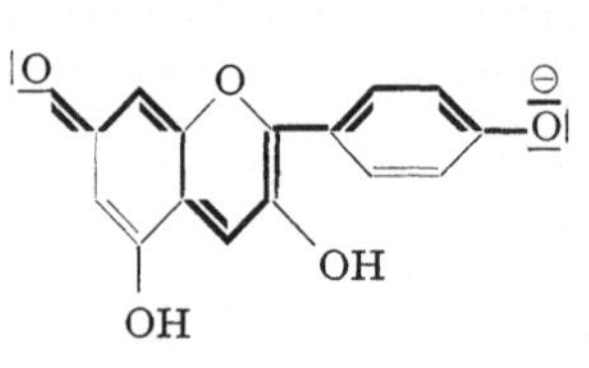   (IXb.)

---

* The wave length of maximum absorption of (VI) is located at 603 mμ (38). Substance (III) has not yet been prepared, but since the lower members of its homologous series, $(H_3C)_2\overset{\ominus}{N}$—$(CH{=}CH)_{j-1}$—$CH{=}\overset{\oplus}{N}(CH_3)_2$, with $j = 2$, 3 and 4 show maxima at 309 mμ, 409 mμ and 511 mμ, respectively (80), we may conclude that in the case of (III) ($j = 5$) λ_{max} will be found at $511 + 100 = 611$ mμ, which is a value comparable with that of (VI), viz. 603 mμ. A similar situation obtains for (VII); here $\lambda_{max} = 555$ mμ (19); the maximum of the corresponding simple compound $|\overset{\ominus}{\underline{O}}$—$(CH{=}CH)_4$—$CH{=}\underline{O}|$ which has not yet been prepared lies probably at 564 mμ, since the lower homolog containing two conjugated double bonds less, viz. the compound $|\overset{\ominus}{\underline{O}}$—$(CH{=}CH)_2$—$CH{=}\underline{O}|$ shows the maximum at 364 mμ (78).

indicated by heavy lines, contains an additional double bond, and, consequently, during the progression (VII) → (IX) the maximum is shifted by about 100 mμ towards the longer wave lengths. Since the maximum of (VII) is located at 555 mμ, (IX) should absorb at 655 mμ.

Similarly, since the violet neutral form of pelargonidine (IX) (*84*) corresponds to the neutral form of benzaurine (VIII) a very similar spectral shift is expected in both instances for the shift, basic form → neutral form. The calculated wave length position for the neutral form of pelargonidine is 655 — 120 = 535 mμ.

2. Aza Derivatives of Symmetrical Cyanine Type Compounds.

When replacing the CH group in the middle of a cyanine chain by a nitrogen atom, one generally finds that the absorption band shifts

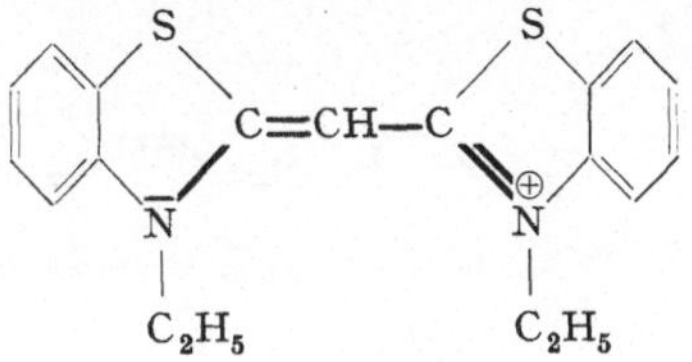

(X.), λ_{max} at 422 mμ (*39*).

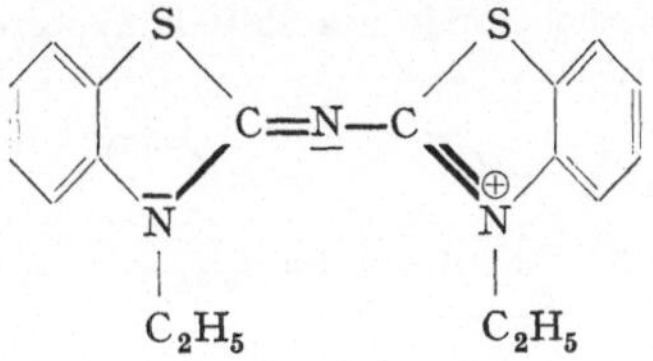

(XI.), λ_{max} at 368 mμ (*39*).

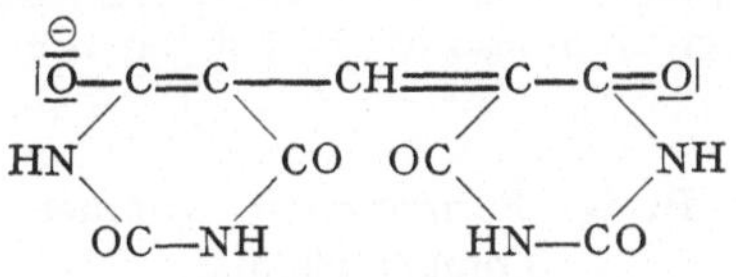

(XII.), λ_{max} at 413 mμ (*30, 79*).

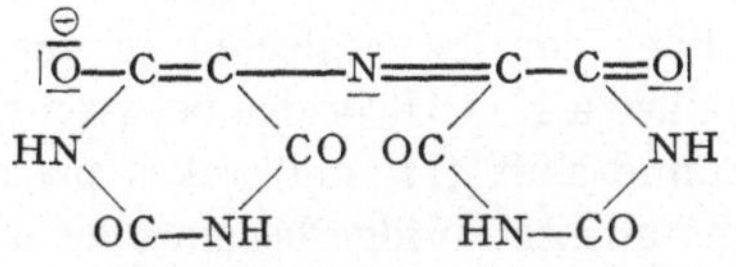

(XIII.), λ_{max} at 522 mμ (*30, 79*).

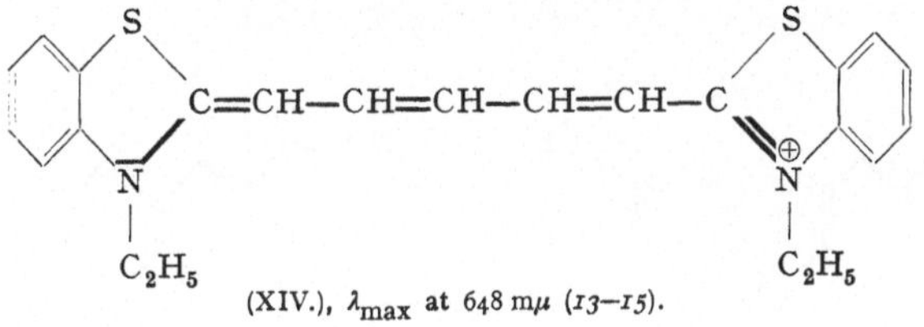

(XIV.), λ_{max} at 648 mμ (*13–15*).

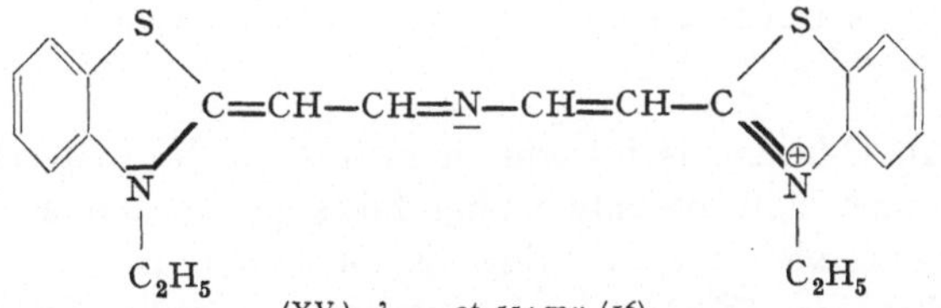

(XV.), λ_{max} at 554 mμ (*16*).

Chart 2. Symmetrical Cyanine Type Compounds: Aza Derivatives.

by 60 to 120 mμ towards the shorter wave lengths, if j (the number of conjugated double bonds in the resonating chain) is an even number. However, a similar shift (50–120 mμ) towards longer wave lengths takes place if j is an odd number. This behavior can be illustrated by some compounds with $j = 2$ (X, XI), $j = 3$ (XII, XIII), and $j = 4$ (XIV, XV); the shifts are, $368 - 422 = - 54$ mμ, $522 - 413 = + 109$ mμ, and $554 - 648 = - 94$ mμ, respectively. *(Chart 2.)*

(One of the two mirror image resonance structures is omitted from the foregoing and the following formulas.)

The aza derivatives of (VI) and (VII) are, respectively, BINDSCHEDLER's green (XVI) and the blue basic form of indophenol (XVII). In both instances a bathochromic aza shift is expected, since $j = 5$.

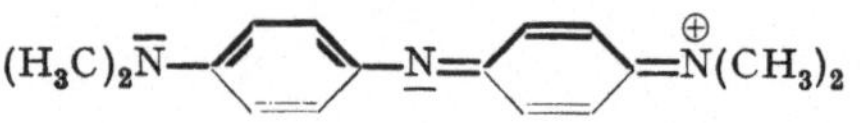

(XVI.) An indamine dye (λ_{max} at 725 mμ) (38). 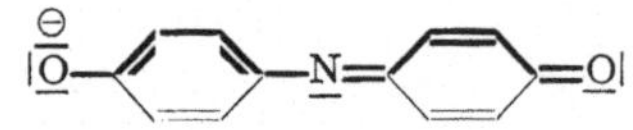   (XVII.) Indophenol (λ_{max} at 630 mμ) (22).

In the first case the shift is $725 - 603 = + 122$ mμ, and in the second one it is $630 - 555 = + 75$ mμ. Structure (VII, p. 174) contains in the center of the molecule a CC_6H_5 group instead of CH, but this is of no significance in this connection.

3. Acridine Type Bridge Formation in Symmetrical Cyanine Type Compounds (65).

As mentioned, in compound (VI) the part of the mesomeric portion indicated by thin lines can be neglected in the present discussion. However, if we introduce a $N \cdot CH_3$ bridge between the positions 4 and 8, a marked hypsochromic shift (112 mμ) takes place, the greenish blue color turns to orange, and the bridge becomes an important part of the resonating portion of the molecule (XVIII).

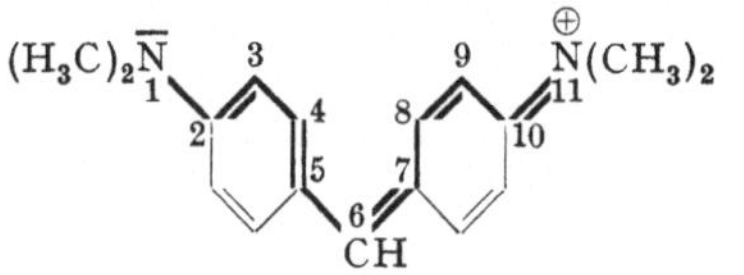

(VI.) A diphenylmethane dye (greenish blue,   (XVIII.) An acridine dye (orange,
λ_{max} at 603 mμ) (38). λ_{max} at 491 mμ) (23).

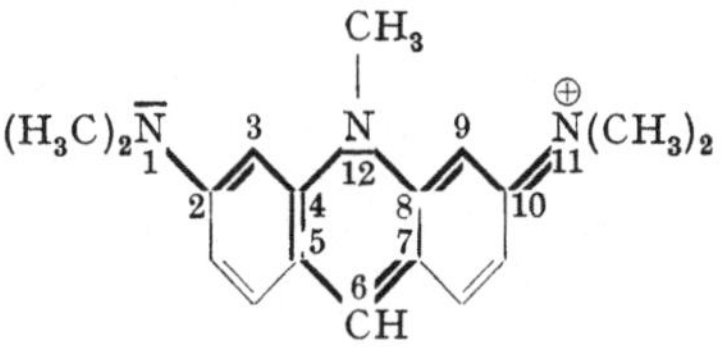

When an O-bridge is introduced instead of $N \cdot CH_3$, (VI → XIX), a hypsochromic shift of only 53 mμ takes place; and a similar shift (66 mμ) is observed if an O-bridge is introduced into the basic form of phenolphthalein (XX), a molecule which is only slightly different from (VII, p. 174).

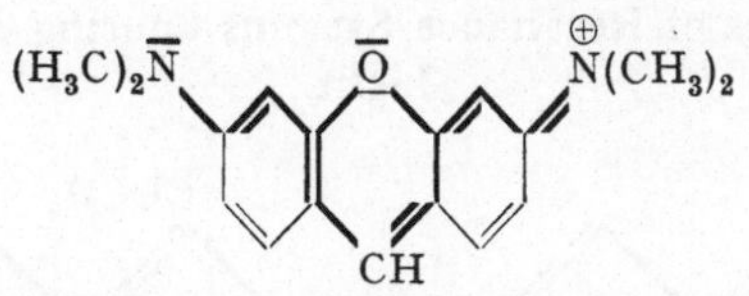

(XIX.) A pyronine dye (red, λ_{max} at 550 mμ) (*13–15*).

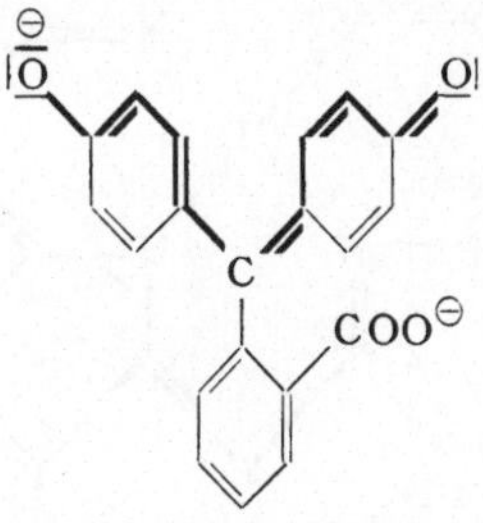

(XX.) Phenolphthalein (red, λ_{max} at 555 mμ) (*28*).

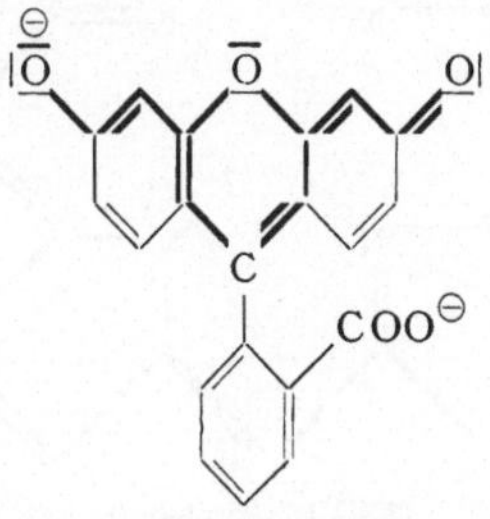

(XXI.) Fluorescein (yellow, λ_{max} 489 mμ) (*22*).

If the CH or C—$C_6H_4COO^{\ominus}$ group at position 6 is replaced by —N=, the bathochromic shift discussed above in Section 2 (p. 175) still appears. Thus, we find a bathochromic shift of 74 mμ when proceeding from acridine orange (XVIII) to the diazine dye (XXII), a bathochromic

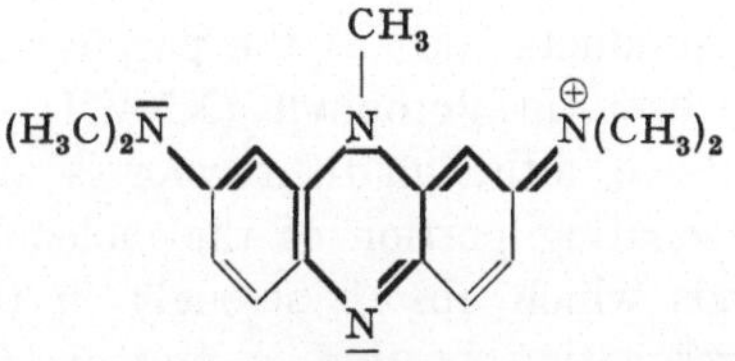

(XXII.)
A diazine dye (violet, λ_{max} at 565 mμ) (*23*).

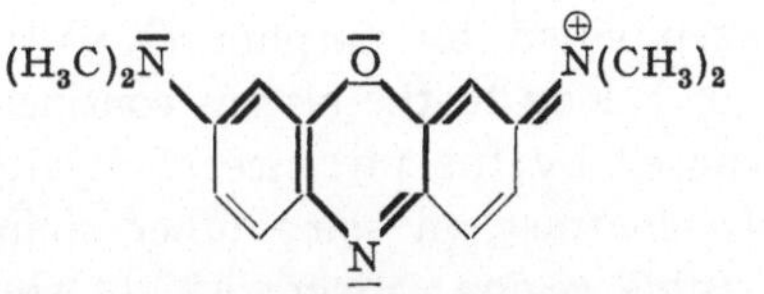

(XXIII.)
An oxazine dye (blue, λ_{max} at 666 mμ) (*13–15*).

shift of 116 mμ if we proceed from the xanthylium dye (XIX) to the oxazine dye (XXIII), and a bathochromic shift of 93 mμ during the change, eosine (XXIV) → iris blue (XXV).

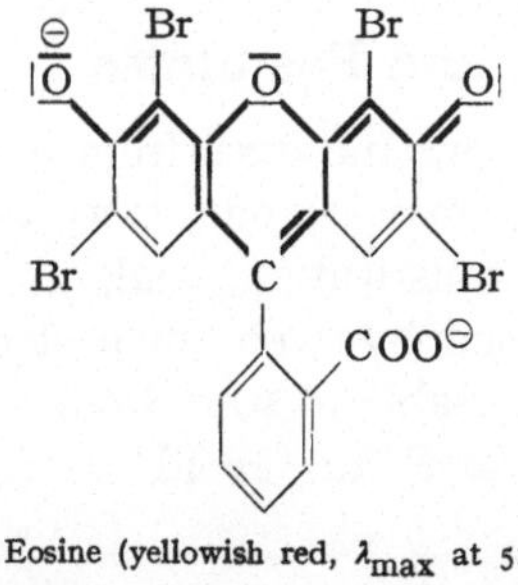

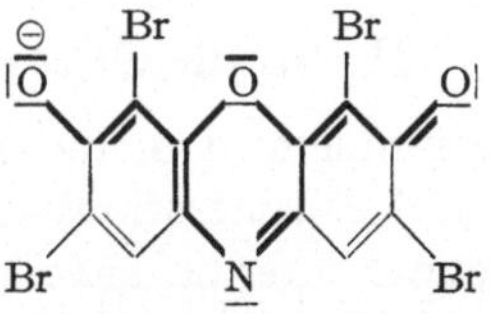

(XXV.) Iris blue (blue, λ_{max} at 609 mμ) (*28*).

(XXIV.) Eosine (yellowish red, λ_{max} at 516 mμ) (*28, 22*).

4. Some Other Types of Resonance Systems Causing Absorption of Visible Light.

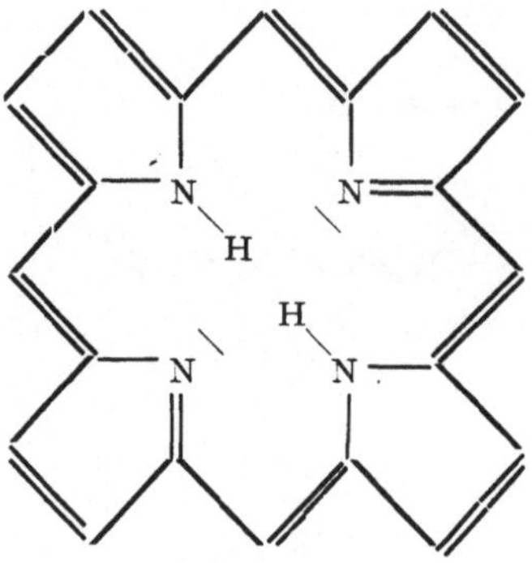

(XXVI.) Porphin.

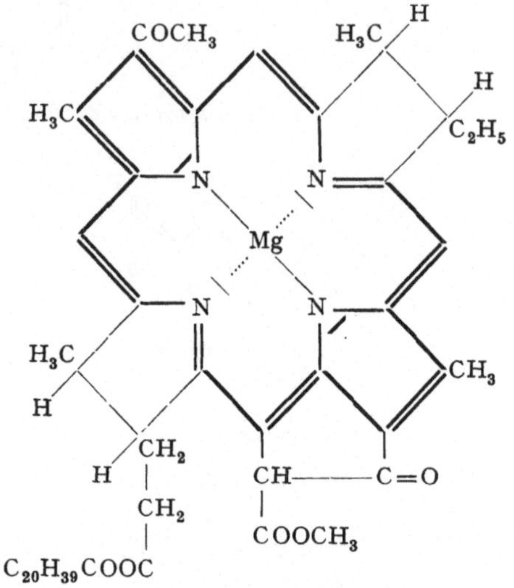

(XXVII.) Bacteriochlorophyll.

The types of resonance systems discussed in Sections 1–3 are present in many well-known dyes. However, numerous other colored substances, among them some important natural products such as the porphyrins, exemplified by porphin (XXVI) or bacteriochlorophyll (XXVII) do not belong to the classes considered above, although their color is also caused by the presence of a large resonating portion of the molecule. In contrast, in some other compounds which absorb strongly in the visible region, a surprisingly small resonating portion is present, for example in WURSTER's blue (XXVIII) or in the blue azulene (XXIX).

(XXVIII.) Wurster's blue.

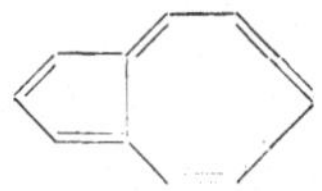

(XXIX.) Azulene.

II. Some General Principles and Postulates.

The problem of predicting the color of organic dyes from quantum mechanical considerations is necessarily an intricate one, since all dyes of practical interest have a complicated constitution, and, as shown in Chapter I, the color is strongly influenced by even small structural changes. To solve this problem it seems advisable to start from a highly simplified quantum mechanical model and to avoid a rigorous mathematical treatment.

Such a simple model is the *free electron gas model*, proposed by SOMMER-FELD *(83)* to describe the metallic state. A similar model was used to predict the color of simple organic dyes *(44, 45)*, [SIMPSON *(80)*], and this model was refined and extended to other dye classes *(46–61, 32–34, 81, 73)*. SCHMIDT *(75)* and PLATT *(72, 73)* have used free electron treatments for a discussion of aromatic hydrocarbons. BAYLISS *(4–7)* and recently, DALE *(21)* as well as LABHART *(62, 63)* have treated the absorption spectra of polyenes on a free electron basis. In the present article we will discuss those aspects of the free electron gas model which allow to predict the color of organic dyes. It will be shown in Chapter III (p. 190) as well as in a forthcoming paper (cf. p. 1(9) that the spectral features discussed in Chapter I can be explained on this basis.

1. Light Quanta and the EINSTEIN-BOHR Frequency Relation.

As is well known, light of frequency ν or of the vacuum wave length $\lambda = c/\nu$ behaves in a certain sense as consisting of energy quanta $h\nu = = hc/\lambda$, where $h = 6.624 \times 10^{-27}$ erg sec is PLANCK's constant and $c = 2.998 \times 10^{10}$ cm. sec^{-1} is the velocity of light. If a hydrogen atom is irradiated by ultraviolet light, a light quantum may be absorbed and the electron jumps from its ground state to an excited state *(Fig. 2)*. The energy, $h\nu = hc/\lambda$ of the light quantum equals the excitation

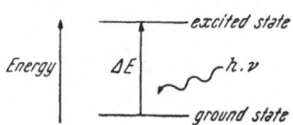

Fig. 2. EINSTEIN-BOHR frequency relation.

energy ΔE, i. e. the energy difference between the ground state and the excited state. Since the H atom can only exist in discrete excited states of given energy levels, narrow absorption lines appear in the ultra-violet spectrum.

If a dye solution is irradiated with visible light, a given dye molecule will absorb a light quantum and thus be raised from its ground state to an excited state. The stability of the excited state is limited, and after a very short time, e. g. less than 10^{-8} sec., the excited molecule will again dissipate its excitation energy.

Each pigment molecule can exist in many excited states which do not differ much in energy and which correspond to different states of rotation and oscillation of the electronically excited molecule; thus a broad absorption band and not a sharp line will be observed. Each of the neighboring excited states can be considered as corresponding to a certain jump of an electron from one state to another. The energy difference ΔE between the two electronic states is then related to the λ_{max} of the absorption band by equation (1). Consequently, according to equation (1),

$$\Delta E = hc/\lambda_{max} \tag{1}$$

since the excitation energy ΔE is known, the wave length λ_{max} of the absorption maximum can be given. Conversely, the spectrum of a dye solution approximately determines ΔE (see Table 1, p. 170).

Since, in order to find the excitation energy ΔE one must know the states of the electron before and after its excitation, we shall first mention the fundamental experiment by means of which the wave nature of electrons has been demonstrated.

2. Wave Particle Duality of the Electron and the DE BROGLIE Relationship.

Fig. 3 shows how S, symbolizing a point source of electrons of defined velocity v and an edge E produces on screen P a shadow of the

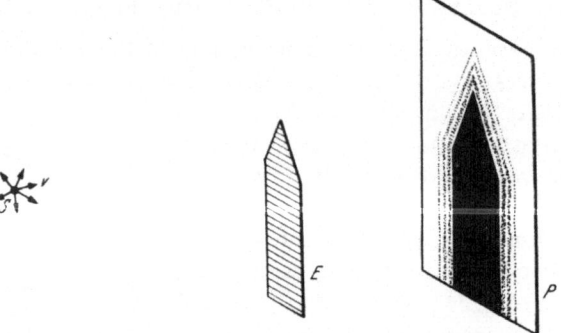

Fig. 3. Wave particle duality of electron. S point source of electrons of velocity v; E edge; P screen.

electron beam. This shadow is not sharp but contains alternating light and dark bands (64). The latter had to be interpreted as interference fringes and for the quantitative evaluation of this experiment it was necessary to postulate (a) that electron waves of the DE BROGLIE wave length,

$$\Lambda = \frac{h}{m\,v}, \tag{2}$$

(where $\Lambda =$ wave length of electron) are emitted, and (b) that the probability of finding an electron at a given spot on the screen is proportional to the square of the amplitude ψ of the wave at the spot. Here $m = 9.107 \times 10^{-28}$ g is the mass of the electron, and v is its velocity.

3. Standing Electron Waves. The PAULI Principle.
(Cf. e. g., 26, 37, 69, 71.)

We next consider an electron between two parallel walls that moves with velocity v perpendicular to the walls, at which the electron waves will be reflected. In general, the direct wave and the reflected wave cancel each other out, and a stationary electronic state is then impossible. No such cancelling effect is, however, realized, when the wave length has certain discrete values. In this case standing electron waves are produced exactly as in the vibrating string analog *(Fig. 4)*. The half wave length $\Lambda/2$ equals either the distance L of the walls (Fig. 4b), or $L/2$ (Fig. 4c), or $L/3$ (Fig. 4d); in general it equals L/n, n being an integer (zero excluded). Thus, the condition for standing waves is

$$\Lambda = \frac{2L}{n}, \qquad (3)$$

where $n = 1, 2, 3 \ldots$

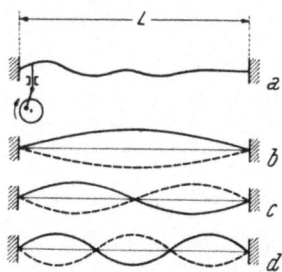

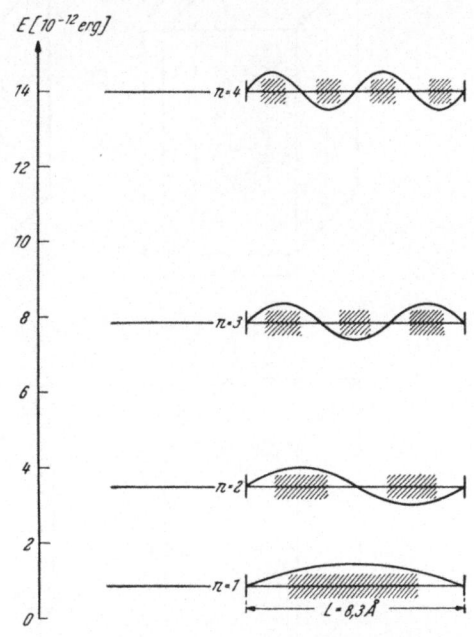

Fig. 4. Stationary oscillations b, c, d of a string fixed at the ends, and standing waves of electron between two parallel walls.

Fig. 5. Electron between two parallel walls. Distance $L = 8.3$ A. Energy levels, wave functions and electron cloud accumulations.

On the other hand, the wave length Λ of the electron is related to the velocity v by the equation (2), hence,

$$\Lambda = \frac{2L}{n} = \frac{h}{mv} \text{ or } v = \frac{hn}{2Lm}.$$

This means that the velocity of the electron can assume only discrete absolute values. The kinetic energy, $\frac{m}{2}v^2$, is thus restricted to the values:

$$E = \frac{m}{2}v^2 = \frac{m}{2}\left(\frac{hn}{2Lm}\right)^2 = \frac{h^2 n^2}{8mL^2}. \qquad (4)$$

Fig. 5 shows the energy levels for the states $n = 1$; 2; 3; and 4, obtained from equation (4) in the particular case, $L = 8.3$ Å.

According to postulate (b) the probability of finding the electron in a given region is proportional to the square of the amplitude of the wave in that region. Especially, it is very probable to find an electron

at the antinodes; the regions of high probability for finding the electron
are shaded in Fig. 5.

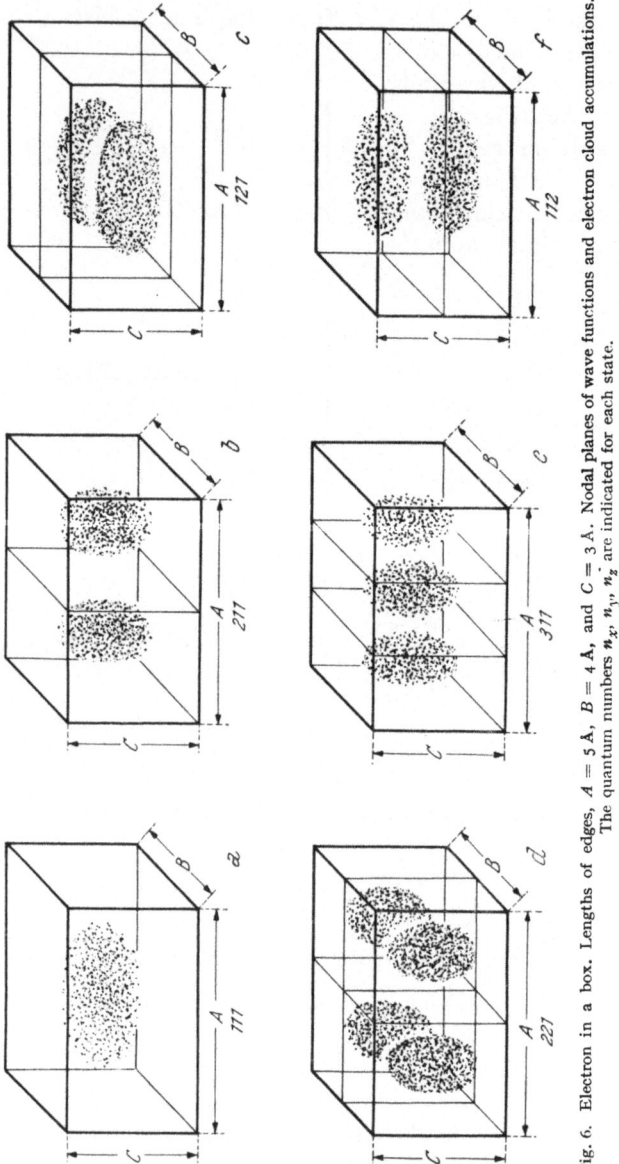

Fig. 6. Electron in a box. Lengths of edges, $A = 5\,\text{A}$, $B = 4\,\text{A}$, and $C = 3\,\text{A}$. Nodal planes of wave functions and electron cloud accumulations. The quantum numbers n_x, n_y, n_z are indicated for each state.

Let us now consider an electron forced to stay inside a rectangular
box with edge lengths A, B, and C. Standing waves are obtained only

under certain conditions, as in the case considered before; the kinetic energy, $E = \frac{m}{2} [v_x^2 + v_y^2 + v_z^2]$ (v_x, v_y and v_z are the respective components of the velocity parallel to A, B and C), is restricted to the discrete values:

$$E = \frac{m}{2} \left[\left(\frac{h\,n_x}{2\,A\,m} \right)^2 + \left(\frac{h\,n_y}{2\,B\,m} \right)^2 + \left(\frac{h\,n_z}{2\,C\,m} \right)^2 \right] = \frac{h^2}{8\,m} \left[\frac{n_x^2}{A^2} + \frac{n_y^2}{B^2} + \frac{n_z^2}{C^2} \right], \quad (5)$$

where n_x, n_y, n_z are integers (zero excluded). The reader will note the close analogy of the equations (4) and (5). Evidently, in each state the walls are nodal planes of the standing electron waves. We find no other nodal planes in the lowest electron state *(Fig. 6a)*, where the antinode is in the center of the box; hence, it is very probable to find the electron in the central region of the box, i. e. the electron cloud is accumulated about the middle of the box as pictured in Fig. 6a. In the next state (Fig. 6b) we find a nodal plane in the center, perpendicular to the longest edge, and thus two electron cloud accumulations. Figs. 6c–6f show the nodal planes and cloud accumulations in some further states. In *Fig. 7* the eight lowest energy levels are given and indicated by the quantum numbers n_x, n_y, n_z for the particular case $A = 5$ Å, $B = 4$ Å, $C = 3$ Å.

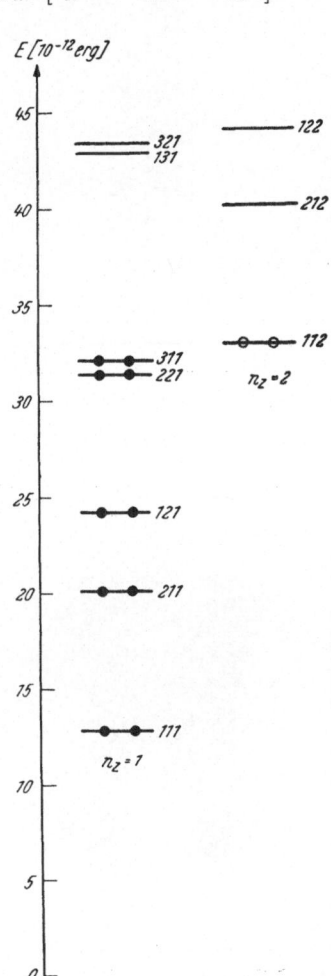

Fig. 7. Energy levels of electron in the box of Fig. 6 with lengths of edges $A = 5$ A, $B = 4$ A, and $C = 3$ A.

Let us now assume the presence of several electrons in the box and neglect their mutual repulsion. Not more than two of these electrons can be simultaneously in the same state (PAULI principle). Hence, when an even number of electrons and the ground state of the system are considered, it is found that all electron states up to a certain level are filled by two electrons each, and all other states are empty. If we consider ten electrons in the box (indicated in Fig. 7 by dots), we find that all occupied states have a quantum number $n_z = 1$ and that the total electron cloud (obtained by adding the distributions of Fig. 6a–6e) has five accumulations as indicated

in *Fig. 8a*. Two additional electrons (marked in Fig. 7 by circles) occupy a
state with $n_z = 2$ and with the cloud accumulations shown in Fig. 8b.

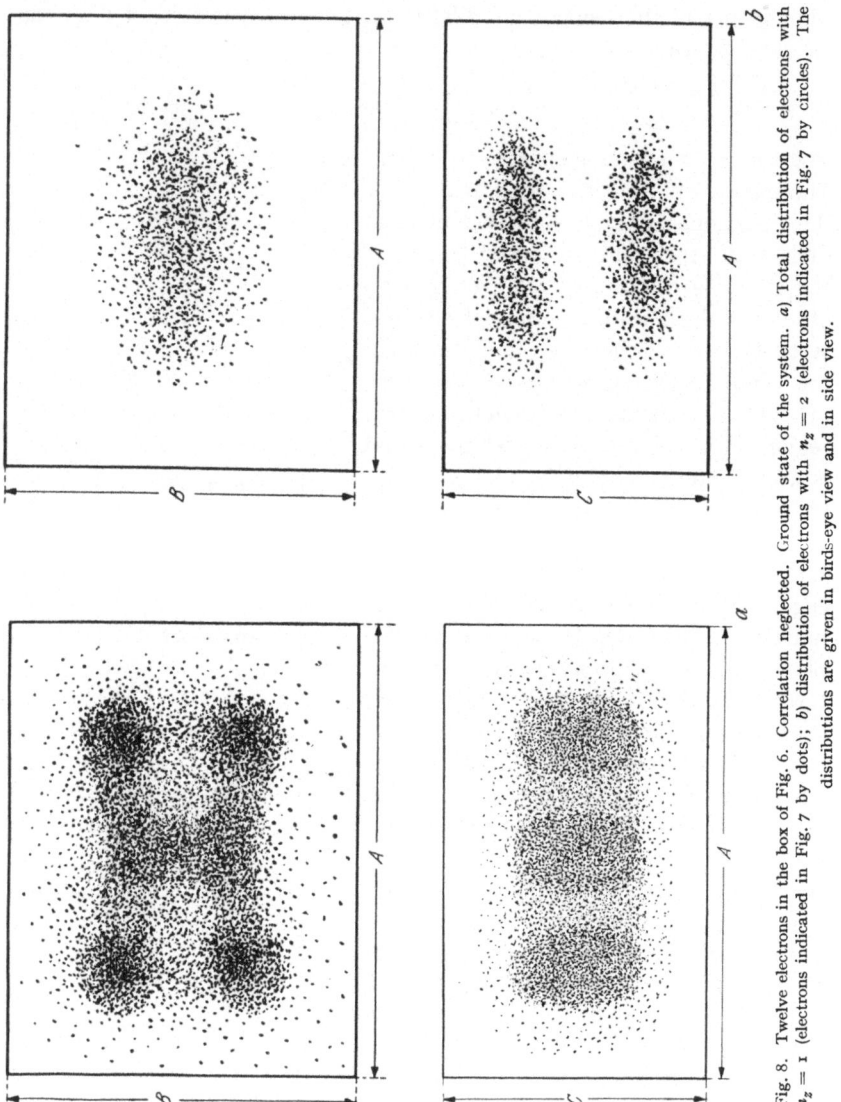

Fig. 8. Twelve electrons in the box of Fig. 6. Correlation neglected. Ground state of the system. *a*) Total distribution of electrons with $n_z = 1$ (electrons indicated in Fig. 7 by dots); *b*) distribution of electrons with $n_z = 2$ (electrons indicated in Fig. 7 by circles). The distributions are given in birds-eye view and in side view.

4. The Chemical Bond; σ and π Electron States.

We intend to give here only a crude picture of the chemical bond
and for a more profound study we refer to textbooks of quantum
chemistry (*26*, *37*, *69*, *71*).

Let us first consider a hydrogen atom. Like the electron in a cubic box, the electron in a hydrogen atom is forced (by the COULOMB field of the proton) to remain within a certain space. Thus, in both instances we find roughly the same shape of the electron cloud in the ground state.

The size of the electron cloud in this state of H is determined by the fact that of all imaginable sizes and shapes of clouds nature realizes that one which corresponds to the lowest energy level (variational principle).

The energy is the sum of the average potential energy $\overline{V}$ and the average kinetic energy $\overline{T}$ of the electron. It immediately follows from COULOMB's law that $\overline{V} = -a/\overline{r}$, where a is a constant and $\overline{r}$ is the average distance of the electron from the nucleus, i. e. $\overline{r}$ is a measure of the size of the cloud*. According to postulate (b) (p. 180), the size of the cloud is a measure of the wave length Λ, thus $\overline{r} \sim \Lambda$, and according to postulate (a), $1/\Lambda$ is proportional to the average velocity of the electron; $\overline{T} = m\,\overline{v^2}/2 \sim (1/\Lambda)^2$ or $\overline{T} = b/\overline{r}^2$, where b is a constant.

If we imagine that the electron cloud of the hydrogen atom could be compressed to a smaller volume than it actually occupies, then the average kinetic energy $\overline{T}$ of the electron would increase more rapidly than the average potential energy $\overline{V}$ would decrease; in contrast, if the cloud would expand, $\overline{V}$ would increase more rapidly than $\overline{T}$ would decrease. The actual size of the electron cloud is thus the result of a compromise between the COULOMB attraction $\overline{V}$, which tends to decrease the size of the cloud, and the requirement of the wave particle duality, which tends to increase $\overline{r}$. By minimizing the energy $\overline{V} + \overline{T} = -(a/\overline{r}) + (b/\overline{r}^2)$ the result $-\overline{T} = 1/2\,\overline{V}$ (virial theorem) is obtained and the value $\overline{r} = 2\,b/a \cong 10^{-8}$ cm. $\cong 1$ Å is found, since $a \cong 2 \times 10^{-19}$ erg cm., $b \cong 10^{-27}$ erg cm².

The variational problem here indicated, to find the actual shape of the electron cloud, i. e. the shape corresponding to the lowest energy, is mathematically identical with the problem of finding the eigenfunctions of the SCHRÖDINGER equation. We will consider this equation in a forthcoming paper.

If the electron cloud in H is compared with the cloud of an electron in a cubic box, the best approximation is obtained if the edge length of the box is 3.8 Å (61) (Fig. 9).

The electron cloud in the H_2 molecule can similarly be approximated by the cloud of an electron in the rectangular box of Fig. 10 (61). The two protons are surrounded by the cloud and they are attracted by COULOMB forces towards the center of the cloud. The equilibrium distance of the nuclei is reached if this attraction

---

* In fact, $V = -e_0^2/r$ and thus $\overline{V} = -e_0^2\,(\overline{1/r})$; e_0 is the charge of the electron. The term $(\overline{1/r})$ is proportional to $1/\overline{r}$ for clouds different in size but of the same shape. Thus, $\overline{V} \sim 1/\overline{r}$ or $\overline{V} = -a/\overline{r}$.

of the positive charge of the protons and the negative charge of the electron cloud is compensated by the COULOMB repulsion of the two protons. The chemical bond effective between the two H atoms can be explained roughly on this basis.

Let us now consider an ethylene molecule, $CH_2=CH_2$, which consists of two C nuclei, four protons and sixteen electrons. All nuclei are in the same plane, the "plane of the molecule". Each C nucleus is surrounded by two electrons (the inner shell electrons) which have a cloud similar in shape to the cloud of an H atom, but smaller in size (the linear scale factor is about 1/6, since the nuclear charge of C is six times the proton charge).

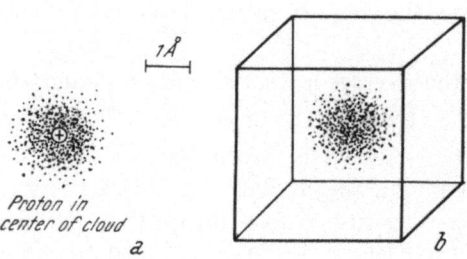

Proton in
center of cloud
a

1 Å

b

Fig. 9. Cloud of electron of the H atom (Fig. 9a) and of electron in a cubic box with edge length of 3.8 Å (Fig. 9b).

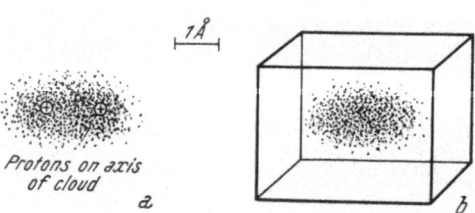

Protons on axis
of cloud
a

1 Å

b

Fig. 10. Cloud of electron of the H_2 molecule (Fig. 10a) and of electron in a rectangular box (edges, $A = 4$ Å, $B = C = 3$ Å) (b).

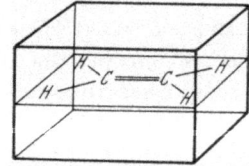

Fig. 11. Box model of ethylene. The twelve valence electrons are treated as electrons in the box as indicated in the figure. This box has the dimensions of that in Figs. 6—8 (pp. 182—184), (edges., $A = 5$ Å, $B = 4$ Å, $C = 3$ Å).

The other twelve electrons (the valence electrons) are located in a complicated COULOMB field which is determined by the coordinates of all nuclei and electrons; however, these electrons can roughly be treated as electrons constrained to the inside of the rectangular box of *Fig. 11*, in a potential which is assumed to be constant within this region*. In this approximation the cloud accumulations and energy levels of the different electronic states are given by Figs. 6 and 7 (pp. 182, 183). The twelve valence electrons occupy the six lowest levels marked in Fig. 7 by dots and circles. The large gap between the highest occupied state and the next state shows that the molecule has a "closed shell" structure and

---

* The approximate valence electron wave functions considered here are not usable in the regions near the C nuclei, since they do not fulfill the orthogonality requirement with the wave functions of the inner shell electrons of C. This is of no importance for the following considerations, since the regions where the inner shell electrons of C are concentrated are small compared with those in which the valence electrons are accumulated. A justification of this statement will be given elsewhere.

thus constitutes a stable planar system. Five pairs of valence electrons (the σ electrons) occupy states with $n_z = 1$, and thus have wave functions with an antinode in the plane of the molecule. One pair of valence electrons (the π electrons) (35, 36) occupies a state with $n_z = 2$; the plane of the molecule is here a nodal plane and the antinodes are above and below it. The total σ electron distribution given in Fig. 8a and the

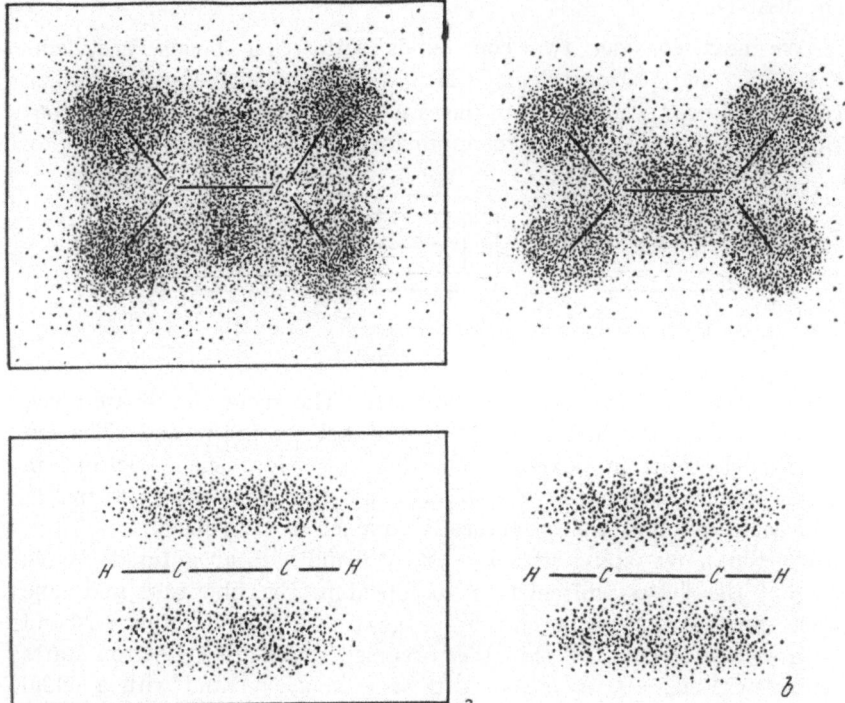

Fig. 12. Ethylene. Cloud accumulations of the σ electrons (birds-eye view) and of the π electrons (side view): a) box model, and b) result of a more profound treatment.

π electron distribution that appears in Fig. 8b are shown again in *Fig. 12a*, and are compared with the distribution in *Fig. 12b* that results from a more profound treatment.

It will be noted that the charge of the valence electrons is concentrated between C and H, and between C and C; and using the same arguments as in the case of the H_2 molecule, we conclude that these atoms are connected by chemical bonds and that the molecule must constitute a stable planar system.

The chemist describes this situation by saying that the C atom first forms a σ bond with each of its three direct neighbors. An electron pair is needed for each σ bond, and thus a σ bond can be considered as an electron pair bond. Both ligands contribute one valence electron each to the mutual σ bond. Thus, each

C atom uses three of its four valence electrons for σ bond formation. The fourth valence electron of each carbon atom forms the π bond between the two C atoms (68),

and each bond line in the formula 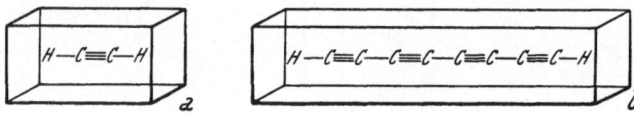 symbolizes a pair of valence electrons.

The single bond lines indicate pairs of σ electrons, the double bond line a pair of σ electrons and a pair of π electrons. The C—C—H and H—C—H bond angles are about 120°.

We next consider two compounds with triple bonds, acetylene, CH≡CH, and tetracetylene, CH≡C—C≡C—C≡C—C≡CH, which constitute linear systems. As in the case of ethylene, the valence electrons can be treated here, in rough approximation, as electrons in the box

Fig. 13. Box models of acetylene and tetracetylene* (edges, $A = 6$ A and $A = 14$ A, respectively, $B = C = 3$ A).

appearing in *Figs. 13a and b*, respectively. The energy levels are given in *Fig. 14a* for the first case and in *Fig. 14b* for the second. The ten valence electrons in acetylene and the thirty-four such electrons in tetracetylene occupy the lowest levels as shown in Figs. 14a and b, and again a closed shell structure is obtained in both instances. Three σ electron states (states with $n_y = n_z = 1$ and with an antinode in the axis of the linear molecule) are occupied in the first case and nine σ electron states in the second case. Again, each C atom forms a σ bond with each neighbor. All the other occupied states are π electron states (states with $n_y = 1$, $n_z = 2$ or with $n_y = 2$, $n_z = 1$ and with a nodal plane containing the axis of the molecule). Since each C atom uses two valence electrons for σ bond formation, the third and the fourth valence electrons contribute to the π electron system. For formal purposes we may distinguish between π_y and π_z electron states, with nodal plane perpendicular to the y and z axis, respectively. Each π electron level is occupied by four electrons (two π_z and two π_y electrons)*.

---

* The length of edge A in the Figures 10, 11, and 13 (pp. 186, 188) has been obtained by assuming that this length stretches 1.5 Å to either side of each terminal H atom. Using the known values of the H—H bond ($\cong 0.7$ Å), the C—H bond ($\cong 1.1$ Å), the C=C bond ($\cong 1.3$ Å) the C≡C bond ($\cong 1.2$ Å), the C—C bond in polyacetylenes ($\cong 1.4$ Å), and the H—C—H bond angle in ethylene (120°), we find in the case of H_2 (Fig. 10): $A = 2 \times 1.5 + 0.7 = 3.7 \cong 4$ Å; in the case of $H_2C=CH_2$ (Fig. 11): $A = 2 \times 1.5 + 2 \times 1.1 \cos 60° + 1.3 = 5.4 \cong 5$ Å; in the case of HC≡CH (Fig. 13): $A = 2 \times 1.5 + 2 \times 1.1 + 1.2 = 6.4$ Å $\cong 6$ Å; and in the case of HC≡C—C≡C—C≡C—C≡CH (Fig. 13): $A = 2 \times 1.5 +$

So far we have considered molecules containing C atoms with two and three neighbors in linear and planar arrangement. If a C atom has four neighbors, all four valence electrons are used to form tetrahedrally arranged σ bonds (68). The

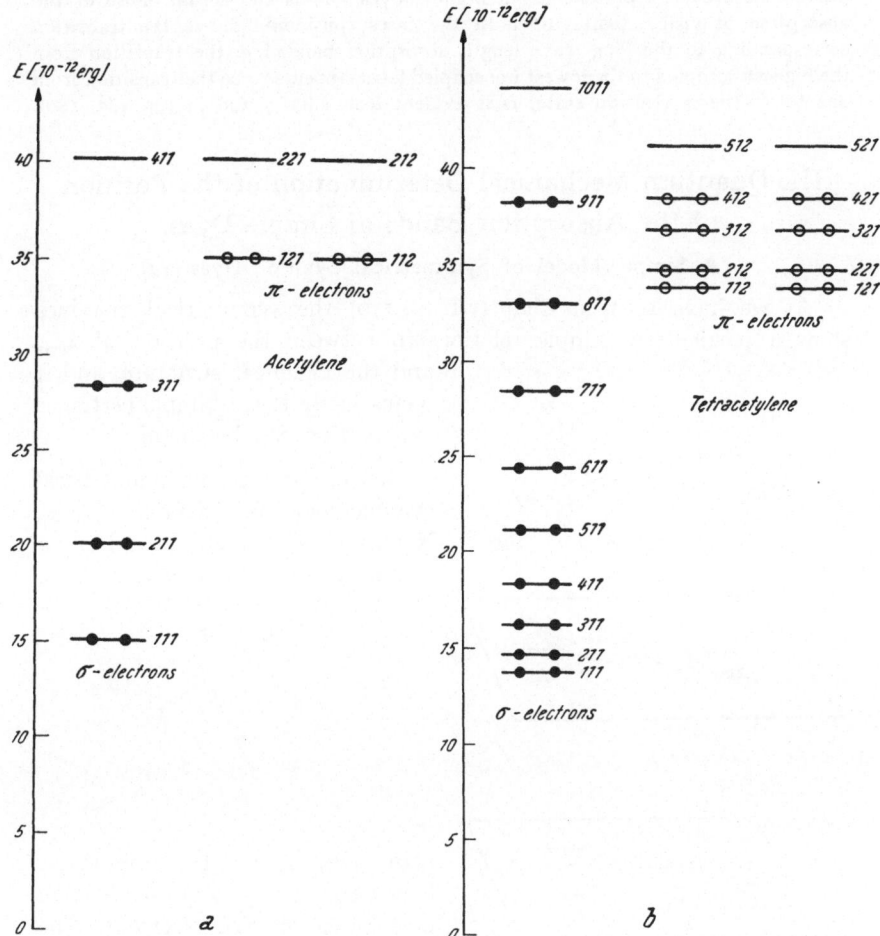

Fig. 14. Box models of acetylene and tetracetylene. Energy levels of valence electron states. Dots: σ electrons; circles: π electrons.

$+ 2 \times 1.1 + 4 \times 1.2 + 3 \times 1.4 = 14.2 \cong 14$ Å. Similarly, it has been assumed that the distance B or C stretches 1.5 Å to either side of the nuclei. Thus, in the case of H_2, acetylene and tetracetylene $B = C = 2 \times 1.5 = 3$ Å, and in the case of ethylene $C = 3$ Å; in the latter instance B was set to equal 4 Å, a value halfway between the limits $2 \times 1.5 + 2 \times 1.1 \sin 60° = 4.9$ Å and $2 \times 1.5 = 3$ Å, obtained by assuming that B stretches 1.5 Å to either side of the H and C nuclei, respectively. The above value, 1.5 Å of the distance of the walls from the H and C nuclei has been justified by a variational treatment of $H_2{}^+$, in which the wave functions of an electron in a box were used as test functions (61).

saturated C atom contributes no electron to a neighboring π electron system, and the π electron cloud does not extend over the region of the saturated atom.

As pointed out in the next Section, a π electron cloud which extends over a number of atoms is present in all organic dyes and is the actual cause of the absorption of visible light. Even in the cases considered above the transition corresponding to the long wave length absorption band, i. e. the transition from the highest occupied to the lowest unoccupied level, appears to be the transition from one to another π electron state, as is evident from Figs. 7 and 14 (pp. 183, 189).

III. Quantum Mechanical Determination of the Position of the Absorption Bands of Simple Dyes.

1. A Simple Model of Symmetrical Cyanine Dyes (44).

As we have seen in Chapter I (p. 170) the symmetrical cyanines show a particularly simple relationship between the position of λ_{max} and the chemical structure, and it seems likely that a simple pertinent explanation can be found.

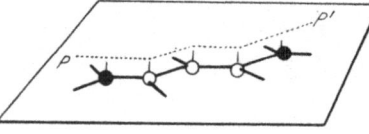

Let us consider the symmetrical cyanine molecule (XXX). The C, N and H atoms are linked by σ bonds and are located in a common

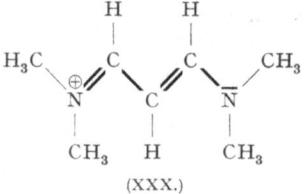

(XXX.)

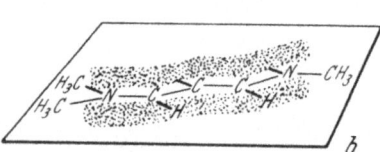

Fig. 15. Symmetrical cyanine (XXX). a) Molecular skeleton, and b) charge cloud of the π electron gas.

plane *(Fig. 15a)*. To form the σ bonds each C atom and each N atom uses three valence electrons. The fourth valence electron of each C atom and the remaining valence electrons of the two terminal N atoms contribute to the π electronic system. These electrons are placed in the electrostatic field of the molecular skeleton. They will be attracted by the positive charges of the C and N atoms, but (as in ethylene) their wave functions must have a node at all points of the plane of the molecule, as required by the Pauli principle. They form an electron gas which in the shape of a charge cloud stretches along the cyanine chain, both above and below the plane of the molecule (Fig. 15b). Since N has five valence electrons and three are used for σ bond formation, the N atom with no charge sign in (XXX) contributes two electrons to the π electron gas, and the N atom with the formal plus charge yields one electron. Since, in addition, each C atom contributes one π electron, we find a total of six π electrons or three π electron pairs in this case. Two of these pairs are indicated in (XXX) by double bonds, the third pair is symbolized by the bar above the N atom without charge sign.

Let us consider a single π electron in the electrostatic field of the rest of the molecule, and let us assume for a moment that it can only move in the direction of the zig-zag line connecting the C and N atoms in the chain, say along the central line of the cloud above the molecular skeleton (dotted line in Fig. 15a). The potential energy of the electron is then roughly constant along the chain, since the electron is practically in the COULOMB field of the nearest C or N atom only, while the field of the more distant C and N atoms is neutralized by other π electrons. Consequently, this

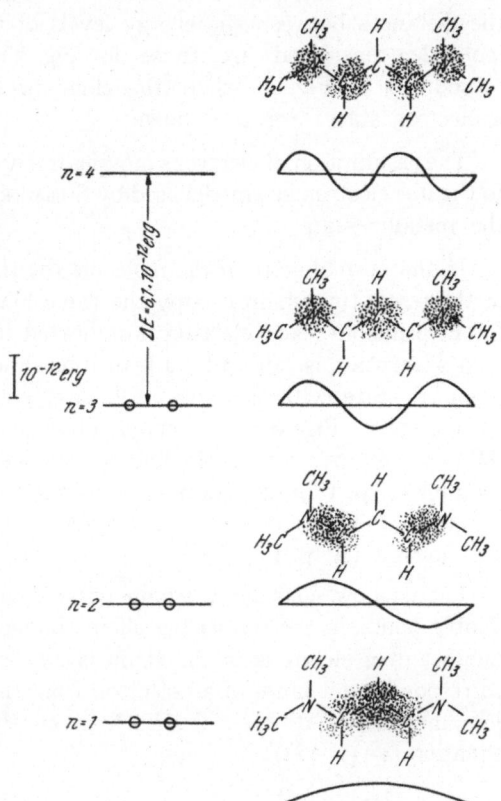

Fig. 16. Symmetrical cyanine (XXX). Charge clouds, standing electron waves and levels of the four most stable π electron states. The distances between the energy levels are the same as in the case of Fig. 5 (p. 181).

electron can move almost freely along the zig-zag line, but beyond the ends of the chain, about at the points P and P' in Fig. 15a, a strong COULOMB attraction force appears. Thus, our electron is about in the same condition as an electron between two walls; and the length L of the zig-zag line corresponds to the distance between the walls. Since L is about 8.3 Å (Fig. 15a), i. e. it equals the distance between the walls of Fig. 5 (p. 181), the states indicated in this Figure are very similar to the possible π electron states of the molecule considered.

It must be noted, however, that the component of the motion of the π electron in the directions perpendicular to the zig-zag line connecting the C and N atoms in the chain has been neglected here. As explained in Chapter II, the contribution of this component to the energy of the electron is constant for each π electron state. Hence, the distances between the energy levels of the π electron states in the molecule considered are those in Fig. 5. These energy levels are reproduced in *Fig. 16* where the cloud accumulations of the different π electron states are also shown.

The assumption of electrons moving freely along the chain is analogous to the free electron gas model used by SOMMERFELD (*83*) when he described the metallic state.

In the normal state of the molecule the three π electron pairs present in the resonating chain occupy the three lowest levels, according to the PAULI principle (each electron is indicated in Fig. 16 by a circle). The light absorption is caused by a transition of an electron from the highest occupied state with $n = 3$ to the next state with $n = 4$. According to Fig. 5 or Fig. 16 the energy difference between these states is $\Delta E = 6.1 \times 10^{-12}$ erg, and thus, the wave length $\lambda_{max} = h\,c/\Delta E$ (cf. Chapter II, p. 179) of maximum absorption is $\lambda_{max} = 3.3 \times 10^{-5}$ cm. $= 330$ mμ. This value is in good agreement with experimental data, viz. 309 mμ (cf. p. 174).

Let us now consider a symmetrical cyanine dye with j conjugated double bonds in the resonating chain connecting the two N atoms. The number of π electrons in the chain is $2j + 2$, and the absorption band corresponds to a jump of an electron from the level $j + 1$ to level $j + 2$. For the energy difference, ΔE, between these levels we obtain from equation (4) (p. 181):

$$\Delta E = \frac{h^2}{8\,m\,L^2}\,[(j + 2)^2 - (j + 1)^2] = \frac{h^2}{8\,m\,L^2}\,(2j + 3). \tag{6}$$

Thus (*44*),

$$\lambda_{max} = \frac{h\,c}{\Delta E} = \frac{8\,m\,c}{h}\,\frac{L^2}{2j + 3}. \tag{7}$$

This remarkable result indicates that (in this approximation) the position of the absorption band is determined by the chain length L and by the number of π electrons ($2j + 2$), since m, h and c are universal constants; λ_{max} does not depend upon any specific properties of the atoms along the chain.

The length of the zig-zag chain which connects the C and N atoms is $j \times 2l$, where l is the bond length of the chain elements ($l = 1.39$ Å, which is the bond length of a C–C one-and-a-half bond as found in benzene). The electron gas stretches by a certain length, αl, to both sides of each terminal N atom, thus

$$L = j \times 2l + 2\alpha l = 2l(j + \alpha);$$

and according to equation (7),

$$\lambda_{max} = \frac{8mc}{h} \frac{4l^2(j+\alpha)^2}{2j+3} = 127 \frac{(j+\alpha)^2}{j+(^3/_2)} \text{ (in m}\mu\text{).} \qquad (8)$$

In Chapter II we have shown that the imaginary walls are in distances of 1.5 Å from the nuclei; from this follows that the length αl on both ends is approximately 1.5 Å and therefore $\alpha \cong 1$. The value $\alpha = 1$

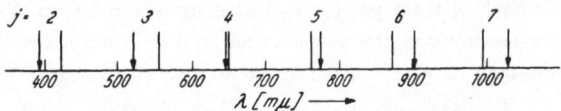

Fig. 17. Symmetrical cyanine type dyes. Homologous series of dyestuffs, with $j = 2, 3, 4, 5, 6, 7$.

Wave length of the absorption maxima: bars, observed values (according to Fig. 1, p. 171); and arrows, calculated values.

has already been used in the treatment of (XXX, p. 190) since $L = 6l = 6 \times 1.39 \text{ Å} = 8.3 \text{ Å}$ (Fig. 15). Generally speaking, the value of α depends on the end groups, and thus in a homologous series of dyes α is a constant for all members of the series. If we proceed from the homologous dye series (III, p. 172) and (XXX) to the series of (IV, p. 172), the value of α increases to 1.3 since the strongly polarizable phenyl end group is introduced instead of methyl. By introducing this value into equation (8) we find the values shown by arrows in *Fig. 17* (44). The experimental values given in Fig. 1d (p. 171) are indicated in Fig. 17 by bars. The excellent agreement between theory and experiment shows that the typical wave length shift $\Delta\lambda_{max}$ of about 100 mμ produced by the inclusion of an additional —CH=CH— group in a cyanine chain can be explained simply. In the case of sufficiently large values of j, equation (8) reduces to the expression $\lambda_{max} = 127(j + \alpha)$, since in this case $\frac{j+\alpha}{j+(^3/_2)} \cong 1$. Thus, beginning with a cyanine of j double bonds we find for the shift $\Delta\lambda_{max}$ produced by an additional —CH=CH— group the value $\Delta\lambda_{max} = 127(j + 1 + \alpha) - 127(j + \alpha) = 127$ mμ. This means that $\Delta\lambda_{max}$ is roughly 100 mμ, independent of j, i. e. independent of the already existing length of the chain.

The light absorption of the dyes (VI), (VII), and (IX) (cf. pp. 173, 174) can be treated by this model if the resonating portion of the molecule as indicated by heavy lines is considered. Actually, the resonating portion is extended over both sides of the benzene nuclei forming a branched π electron gas. The simplified treatment as an unbranched gas, i. e. neglecting the π electrons of the double bonds indicated by thin lines, leads to results very similar to those obtained by a more refined treatment based on a branched electron gas model.

The role of the terminal O atoms in (VII) and (IX) requires some consideration. Each such atom uses one electron for the formation of the σ bond with the neighboring C atom. Two pairs of electrons occupy states with an antinode of the wave function in the plane of the molecule, which are closely analogous to σ electron states in ethylene. These electron pairs clearly do not contribute to the π electron gas. Each of these pairs is indicated by a thin bar in (VII) and (IX). Since O has six electrons in the valence shell and $O^{\ominus}$ seven, one electron in O and two electrons in $O^{\ominus}$ are left and contribute to the π electron gas. The π electrons are indicated by the heavy bar at $O^{\ominus}$ and again by double bonds.

2. A Simple Treatment of the Aza Derivatives of Symmetrical Cyanine Type Compounds (49, 51).

Let us now consider the aza derivative of the symmetrical cyanine (XXX, p. 190), viz. compound (XXXI). The nitrogen atom located in the

(XXXI.)

center of the chain has two neighbors (in contrast to the terminal N atoms which have three neighbors). Two of its five valence electrons are used for σ bond formation and two are in a state similar to that of the σ electrons at the corresponding C in the parent compound (XXX), involved in a C—H bond. These two electrons are indicated in the formula by a thin bar. They have an antinode of the wave function in the plane of the molecule and thus are clearly not π electrons (in contrast to the two electrons indicated by a heavy bar at the terminal N atom without charge sign). Thus the central N atom contributes one electron to the π electron gas of the system, and the situation is very similar to that in (XXX).

It must, however, be considered, that N is more electro-negative, i. e. more electron-hungry, than C. If we introduce N instead of CH at the position indicated in *Fig. 18* by a circle, an electron, e. g., in the state

$n = 3$ of Fig. 18, will gain COULOMB energy and the energy level of this state will decrease. A similar change will take place in the levels of all states which show an accumulation of the charge cloud at the position

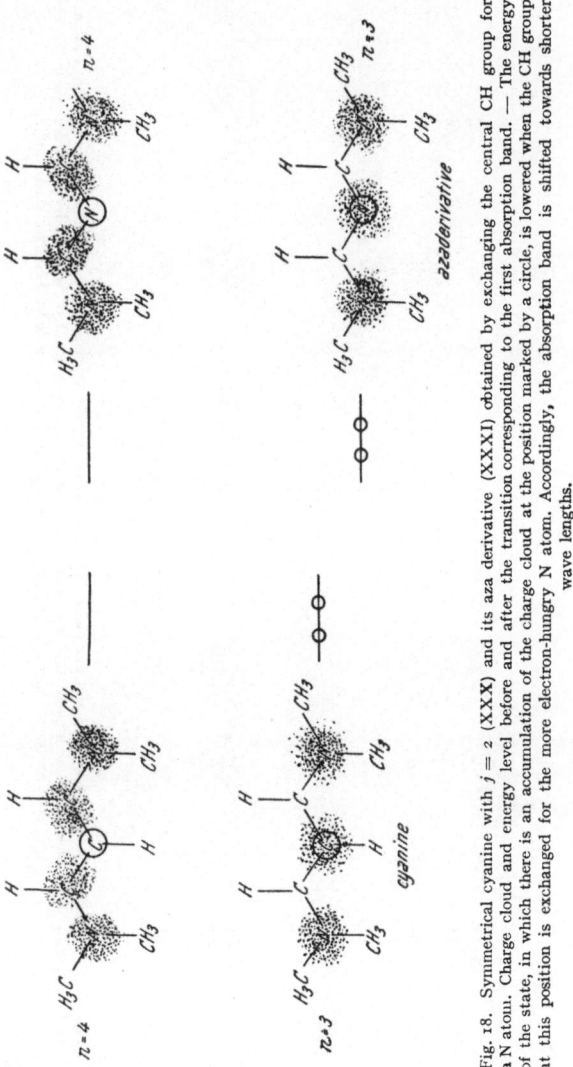

Fig. 18. Symmetrical cyanine with $j = 2$ (XXX) and its aza derivative (XXXI) obtained by exchanging the central CH group for a N atom. Charge cloud and energy level before and after the transition corresponding to the first absorption band. — The energy of the state, in which there is an accumulation of the charge cloud at the position marked by a circle, is lowered when the CH group at this position is exchanged for the more electron-hungry N atom. Accordingly, the absorption band is shifted towards shorter wave lengths.

where the N atom was introduced. However, the energy levels of those states which have a node at that position will be far less affected. In such a distribution of the charge density cloud the electron-hungry N atom does not benefit much. Consequently, the distance between

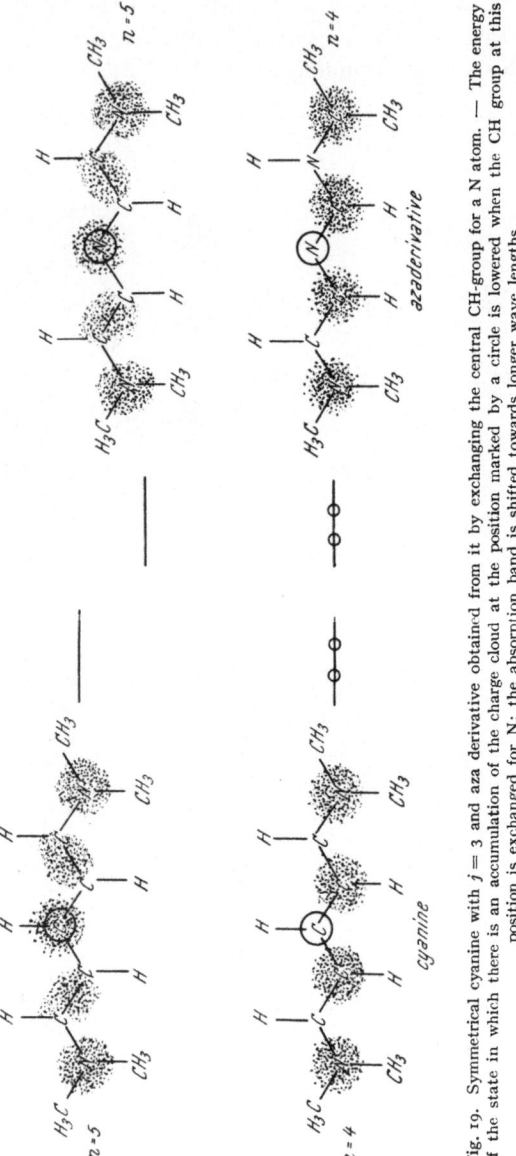

Fig. 19. Symmetrical cyanine with $j = 3$ and aza derivative obtained from it by exchanging the central CH-group for a N atom. — The energy of the state in which there is an accumulation of the charge cloud at the position marked by a circle is lowered when the CH group at this position is exchanged for N; the absorption band is shifted towards longer wave lengths.

the highest occupied state and the next higher state will increase as we go from the cyanine to the aza derivative, and the absorption band will be shifted towards shorter wave lengths (Fig. 18).

If we increase the chain length by the inclusion of an additional —CH=CH— group and replace the central CH by a N atom *(Fig. 19)*, a shift towards the longer wave length region is expected, because the highest occupied state has a node and the next higher state an antinode at the position where the N atom has been introduced (Fig. 19). Thus, the energy level of the highest occupied state will remain unaltered and the level of the next state will decrease; the distance between these will be shortened and a spectral shift towards the red will take place.

In general, when replacing the central CH group by a N atom, a shift towards shorter waves is expected if an even number of double bonds (j) is present between the two terminal N atoms; and an opposite shift will take place in case j is an odd number. This is exactly the observed behavior (cf. p. 175).

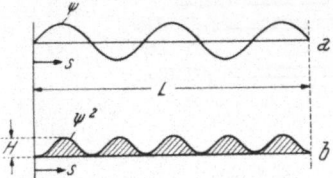

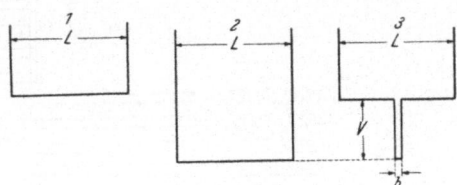

Fig. 20. Sine function (*a*) and square of this function (*b*). The wave function is normalized if the shaded area is unity. Since this area is $HL/2$, the normalization condition requires the value, $H = 2/L$.

Fig. 21.

These considerations can easily be placed on a quantitative basis. A perturbation treatment shows that the displacement ε of a given energy level, caused by exchanging carbon for nitrogen, is given by the expression

$$\varepsilon = - A\,\psi^2,\qquad(9)$$

where A is a constant, characteristic for the hetero atom, its value increasing with the electro-negativity of that atom. In the case of $=N{-}$, $A = 3.9 \times 10^{-20}$ erg cm., ψ is the value of the normalized wave function at the heteroatom. A wave function ψ (*s*) *(Fig. 20a)* is normalized, if the arbitrary amplitude of the wave is determined by the condition, that in a plot of the square of the wave function, the area under the curve (shaded area in Fig. 20b) is unity.

Equation (9) was obtained by assuming that the π electrons can be treated, in the parent molecule (XXX) as electrons in the potential 1 of *Fig. 21*, and in the aza derivative (XXXI) as electrons in potential 3, which has a narrow trough at the position of the N atom. Let us consider for a moment the potential 2 which is similar to 1 but its bottom is at the lowest level of potential 3, i. e. at level $- V$, if the bottom of potential 1 is chosen as zero energy. Clearly, the energy of a given state in potential 2 is $E_n - V$, if E_n is the energy of state n in potential 1; the corresponding expression in potential 3 is $E_n - VP$ where P is the probability of finding the electron in the trough. The shape of the wave function of state n changes somewhat if we proceed from potential 1 or 2 to potential 3; this change is here neglected, which is justified [according to the perturbation theory (*26, 37, 69, 71*)] if Vb is sufficiently small. Furthermore, if b is small compared to the wave length Λ of the electron, the normalized wave function ψ can be considered as being constant inside the trough. Thus, $P = \psi^2 b$ since $\psi^2 ds$ represents the probability of finding the electron in the interval ds. Hence, the energy $E_n + \varepsilon$ of state n in the potential 3 equals $E_n - V b \psi^2$ or $\varepsilon = - V b \psi^2$ which becomes identical with equation (9) if we set $A = V b$.

We will now consider sine wave functions extending over the length $L = 2 (j + \alpha) l$; in this case the value of the square of the normalized wave function at the antinodes is $\psi^2 = H = 2/L$ (see Fig. 20). Thus,

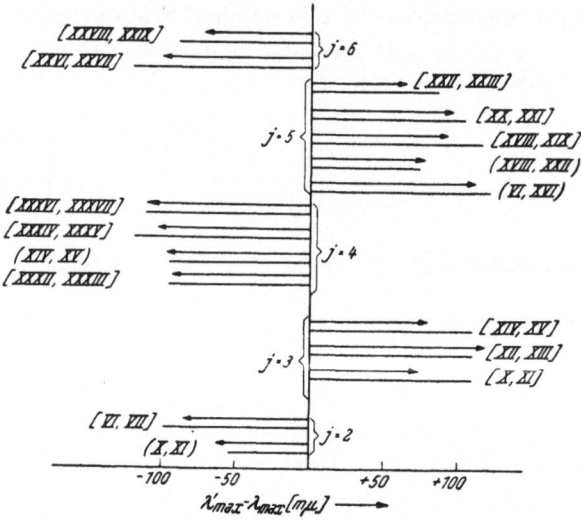

Fig. 22. Aza shift $\lambda'_{max} - \lambda_{max}$ of the absorption maximum. — λ_{max} refers to a symmetrical cyanine type, λ'_{max} to its aza derivative obtained by exchanging the central CH group for N; j is the number of double bonds in the resonating chain between the two terminal N or O atoms. Bar: observed shift; arrow: calculated shift. The roman numbers given in ( ) or [ ] indicate the pair of dyes to which the data refer: ( ) refers to a pair of dyes which we have considered here, and the numbers in ( ) are those assigned to the formulas of the cyanine and of its aza derivative; [ ] refers to a case not further considered here and the numbers in [ ] are those given in reference (51) to the cyanine type of dyes and aza derivatives.

in each state with an antinode in the middle of the chain (i. e. at the central N atom),

$$\varepsilon = - A \, (2/L) = - \frac{2 A}{2 (j + \alpha) l} = - \frac{3.9 \times 10^{-20}}{(j + \alpha) \, 1.39 \times 10^{-8}} = - \frac{2.8 \times 10^{-12}}{j + \alpha}$$

(perturbation energy in erg), and in each state with a node at that position, $\varepsilon = 0$. It is easy to calculate the expected aza shift $\lambda'_{max} - \lambda_{max}$ if we proceed for instance from (VI) (p. 173) to (XVI) (p. 176). In this case $j = 5$. By introducing the experimental value $\lambda_{max} = 603$ mμ of the parent molecule (VI) into equation (8, p. 193), the value $\alpha = 0.55$ is obtained. Consequently, for the highest occupied state $\varepsilon = 0$, and for the next state,

$$\varepsilon = - \frac{2.8 \times 10^{-12}}{5 + 0.55} = - 0.50 \; 10^{-12} \; \text{erg}.$$

The excitation energy $\Delta E'$ of the aza derivative is $\Delta E' = \Delta E + \varepsilon$, where

$$\Delta E = \frac{h\,c}{\lambda_{max}} = \frac{6.62 \times 10^{-27} \times 3.00 \times 10^{-10}}{6.03 \times 10^{-5}} = 3.30 \times 10^{-12} \text{ erg}$$

is the excitation energy of the parent molecule. Thus,

$$\Delta E' = 3.30 \times 10^{-12} - 0.50 \; 10^{-12} = 2.80 \times 10^{-12} \text{ erg,}$$

$$\lambda'_{max} = \frac{h\,c}{\Delta E'} = 7.09 \; 10^{-5} \text{ cm} = 709 \text{ m}\mu,$$

and

$$\lambda'_{max} - \lambda_{max} = 709 - 603 = + 106 \text{ m}\mu.$$

The observed shift (cf. p. 176) is $+ 122$ mμ. In *Fig. 22* the aza shift $\lambda'_{max} - \lambda_{max}$ (arrows) calculated in this way in a number of cases is compared with the observed shift (bars). An excellent verification of

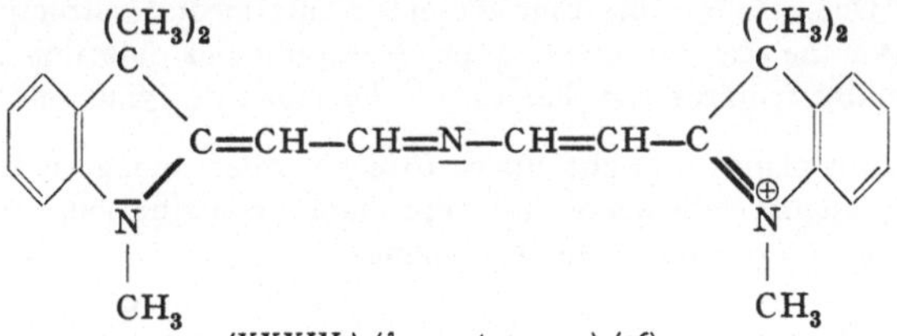

(XXXIII.) (λ_{max} at 544 mμ) (*16*).

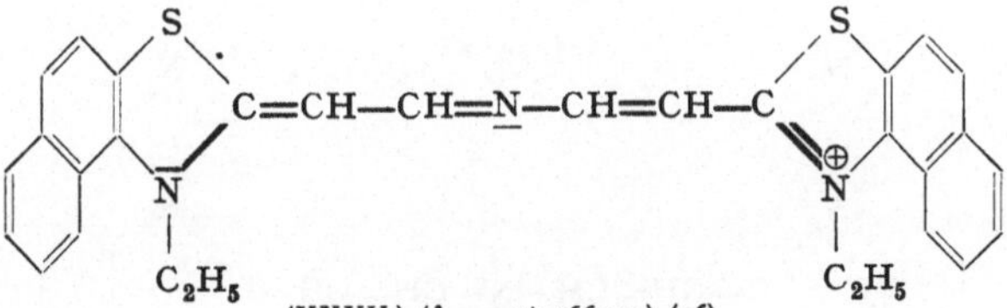

(XXXV.) (λ_{max} at 566 mμ) (*16*).

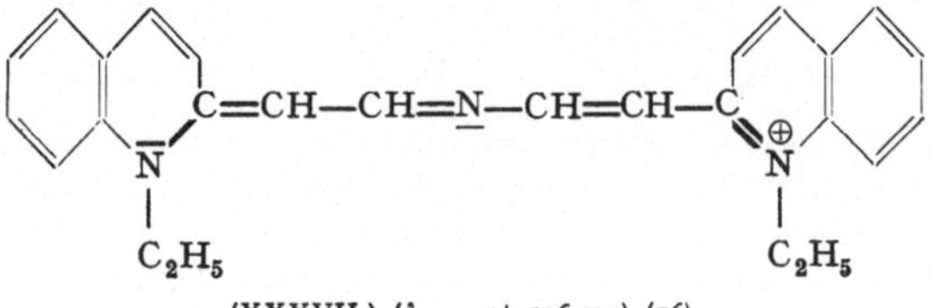

(XXXVII.) (λ_{max} at 596 mμ) (*16*).

Chart 3.

the theory is thus obtained. Although no aza derivative of a dye with $j = 4$ was known when this approximation was first proposed (*49, 51*), recently BROOKER (*16*) has prepared several such dyes and he has kindly given permission to the writer to publish some of his pertinent data.

These compounds are, (XV, p. 175), (XXXIII), (XXXV), and (XXXVII) *(Chart 3).**

The corresponding cyanines are designated by the numbers (XXXII), (XXXIV) and (XXXVI), and their respective maxima are located at 640 mμ, 685 mμ, and 708 mμ (*16*).

In the above discussion we have considered the potential trough of the central N atom in an aza cyanine but we have neglected the troughs of the terminal N atoms in a cyanine or aza cyanine molecule. However, this neglect has no

* *Addendum.* Dr. BROOKER has asked me to present at this point his following interpretation of the aza shift in the case of dyes (XIV; XV), (XXXII; XXXIII), (XXXIV; XXXV) and (XXXVI; XXXVII), which in some fundamental points differs from that given above:

"*Theoretical treatment of the absorption of meso-aza-dicarbocyanines.* Replacement of the pentamethine chain of a symmetrical dicarbocyanine by $=$CH—CH$=$N—CH$=$CH— brings about a considerable hypsochromic shift in each of the four cases studied. At first sight it seems possible to account for this in two ways: One can imagine that either an intermediate structure of type A is responsible for the effect or one of type B. An obvious objection of structure A is that it formally requires the 'halves' of the molecule lying on either side of $\overset{\oplus}{=}$N$=$ to occupy planes at right angles to each other, though it is possible that this structure might still make an appreciable contribution, even with the heterocyclic rings of the dye held in a plane.

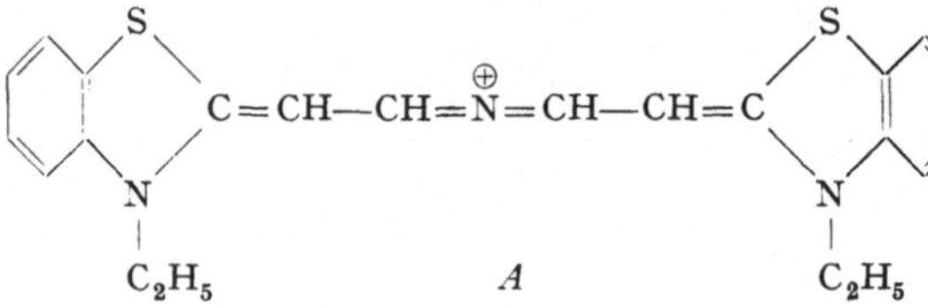

Aside from this, however, A structure would be expected to become increasingly important the *lower* the basicity of the rings, while the B structure should become more important with increasing basicity of the rings. Since the hypsochromic effects would be expected to be proportional to the significance of the intermediate structures, whether these are of type A or B, comparison of the hypsochromic shifts in a series of symmetrical dyes, where the nuclei differ in basicity, might enable a choice to be made between A and B structures.

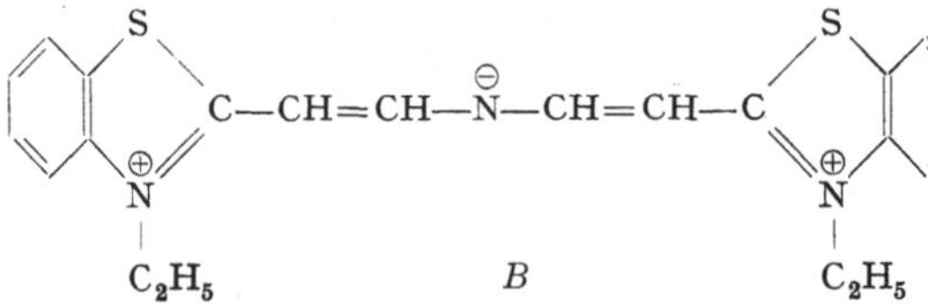

If the intermediate structures of type B were alone significant, the hypsochromic shifts would be expected to increase steadily from indolenine through to β-naphthothiazole (or-2-quinoline), but actually this increase in shift is relatively small. The results seem to be consistent with contributions from structures of both types A and B, the relative participation of B increasing slightly with increasing basicity of the heterocyclic nuclei of the dye."

appreciable effect on the value of the excitation energy, since the wave functions of an electron, both in the highest occupied state and in the next state, have an electron cloud accumulation at each terminal N nucleus. Thus, when an electron jumps from one state to the next higher one, it neither gains nor looses much COULOMB energy of attraction towards these nuclei.

We intend to consider in the forthcoming article various hetero atom groups.

Constant A in equation (9, p. 197) is greater for $=\overset{\oplus}{N}-$ than for $=\underline{N}-$, since the π
$\qquad\qquad\qquad\qquad\qquad\qquad\qquad\qquad\qquad\qquad\qquad |$
$\qquad\qquad\qquad\qquad\qquad\qquad\qquad\qquad\qquad\qquad\quad R$

electrons are located in the field of $\overset{\diagdown\overset{\oplus\oplus}{}}{\underset{\diagup}{N}}-$, while in $=\underline{N}-$ they are in the field

of $-\overset{\oplus}{\underset{\underline{=}}{N}}-$ which is obviously smaller. Again, a larger A value is found in the case of $-\underline{\underline{O}}-$, since oxygen is more electro-negative than nitrogen. The following value is found for first row elements,

$$A = (Z_{eff;\ heteroatom} - 3.25) \times 6.0 \times 10^{-20}\ \text{erg cm.} \tag{10}$$

whereby $Z_{eff;\ heteroatom}$ is the SLATER effective charge (82) of the heteroatom ($Z_{eff;\ N^+} = 3.90$; $Z_{eff;\ N^{++}} = 4.25$; $Z_{eff;\ O^+} = 4.55$; $Z_{eff;\ O^{++}} = 4.90$), and 3.25 is the SLATER effective charge of carbon.

Thus, from equation (10) the following A values are found:

for $-\underline{N}=$ or $-\overset{\ominus}{\underline{N}}-$: $A = 3.9 \times 10^{-20}$ erg cm. $\qquad$ (10a)

for $-\overset{\oplus}{\underline{N}}=$ or $-\underline{N}-$: $A = 6.0 \times 10^{-20}$ erg cm. $\qquad$ (10b)
$\quad\ \ R \qquad\quad\ \ R$

for $|\underline{O}=$ or $|\overset{\ominus}{\underline{O}}-$: $A = 7.8 \times 10^{-20}$ erg cm. $\qquad$ (10c)

for $-\overset{\oplus}{\underline{O}}=$ or $-\underline{O}-$: $A = 9.9 \times 10^{-20}$ erg cm. $\qquad$ (10d)

References.

1. ARENS, H.: Farbmetrik. Berlin: Akad. Verlagsges. 1941.
2. ARMITAGE, J. B., C. L. COOK, E. R. H. JONES and M. C. WHITING: Researches on Acetylenic Compounds. Part XXXVI. The Synthesis of Symmetrical Conjugated Triacetylenic Compounds. J. Chem. Soc. (London) **1952**, 2010.
3. ARMITAGE, J. B., E. R. H. JONES and M. C. WHITING: Researches on Acetylenic Compounds. Part XXXVII. The Synthesis of Conjugated Tetra-acetylenic Compounds. J. Chem. Soc. (London) **1952**, 2014.
4. BAYLISS, N. S.: A "Metallic" Model for the Spectra of Conjugated Polyenes. J. Chem. Physics 16, 287 (1948).
5. — The Potential Energy in Conjugated Polyenes and the Effective Nuclear Charge of the Carbon Atom. J. Chem. Physics 17, 1353 (1949).
6. — Conjugated Compounds. II. Simple Potential-energy Functions, Absorption Spectra, and Ionization in Linear Polyenes. Austral. J. Sci. Res., Ser. A 3, 109 (1950).
7. — The Free-Electron Approximation for Conjugated Compounds. Quart. Rev. Chem. Soc. (London) 6, 319 (1952).
8. BEILENSON, B., N. I. FISHER and F. M. HAMER: A Comparison of the Absorption Spectra of some Typical Unsymmetrical Cyanine Dyes. Proc. Roy. Soc. (London), Ser. A 163, 138 (1937).

9. BOHLMANN, F.: Konstitution und Lichtabsorption, VI. Mitt.: Zur Deutung von Polyacetylen-Spektren, sowie Darstellung von Bis-*tert*.-butyl-decapentain-(1.3.5.7.9). Chem. Ber. **86**, 63 (1953).

10. — Polyacetylene, IV. Mitt.: Darstellung von Di-*tert*.-butyl-polyacetylenen. Chem. Ber. **86**, 657 (1953).

11. BOHLMANN, F. und H. J. MANNHARDT: Konstitution und Lichtabsorption, VIII. Mitt.: Darstellung und Lichtabsorption von Dimethyl-polyenen. Chem. Ber. **89**, 1307 (1956).

12. BOUMA, P. J.: Farbe und Farbwahrnehmung. Eindhoven: Philips. 1951.

13. BROOKER, L. G. S.: Sensitizing and Desensitizing Dyes. In: C. E. K. MEES, The Theory of the Photographic Process, 2nd Ed., Chapter 11, p. 371. New York: Macmillan. 1954.

14. — Spectra of Dye Molecules. Absorption and Resonance in Dyes. Rev. Mod. Physics **14**, 275 (1942) (cf. especially p. 289).

15. — Resonance in Organic Chemistry. In: Advances in Nuclear Chemistry and Theoretical Organic Chemistry **4**, 130. New York: Interscience Publ. 1945.

16. — Private communication.

17. BROOKER, L. G. S., G. H. KEYES and W. W. WILLIAMS: Color and Constitution. V. The Absorption of Unsymmetrical Cyanines. Resonance as a Basis for a Classification of Dyes. J. Amer. Chem. Soc. **64**, 199 (1942).

18. BROOKER, L. G. S., R. H. SPRAGUE, C. P. SMYTH and G. L. LEWIS: Color and Constitution. I. Halochromism of Anhydronium Bases Related to the Cyanine Dyes. J. Amer. Chem. Soc. **62**, 1116 (1940).

19. BURAWOY, A.: Licht-Absorption und Konstitution, II. Mitt.: Heteropolare organische Verbindungen. Ber. dtsch. chem. Ges. **64**, 462 (1931).

20. COOK, C. L., E. R. H. JONES and M. C. WHITING: Researches on Acetylenic Compounds. Part XXXIX. General Routes to Aliphatic Polyacetylenic Hydrocarbons and Glycols. J. Chem. Soc. (London) **1952**, 2883.

21. DALE, J.: The Free-Electron Model, "Overtone" Bands, and Vibrational Structure in Absorption Spectra of Polyenes and Polyenynes. Acta Chem. Scand. **11**, 265 (1957). — For connected papers by the same author cf. Acta Chem. Scand. **8**, 1235 (1954); **11**, 640, 650 and 971 (1957).

22. DEWAR, M. J. S.: The Electronic Theory of Organic Chemistry, pp. 311–312. Oxford: Clarendon Press. 1949.

23. — Colour and Constitution. Part I. Basic Dyes. J. Chem. Soc. (London) **1950**, 2329.

24. DILTHEY, W.: Beiträge zur Kenntnis der Triphenylmethanfarbstoffe. J. prakt. Chem. [2] **109**, 273 (1925).

25. DILTHEY, W. und R. WIZINGER: Über eine Erweiterung der WITTSchen Farbtheorie auf koordinationschemischer Grundlage. J. prakt. Chem. [2] **118**, 321 (1928).

26. EYRING, H., J. WALTER and G. E. KIMBALL: Quantum Chemistry. New York: J. Wiley & Sons. 1944.

27. FERGUSON, L. N.: Relationships between Absorption Spectra and Chemical Constitution of Organic Molecules. Chem. Rev. **43**, 385 (1948).

28. FIERZ-DAVID, H. E.: Künstliche organische Farbstoffe. Berlin: Springer. 1926.

29. FISHER, N. I. and F. M. HAMER: A Comparison of the Absorption Spectra of Some Typical Symmetrical Cyanine Dyes. Proc. Roy. Soc. (London), Ser. A **154**, 703 (1936).

30. GYSLING, H. und G. SCHWARZENBACH: Metallindikatoren II. Beziehungen zwischen Struktur und Komplexbildungsvermögen bei Verwandten des Murexids. Helv. Chim. Acta **32**, 1484 (1949).

31. HAMER, F. M.: The Cyanine Dyes. Quart. Rev. Chem. Soc. (London) **4**, 327 (1950).

32. HORNIG, J. F., WALTER HUBER and H. KUHN: Nature of the Free Electron Approximation: The Simple Example of the H_2^+ Ion. J. Chem. Physics **25**, 1296 (1956).

33. HUBER, WALTER, J. F. HORNIG und H. KUHN: Über den Potentialverlauf entlang der Molekülkette im verfeinerten eindimensionalen Elektronengasmodell. Untersuchungen am Beispiel des Wasserstoffmolekülions. Z. physik. Chemie. Neue Folge **9**, 1 (1956).

34. HUBER, WERNHARD, H. KUHN und WALTER HUBER: Elektronengasmodell zur quantitativen Deutung der Lichtabsorption von organischen Farbstoffen. II. Teil C. Farbstoffe vom Acridintypus. Helv. Chim. Acta **36**, 1597 (1953).

35. HÜCKEL, E.: Zur Quantentheorie der Doppelbindung. Z. Physik **60**, 423 (1930).

36. — Zur modernen Theorie ungesättigter und aromatischer Verbindungen. Z. Elektrochem. **61**, 866 (1957).

37. KAUZMANN, W.: Quantum Chemistry. An Introduction. New York: Academic Press. 1957.

38. KNOTT, E. B.: The Colour of Organic Compounds. Part I. A General Colour Rule. J. Chem. Soc. (London) **1951**, 1024.

39. KNOTT, E. B. and L. A. WILLIAMS: The Colour of Organic Compounds. Part III. A New Method of Assessing the $\pm M$ Effect of Heterocyclic Nuclei. J. Chem. Soc. (London) **1951**, 1586.

40. KÖNIG, W.: Über den Begriff der „Polymethinfarbstoffe" und eine davon ableitbare allgemeine Farbstoff-Formel als Grundlage einer neuen Systematik der Farbenchemie. J. prakt. Chem. [2] **112**, 1. (1926).

41. KÖNIG, W., K. HEY, FR. SCHULZE, E. SILBERKWEIT und K. TRAUTMANN: Über Strepto- und Heterocyclo-Polymethin-Farbstoffe aus Furfurol und dessen Vinylen-Homologen. Ber. dtsch. chem. Ges. **67**, 1274 (1934).

42. KÖNIG, W. und W. MEIER: Über Thio- und Oxocyanine. (6. Mitt. über Cyaninfarbstoffe.) J. prakt. Chem. [2] **109**, 324 (1925).

43. KÖNIG, W. und K. SEIFERT: Über das Vinylen-Homologe des „MICHLERschen Hydrolblaus". Ber. dtsch. chem. Ges. **67**, 2112 (1934).

44. KUHN, H.: Elektronengasmodell zur quantitativen Deutung der Lichtabsorption von organischen Farbstoffen. I. Helv. Chim. Acta **31**, 1441 (1948).

45. — Free Electron Model for Absorption Spectra of Organic Dyes. J. Chem. Physics **16**, 840 (1948).

46. — A Quantum-Mechanical Theory of Light Absorption of Organic Dyes and Similar Compounds. J. Chem. Physics **17**, 1198 (1949).

47. — Theoretische Deutung der Lichtabsorption organischer Farbstoffe. Z. Elektrochem. **53**, 165 (1949).

48. — Quantenmechanische Behandlung von Farbstoffen mit verzweigtem Elektronengas. Helv. Chim. Acta **32**, 2247 (1949).

49. — Lichtabsorption organischer Farbstoffe. Chimia **4**, 203 (1950).

50. — Elektronengasmodell zur quantitativen Deutung der Lichtabsorption von organischen Farbstoffen. II. Teil A. Ermittlung der Intensität von Absorptionsbanden. Helv. Chim. Acta **34**, 1308 (1951).

51. — Elektronengasmodell zur quantitativen Deutung der Lichtabsorption von organischen Farbstoffen. II. Teil B. Störung des Elektronengases durch Heteroatome. Helv. Chim. Acta **34**, 2371 (1951).

52. — Chemische Bindung und Zustände von Elektronen in Molekülen. Experientia **9**, 41 (1953).

53. — Lichtabsorption organischer Farbstoffe. (Neuere Ergebnisse der Elektronengasmethode.) Chimia **9**, 237 (1955).

54. Kuhn, H.: Note on the Branching Condition in the One-Dimensional Free Electron Gas Model. J. Chem. Physics **22**, 2098 (1954).

55. — Verfeinertes eindimensionales Elektronengasmodell. Verzweigungsbedingung und Orthogonalitätsrelation. Z. Naturforsch. **9** a, 989 (1954).

56. — Physical Basis of the Free-Electron Gas Model of Branched Molecules. J. Chem. Physics **25**, 293 (1956).

57. — Die Verzweigungsbedingung in der Elektronengasmethode. Z. Elektrochem. **58**, 219 (1954).

58. — Zweidimensionales Elektronengasmodell organischer Farbstoffe. Angew. Chem. **69**, 239 (1957).

59. — Neuere Untersuchungen über das Elektronengasmodell organischer Farbstoffe. Angew. Chem. **70** (1958) [in print].

60. Kuhn, H., Walter Huber et F. Bär: Modèle de l'électron libre amélioré à une dimension. Position et structure des bandes d'absorption des poly-ynes et polyènes. Longueurs des liaisons. Proc. Intern. Conf., Calcul des fonctions d'onde moléculaires. Paris, 1957 (in print).

61. Kuhn, H. und Wernhard Huber: Kastentestfunktion als Näherung für die Wellenfunktion des Elektrons im Wasserstoffatom und im Wasserstoffmolekelion. Helv. Chim. Acta **35**, 1155 (1952).

62. Labhart, H.: FE Theory Including an Elastic σ Skeleton. I. Spectra and Bond Lengths in Long Polyenes. J. Chem. Physics **27**, 957 (1957).

63. — FE Theory Including an Elastic σ Skeleton. II. Changes of Molecule Dimensions due to the Optical Excitation. J. Chem. Physics **27**, 963 (1957).

64. Laue, M. v.: Materiewellen und ihre Interferenzen. Leipzig: Becker und Erler. 1944.

65. Lewis, G. N.: Rules for the Absorption Spectra of Dyes. J. Amer. Chem. Soc. **67**, 770 (1945).

66. Maccoll, A.: Colour and Constitution. Quart Rev. Chem. Soc. (London) **1**, 16 (1947).

67. Orndorff, W. R., R. C. Gibbs, S. A. McNulty and C. V. Shapiro: The Absorption Spectra of Fuchsone, Benzaurin and Aurin. J. Amer. Chem. Soc. **49**, 1545 (1927).

68. Pauling, L.: The Nature of the Chemical Bond and the Structure of Molecules and Crystals. Ithaca, N. Y.: Cornell Univ. Press. 1945.

69. Pauling, L. and E. B. Wilson, Jr.: Introduction to Quantum Mechanics. New York: McGraw-Hill. 1935.

70. Pestemer, M. und D. Brück: Absorptions-Spektroskopie im Sichtbaren und Ultraviolett. In: Methoden der organischen Chemie (Houben-Weyl), 4. Aufl., Bd. III, Teil 2, S. 597. Stuttgart: G. Thieme. 1955.

71. Pitzer, K. S.: Quantum Chemistry. New York: Prentice-Hall. 1953.

72. Platt, J. R.: Classification of Spectra of cata-Condensed Hydrocarbons. J. Chem. Physics **17**, 484 (1949).

73. — Electronic Structure and Excitation of Polyenes and Porphyrins. In: A. Hollaender, Radiation Biology, Vol. III, p. 71. New York: McGraw-Hill. 1956.

74. — Wavelength Formulas and Configuration Interaction in Brooker Dyes and Chain Molecules. J. Chem. Physics **25**, 80 (1956).

75. Schmidt, O.: Die Beziehungen zwischen Dichteverteilung bestimmter Valenzelektronen (*B*-Elektronen) und Reaktivität bei aromatischen Kohlenwasserstoffen. — Die Charakterisierung der einfachen und Krebs erzeugenden aromatischen Kohlenwasserstoffe durch die Dichteverteilung bestimmter Valenzelektronen (*B*-Elektronen) (2. Mitt. über Dichteverteilung der *B*-Elektronen). —

Weitere Untersuchungen zum Kastenmodell (Zylinderring, Kompression der B-Elektronen) (3. Mitt. über Dichteverteilung und Energiespektrum der B-Elektronen). — Beiträge zum Mechanismus der Anregungsvorgänge in der krebskranken und gesunden Zelle (4. Mitt. über Dichteverteilung und Energiespektrum der B-Elektronen). — Die Berechnung der diamagnetischen Anisotropie der Aromaten aus der vom Kastenmodell gelieferten Dichteverteilung der B-Elektronen (5. Mitt. über die Dichteverteilung und das Energiespektrum der B-Elektronen). Z. physik. Chem., Abt. B 39, 59 (1938); 42, 83 (1939); 44, 185, 194 (1939); 47, 1 (1940).

76. SCHULTZE, W.: Farbenphotographie und Farbfilm. Berlin: Springer. 1953.

77. SCHWARZENBACH, G.: Aciditätskonstanten, Resonanzenergien und Lichtabsorption bei einfachen Farbstoffen. Z. Elektrochem. 47, 40 (1941).

78. SCHWARZENBACH, G., K. LUTZ und E. FELDER: Die Absorptionsspektren der allereinfachsten „Farbstoffe". Helv. Chim. Acta 27, 576 (1944).

79. SCHWARZENBACH, G. und H. GYSLING: Metallindikatoren I. Murexid als Indikator auf Calcium und andere Metall-Ionen. Komplexbildung und Lichtabsorption. Helv. Chim. Acta 32, 1314 (1949).

80. SIMPSON, W. T.: Electronic States of Organic Molecules. J. Chem. Physics 16, 1124 (1948).

81. — On the Theory of the π-Electron System in Porphines. J. Chem. Physics 17, 1218 (1949).

82. SLATER, J. C.: Atomic Shielding Constants. Physic. Rev. 36, 57 (1930).

83. SOMMERFELD, A. und H. BETHE: Elektronentheorie der Metalle. In: Handbuch der Physik, 2. Aufl., Bd. 24, 2. Teil, S. 333. Berlin: J. Springer. 1933.

84. WILLSTÄTTER, R. und E. K. BOLTON: Untersuchungen über die Anthocyane. IV. Über den Farbstoff der Scharlachpelargonie. Liebigs Ann. Chem. 408, 42 (1915) (s. insbes. S. 60).

85. WITT, O. N.: Zur Kenntniss des Baues und der Bildung färbender Kohlenstoffverbindungen. Ber. dtsch. chem. Ges. 9, 522 (1876).

86. — Über Derivate des α-Naphthols. Ber. dtsch. chem. Ges. 21, 321 (1888) (siehe insbes. S. 325).

87. WIZINGER, R.: Über das Wesen der Auxochrome und Antiauxochrome. Angew. Chem. 39, 564 (1926).

88. — Die chemische und optische Wirkung ionoider Atome. J. prakt. Chem. 157, 129 (1941).

89. — Organische Farbstoffe. Berlin und Bonn: Dümmler. 1933.

(Received, April 1, 1958.)

Namenverzeichnis. Index of Names. Index des Autors.

Sachverzeichnis. Index of Subjects. Index des Matières.

SPRINGER-VERLAG IN WIEN

Fortschritte der Chemie organischer Naturstoffe. Progress in the Chemistry of Organic Natural Products. Progrès dans la chimie des substances organiques naturelles. Herausgegeben von L. Zechmeister, California Institute of Technology, Pasadena, California, U. S. A.

Bisher erschienen:

Erster Band: Mit 41 Abbildungen im Text. VI, 371 Seiten. Gr.-8⁰. 1938.
Ganzleinen S 348.—, DM 72.25, $ 17.20, sfr. 74.—

Zweiter Band: Mit 24 Abbildungen im Text. VII, 366 Seiten. Gr.-8⁰. 1939.
Ganzleinen S 348.—, DM 72.25, $ 17.20, sfr. 74.—

Dritter Band: Mit 10 Abbildungen im Text. VI, 252 Seiten. Gr.-8⁰. 1939.
Ganzleinen S 264.—, DM 55.45, $ 13.20, sfr. 56.80

Über den Inhalt der drei Bände erteilt der Verlag bereitwilligst Auskunft.

Vierter Band: Mit 47 Abbildungen im Text. VIII, 499 Seiten. Gr.-8⁰. 1945.
Ganzleinen S 474.—, DM 99.10, $ 23.60, sfr. 101.50

Inhalt: **Tschesche, R.** Die Chemie der pflanzlichen Herzgifte, Krötengifte, Saponine und Alkaloide der Steroidgruppe. — **Wieland, Th.** und **Irmentraut Löw.** Zur Biochemie der Vitamin-B-Gruppe (Pantothensäure und Vitamin B₆). — **Purrmann, R.** Pterine. — **Schramm, G.** Die Biochemie der Virusarten. — **Bernhard. K.** und **H. Lincke.** Biologische Oxydationen. — **Trurnit, H. J.** Über monomolekulare Filme an Wassergrenzflächen und über Schichtfilme.

Fünfter Band: Mit 34 Abbildungen. VIII, 417 Seiten. Gr.-8⁰. 1948.
Ganzleinen S 305.—, DM 50.40, $ 12.—, sfr. 52.20

Inhalt: **Karrer, P.** Carotinoid-Epoxyde und furanoide Oxyde von Carotinoidfarbstoffen. — **Fox, D. L.** Some Biochemical Aspects of Marine Carotenoids. — **Haagen-Smit, A. J.** Azulenes. — **Hilditch, T. P.** Recent Advances in the Study of Component Acids and Component Glycerides of Natural Fats. — **Hassid, W. Z.** and **M. Doudoroff.** Enzymatically Synthesized Polysaccharides and Disaccharides. — **Pacsu, E.** Recent Development in the Structural Problem of Cellulose. — **Brauns, F. E.** Lignin. — **Deulofeu, V.** The Chemistry of the Constituents of Toad Venoms. — **Geiger, E.** Biochemistry of Fish Proteins. — **Beadle, G. W.** Some Recent Developments in Chemical Genetics. — **Rasmussen, R. S.** Infrared Spectroscopy in Structure Determination and its Application to Penicillin.

Sechster Band: Mit 32 Abbildungen. VIII, 392 Seiten. Gr.-8⁰. 1950.
Ganzleinen S 338.—, DM 55.80, $ 13.30, sfr. 57.80

Inhalt: **Deuel, H. J. jr.** and **S. M. Greenberg.** Some Biochemical and Nutritional Aspects in Fat Chemistry. — **Lederer, E.** Odeurs et parfums des animaux. — **Hoffmann-Ostenhof, O.** Vorkommen und biochemisches Verhalten der Chinone. — **Reti, L.** Cactus Alkaloids and Some Related Compounds. — **Bonner, J.** Plant Proteins. — **Dhéré, Ch.** Progrès récents en spectrochimie de fluorescence des produits biologiques.

Siebenter Band: Mit 12 Abbildungen. VII, 330 Seiten. Gr.-8⁰. 1950.
Ganzleinen S 325.—, DM 53.70, $ 12.80, sfr. 55.50

Inhalt: **Jeger, O.** Über die Konstitution der Triterpene. — **Heusser, H.** Konstitution, Konfiguration und Synthese digitaloider Aglykone und Glykoside. — **Niemann, C.** Thyroxine and Related Compounds. — **Cook, A. H.** Penicillin and its Place in Science. — **Stoll, A.** and **B. Becker.** Sennosides A and B, the Active Principles of Senna. — **Williams, J. W.** Some Recent Developments in the Chemistry of Antibodies.

Achter Band: Mit 47 Abbildungen. XI, 400 Seiten. Gr.-8⁰. 1951.
Ganzleinen S 427.—, DM 70.50, $ 16.80, sfr. 72.20

Inhalt: **Frey-Wyssling, A.** and **K. Mühlethaler.** The Fine Structure of Cellulose. — **Stacey, M.** and **C. R. Ricketts.** Bacterial Dextrans. — **Leloir, L. F.** Sugar Phosphates. — **Kenner, G. W.** The Chemistry of Nucleotides. — **Schinz, H.** Die Veilchenriechstoffe. — **Asahina, Y.** Neuere Entwicklungen auf dem Gebiete der Flechtenstoffe. — **Galinovsky, F.** Lupinen-Alkaloide und verwandte Verbindungen. — **Pailer, M.** Brechwurzel-Alkaloide. — **Corey, R. B.** X-Ray Diffraction Studies of Crystalline Amino Acids and Peptides. — **Zechmeister, L.** and **M. Rohdewald.** Some Aspects of Enzyme Chromatography.

Weitere Bände siehe nächste Seite!

Zu beziehen durch Ihre Buchhandlung

SPRINGER-VERLAG IN WIEN

Fortsetzung von vorhergehender Seite

Neunter Band: Mit 20 Abbildungen. XI, 535 Seiten. Gr.-8⁰. 1952.
Ganzleinen S 498.—, DM 82.50, $ 19.60, sfr. 84.50

Inhalt: **Inhoffen, H. H.** und **H. Siemer.** Synthetische Chemie der Carotinoide. — **Baxter, J. G.** Synthesis and Properties of Vitamin A and Some Related Compounds. — **Meunier, P.** Les Antivitamines. — **Stoll, A.** Recent Investigations on Ergot Alkaloids. — **Tomita, M.** Die Alkaloide der Menispermaceae-Pflanzen. — **Dean, F. M.** Naturally Occurring Coumarins. — **Borsook, H.** The Biosynthesis of Proteins and Peptides, including Isotopic Tracer Studies. — **Kalckar, H. M.** The Enzymes of Nucleoside Metabolism. — **McNutt, W. S.** Nucleosides and Nucleotides as Growth Substances for Microorganisms. — **Campbell, D. H.** and **N. Bulman.** Some Current Concepts of the Chemical Nature of Antigens and Antibodies.

Zehnter Band: Mit 19 Abbildungen. IX, 529 Seiten. Gr.-8⁰. 1953.
Ganzleinen S 498.—, DM 83.—, $ 19.80, sfr. 85.—

Inhalt: **Alder, K.** und **Marianne Schumacher.** Anwendungen der Dien-Synthese für die Erforschung von Naturstoffen. — **Mark, H.** Physical Chemistry of Rubbers. — **Asselineau, J.** et **E. Lederer.** Chimie des lipides bactériens. — **Rosenkranz, G.** and **F. Sondheimer.** Syntheses of Cortisone. — **Chatterjee, A.** Rauwolfia Alkaloids. — **Feinstein, L.** and **M. Jacobson.** Insecticides Occurring in Higher Plants.

Elfter Band: Mit 67 Abbildungen. VIII, 457 Seiten. Gr.-8⁰. 1954.
Ganzleinen S 448.—, DM 74.80, $ 18.—, sfr. 77.40

Inhalt: **Peat, S.** Starch: Its Constitution, Enzymic Synthesis and Degradation. — **Freudenberg, K.** Neuere Ergebnisse auf dem Gebiete des Lignins und der Verholzung. — **Inhoffen, H. H.** und **K. Brückner.** Probleme und neuere Ergebnisse in der Vitamin D-Chemie. — **Schmid, H.** Natürlich vorkommende Chromone. — **Pauling, L.** and **R. B. Corey.** The Configuration of Polypeptide Chains in Proteins. — **Schroeder, W. A.** Column Chromatography in the Study of the Structure of Peptides and Proteins. — **Lemberg, R.** Porphyrins in Nature. — **Albert, A.** The Pteridines.

Zwölfter Band: Mit 15 Abbildungen. X, 550 Seiten. Gr.-8⁰. 1955.
Ganzleinen S 497.—, DM 82.80, $ 19.80, sfr. 85.10

Inhalt: **Haagen-Smit, A. J.** Sesquiterpenes and Diterpenes. — **Jones, E. R. H.** and **T. G. Halsall.** Tetracyclic Triterpenes. — **Tschesche, R.** Neuere Vorstellungen auf dem Gebiete der Biosynthese der Steroide und verwandter Naturstoffe. — **Haxo, F. T.** Some Biochemical Aspects of Fungal Carotenoids. — **Warren, F. L.** The Pyrrolizidine Alkaloids. — **Thompson, E. O. P.** and **A. R. Thompson.** Paper Chromatography in the Study of the Structure of Peptides and Proteins. — **Roche, J.** et **R. Michel.** Acides aminés iodés et iodoprotéines. — **Slotta, K.** Chemistry and Biochemistry of Snake Venoms. — **Beadle, G. W.** Gene Structure and Gene Action.

Dreizehnter Band: Mit 48 Abbildungen. XII, 624 Seiten. Gr.-8⁰. 1956.
Ganzleinen S 645.—, DM 107.50, $ 25.60, sfr. 110.10

Inhalt: **Cole, A. R. H.** Infrared Spectra of Natural Products. — **Schmidt, O. Th.** Gallotannine und Ellagen-Gerbstoffe. — **Tamm, Ch.** Neuere Ergebnisse auf dem Gebiete der glykosidischen Herzgifte: Grundlagen und die Aglykone. — **Nozoe, T.** Natural Tropolones and Some Related Troponoids. — **Price, J. R.** Alkaloids Related to Anthranilic Acid. — **Chatterjee, A.,** **S. C. Pakrashi** and **G. Werner.** Recent Developments in the Chemistry and Pharmacology of Rauwolfia Alkaloids. — **Graßmann, W.** und **E. Wünsch.** Synthese von Peptiden.

Vierzehnter Band: Mit 38 Abbildungen. VIII, 377 Seiten. Gr.-8°. 1957.
Ganzleinen S 450.—. DM 75.—, $ 17.85, sfr. 76.80

Inhalt: **Bohlmann, F.** und **H. J. Mannhardt.** Acetylenverbindungen im Pflanzenreich. — **Tamm, Ch.** Neuere Ergebnisse auf dem Gebiete der glykosidischen Herzgifte: Zucker und Glykoside. — **Brockmann, H.** Photodynamisch wirksame Pflanzenfarbstoffe. — **Birch, A. J.** Biosynthetic Relations of Some Natural Phenolic and Enolic Compounds. — **Sobotka, H., N. Barsel** and **J. D. Chanley.** The Aminochromes. — **Morton, R. A.** and **G. A. J. Pitt.** Visual Pigments. — **Brown, H.** The Carbon Cycle in Nature.

Fünfzehnter Band: Mit 81 Abbildungen. VI, 244 Seiten. Gr.-8⁰. 1958.
Ganzleinen S 246.—, DM 41.—, $ 9.75, sfr. 42.—

Inhalt: **Schlubach, H. H.** Der Kohlenhydratstoffwechsel der Gräser. — **Zechmeister, L.** Some in vitro Conversions of Naturally Occurring Carotenoids. — **Hartwell, J. L.** and **A. W. Schrecker.** The Chemistry of Podophyllum. — **Dorothy Crowfoot Hodgkin.** X-ray Analysis and the Structure of Vitamin B_{12}.